E'7
12.00

An Introduction to

ATMOSPHERIC
PHYSICS

International Geophysics Series

Edited by

J. VAN MIEGHEM

Royal Belgian Meteorological Institute
Uccle, Belgium

An Introduction to

ATMOSPHERIC PHYSICS

ROBERT G. FLEAGLE
JOOST A. BUSINGER
University of Washington
Seattle, Washington

1963

 ACADEMIC PRESS New York and London

ACADEMIC PRESS, INC.
111 Fifth Avenue, New York, New York 10003

United Kingdom Edition published by
ACADEMIC PRESS, INC. (LONDON) LTD.
Berkeley Square House, London W1X 6BA

LIBRARY OF CONGRESS CATALOG CARD NUMBER: 63-12922

Third Printing, 1970

PRINTED IN THE UNITED STATES OF AMERICA

Preface

This book is addressed to those who wish to understand the relationship between atmospheric phenomena and the nature of matter as expressed in the principles of physics. The interesting atmospheric phenomena are more than applications of gravitation, of thermodynamics, of hydrodynamics, or of electrodynamics; and mastery of the results of controlled experiment and of the related theory alone does not imply an understanding of atmospheric phenomena. This distinction arises because the extent and the complexity of the atmosphere permit effects and interactions which are entirely negligible in the laboratory or are deliberately excluded from it. The objective of laboratory physics is, by isolating the relevant variables, to reveal the fundamental properties of matter; whereas the objective of atmospheric physics, or of any observational science, is to understand those phenomena which are characteristic of the whole system. For these reasons the exposition of atmospheric physics requires substantial extensions of classical physics. It also requires that understanding be based on a coherent "way of seeing" the ensemble of atmospheric phenomena. Only then is understanding likely to stimulate still more general insights.

In this book the physical properties of the atmosphere are discussed. Atmospheric motions, which are part of atmospheric physics and which logically follow discussion of physical properties, have not been included because they require more advanced mathematical methods than those used here. We hope to treat atmospheric motions in a later volume.

The content of the book has been used as a text for a year's course for upper division and beginning graduate students in the atmospheric sciences. In the course of time, it has filtered through the minds of more than a dozen groups of these critics. It also should be of useful interest to students of other branches of geophysics and to students of physics who want to extend their horizons beyond the laboratory. And, finally, we believe that the tough-minded amateur of science may find pleasure in seeking in these pages a deeper appreciation of natural phenomena.

Although an understanding of the calculus and of the principles of physics is assumed, these fundamentals are restated where they are relevant; and the book is self-contained for the most part. Compromise has been necessary in a few cases, particularly in Chapter VII, where full development would have required extended discussion of advanced and specialized material.

The text is not intended for use as a reference; original sources must be sought from the annotated general references listed at the end of each chapter. Exceptions have been made in the case of publications which are so recent that they have not been included in standard references and in a few cases where important papers are not widely known. In the case of new or controversial material, an effort has been made to cut through irrelevant detail and to present a clear, coherent account. We have felt no obligation to completeness in discussing details of research or in recounting conflicting views. Where it has not been possible for us to make sound judgments, we have tried to summarize the problem in balanced fashion, but we have no illusions that we have always been right in these matters. We accept the inevitable fact that errors remain imbedded in the book, and we challenge the reader to find them. In this way, even our frailties may make a positive contribution.

We are indebted for numerous suggestions to our colleagues who have taken their turns at teaching the introductory courses in Atmospheric Physics: Professors F. I. Badgley, K. J. K. Buettner, D. O. Staley, and Mr. H. S. Muench. Parts of the manuscript have profited from the critical comments of Professors E. F. Danielsen, B. Haurwitz, J. S. Kim, J. E. McDonald, B. J. Mason, H. A. Panofsky, T. W. Ruijgrok, and F. L. Scarf. Last, and most important of all, we acknowledge the silent counsel of the authors whose names appear in the bibliography and in the footnotes. We are keenly aware of the wisdom of the perceptive observer who wrote:

> Von einem gelehrten Buche abgeschrieben ist ein Plagiat,
> Von zwei gelehrten Büchern abgeschrieben ist ein Essay,
> Von drei gelehrten Büchern abgeschrieben ist eine Dissertation,
> Von vier gelehrten Büchern abgeschrieben ist ein fünftes gelehrtes Buch.

Seattle, Washington   R. G. F.
March, 1963   J. A. B.

Table of Contents

CHAPTER III: PROPERTIES AND BEHAVIOR OF CLOUD PARTICLES

CHAPTER IV: SOLAR AND TERRESTRIAL RADIATION

CHAPTER V: TRANSFER PROCESSES AND APPLICATIONS

CHAPTER VI: GEOMAGNETIC PHENOMENA

CHAPTER VII: ATMOSPHERIC SIGNAL PHENOMENA

APPENDIX I: MATHEMATICAL TOPICS

APPENDIX II: PHYSICAL TOPICS

Bibliography

Index

CHAPTER I

Gravitational Effects

"We dance round in a ring and suppose,
But the Secret sits in the middle and knows." ROBERT FROST

VIEWED FROM A DISTANCE of several earth radii our planet appears as a smooth spheroid strongly illuminated by a very distant sun. As seen from above the northern hemisphere the earth rotates about its axis in a counter-clockwise sense once in 23 hours 56 minutes and revolves about the sun in the same sense once a year in a nearly circular orbit. The axis of rotation is inclined to the plane of the earth's orbit at an angle of 66° 33′.

The earth is enveloped in a shell of air which, though transparent for the most part, gives the earth a blue tinge. Opaque patches of cloud float imbedded in a layer of the atmosphere no greater than six to ten kilometers in thickness lying just above the earth's surface. Each of these cloud areas exhibits complex structures of various scales, and each undergoes a characteristic life cycle. Cloud areas range in horizontal scale from a few tens of meters to a few thousand kilometers. Systematic relations can be readily detected between size and structure of the cloud areas and their latitude, season of the year and the underlying earth's surface.

Half of the atmosphere is always confined to the 6-km layer in which most of the clouds are found, and 99% is confined to the lowest thirty kilometers. A revealing impression of the relative vertical and horizontal dimensions of the atmosphere is shown in Fig. 1.1. Although the atmosphere is heavily concentrated close to the earth, air density sufficient to produce observable effects is present at a height of one thousand kilometers, and a very small proportion of the atmosphere extends even farther into space. Therefore, for applications in which mass is the relevant factor, the atmosphere properly is considered a thin shell; for other applications the shell concept is inadequate.

How is the curious vertical distribution of mass in the atmosphere to be explained? Thorough analysis of this problem requires consideration of complex energy transformations which will be discussed at a later stage; we shall then have occasion to refer to this problem of the vertical distribution of density in the atmosphere. At this stage, however, let us fix attention on the most important aspect: Why does the earth have an atmosphere at all? The answer lies in an understanding of gravitation, but first we must establish the ground rules.

FIG. 1.1. The atmosphere as seen by John Glenn from a height of about 200 kilometers over the Indian Ocean after sunset 20 February 1962. The photograph covers about 10° of latitude (by courtesy of Project Mercury Weather Support Group, U. S. Weather Bureau).

1.1 Fundamental Concepts

The fundamental concepts of physics, those defined only in terms of intuitive experience, may be considered to be space, time, and force. We utilize these concepts without precise definition and with the faith that they have the same meaning to all who use them. The choice of fundamental concepts is somewhat arbitrary; it is possible to choose other quantities as fundamental, and to define space, time, and force in terms of these alternate choices. For instance, space, time, and mass are sometimes taken as fundamental concepts. Our reason for choosing force as fundamental is that electrical charge can then be defined by Coulomb's Law without introduction of another fundamental quantity.

The events which are important in ordinary or Newtonian mechanics are considered to occur within a frame of reference having three spatial coordinates. These define the position of the event in terms of its distance from an arbitrary reference point; the distance may be measured along mutually perpendicular straight lines (Cartesian coordinates), or along any of a number of lines associated with other coordinate systems. Time is expressed in terms of the simultaneous occurrence of an arbitrary familiar event, for instance, the rotation of the earth about its axis or the characteristic frequency of an electromagnetic wave emitted by an atom.

1.2   Law of Universal Gravitation

Sir Isaac Newton first recognized that the motions of the planets as well as many terrestrial phenomena (among which we may include the free-fall of apples) rest on a single precise and universal statement relating the separation of two bodies, their masses, and an attractive force between them. This Law of Universal Gravitation states that every particle of matter in the universe attracts every other particle with a force directly proportional to the product of the two masses and inversely proportional to the square of their separation. It may be written in vector form

$$\mathbf{F} = -G \frac{Mm}{r^2} \frac{\mathbf{r}}{r} \tag{1.1}$$

where $\mathbf{F}$ represents the force exerted on the mass m at a distance $\mathbf{r}$ from the mass M, and G represents the universal gravitational constant. Vector notation and elementary vector operations are summarized in Appendix I.B. The scalar equivalent of Eq. (1.1), $F = GMm/r^2$, gives less information than the vector form, and in this case the direction of the gravitational force must be specified by an additional statement. The universal gravitational constant has the value 6.67×10^{-8} dyn cm^{2}g^{-2}. The systems of units, dimensions, and physical constants used throughout the book are summarized in Appendix II.

The Law of Universal Gravitation is fundamental in the sense that it rests on insight into the structure of the universe and cannot be derived by logical deduction from more fundamental principles. It is an example of the inverse square relation which makes its appearance over and over in fundamental principles and in applications. Relativity theory introduces a generalization of the law of universal gravitation, but the change from Eq. (1.1) is so small that it may be neglected for most geophysical purposes.

1.3   Newton's Laws of Motion

Publication in 1686 of Newton's "Philosophiae Naturalis Principia Mathematica," or the "Principia" as it has come to be known, may be recognized as the most important single event in the history of science. For in one published work Newton stated both the Law of Universal Gravitation and the Laws of Motion; together these laws provided the base for all later developments in mechanics until the twentieth century.

Newton's First Law, which states that a body at rest remains at rest and a body in motion remains in motion until acted on by a force, is actually a form of the more general Second Law. The Second Law states that *in an unaccelerated coordinate system the resultant force acting on a body equals the time rate of change of momentum of the body.* It may be written in vector form

$$\mathbf{F} = \frac{d}{dt} (m\mathbf{v})$$

(1.2)

where $\mathbf{v}$ represents velocity and t, time. Equation (1.2) is the central equation in development of the physics of atmospheric motions. It, like the Law of Universal Gravitation, is fundamental in the sense that it cannot be derived from more fundamental statements; it states a property of the universe. Newton's Second Law can be transformed into the equation of conservation of angular momentum. This is left as problem 1 at the end of the chapter.

Forces occur in pairs. If I push against the wall, I feel the wall pushing against my hand; the two forces are equal in magnitude and opposite in direction. If I push against a movable body, for example, in throwing a ball, I experience a force which is indistinguishable from that exerted by the wall in the first example. Newton's Third Law of Motion expresses these observations in the form: *for every force exerted on a body A by body B there is an equal and opposite force exerted on body B by body A.* This may be expressed by

$$\mathbf{F}_{A,B} = -\mathbf{F}_{B,A}$$

(1.3)

Equation (1.3) is a third fundamental equation.

Notice that the forces referred to in discussing the Second Law and the Law of Universal Gravitation are forces acting *on* the body under consideration. Only the forces acting *on* a body can produce an effect on the body.

1.4   The Earth's Gravitational Field

The gravitational force with which we are most familiar is that exerted by the earth on a much smaller object: a book, a stone, or our bodies. We call this (with a correction to be explained in Sections 1.5 and 1.8) the weight of the object. If we simplify by considering an object to be represented by a point mass, and if r represents the separation of the point mass from the center of the homogeneous spherical earth of mass M, the gravitational force exerted by the earth is also expressed by Eq. (1.1). This result may be verified by solving problem 2; problem 3 provides an extension. A gravitational field is said to be associated with the earth; a mass experiences a force at any point in this field.

For geophysical applications it is convenient to write Eq. (1.1) in the form

$$\mathbf{F} = m\mathbf{g}^* \tag{1.4}$$

where

$$\mathbf{g}^* \equiv -G\frac{M}{r^2}\frac{\mathbf{r}}{r} \tag{1.5}$$

The vector $\mathbf{g}^*$ is called the gravitational force per unit mass or the gravitational acceleration, and it is directed vertically downward toward the center of the earth. The slight flattening of the earth at the poles makes this statement less than exact, but still adequate for nearly all atmospheric problems. In this discussion the contributions to the gravitational field made by the moon, sun, and other extraterrestrial bodies are not considered; these effects are applied in Section 1.10 to the discussion of atmospheric tides. Problem 4 provides an application of Eq. (1.5).

It is useful to express Eq. (1.5) in terms of the radius of the earth, R, and height above the earth, z. The scalar form is

$$g^* \equiv \frac{GM}{R^2(1 + z/R)^2} \equiv \frac{g_0^*}{(1 + z/R)^2} \tag{1.6}$$

The identity on the right-hand side of (1.6) represents the definition of g_0^*, the gravitational force per unit mass at sea level. Its value as computed from Eq. (1.6) varies by 6.6 dyn g^{-1} between the pole where R is 6356.9 km and the equator where R is 6378.4 km. But Eq. (1.6) is not quite correct for the triaxial spheroidal shape of the earth; it is easy to recognize that Eq. (1.6) overestimates g_0^* at the equator because the "extra" mass is distributed symmetrically about the plane of the equator. As a result of these effects g_0^* is 983.2 dyn g^{-1} at the pole and 981.4 dyn g^{-1} at the equator. There are also small local anomalies in g_0^* associated with mountains and the inhomogeneous distribution of density within the earth's crust. For applications in which z is very much less than R, Eq. (1.6) may be expanded in a Taylor series about $z = 0$; if only the first two terms are retained, the result is

$$g^* = g_0^*(1 - 2z/R) \tag{1.7}$$

Equation (1.7) shows that near the earth's surface g^* decreases linearly with height. Even at four hundred kilometers above the earth, where g^* is about 20% less than g_0^*, the first neglected term in the Taylor expansion, $(z/R)^2$, contributes an error of only 1%. Beyond about four hundred kilometers g^* decreases more rapidly with height than the linear rate. A brief account of the Taylor series is given in Appendix I.C.

1.5   The Force of Gravity

The force acting on a static unit mass or the acceleration which we measure in a laboratory fixed on the earth's surface is not exactly that given by Eq. (1.5) because the coordinate system within which the measurements are made is itself an accelerating system. It is most convenient in this case to consider that an "apparent" centrifugal force acts on a body rotating with the earth, even though it is clear that, as viewed from a nonrotating coordinate system in space, no "real" centrifugal force acts on the rotating body. It is the coordinate system fixed on the rotating earth which is accelerated, but by adding the apparent centrifugal force, we can use Newton's Second Law even in the rotating system.

The apparent centrifugal force per unit mass is easily found to be given by $\omega^2 \mathbf{r}$, where ω represents the *angular frequency* (often called angular velocity) of the earth and $\mathbf{r}$ is the radius vector drawn perpendicularly from the earth's axis to the point in rotation as shown in Fig. 1.2. The vector

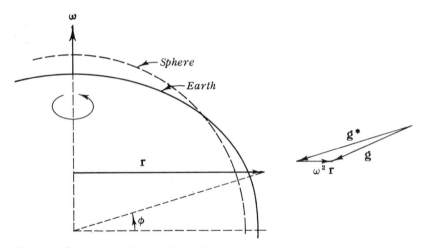

Fig. 1.2. Cross section through the earth showing the vector sum of the gravitational ($\mathbf{g}^*$) and centrifugal ($\omega^2 \mathbf{r}$) forces acting on unit mass fixed with respect to the rotating earth and the resulting distortion of the earth's figure. Latitude is represented by ϕ and the earth's angular frequency by ω.

sum of $\mathbf{g}^*$ and $\omega^2 \mathbf{r}$ represents the force per unit mass or acceleration measured by plumb bob or by a falling weight. It is called the *acceleration of gravity* and is expressed by the vector $\mathbf{g}$. The *force of gravity* is then defined by

$$\mathbf{F}_g \equiv m\mathbf{g} \tag{1.8}$$

The direction of $\mathbf{F}_g$ and of $\mathbf{g}$ is of course normal to horizontal surfaces (sea level, in the absence of currents of air or water, for example). The angle between $\mathbf{g}^*$ and $\mathbf{g}$ clearly vanishes at pole and equator, and it is very small even at 45° of latitude. Although small, this angle accounts for the flattening of the earth at the poles and the bulge at the equator; for the earth is a plastic body and takes on the form which tends to minimize internal stresses. Problem 5 requires calculation of the angle between $\mathbf{g}^*$ and $\mathbf{g}$.

The magnitude of g is easily calculated if we assume that $\mathbf{g}^*$ and $\mathbf{g}$ have the same direction. The component of $\omega^2 \mathbf{r}$ in the direction of $\mathbf{g}^*$ is $\omega^2(R + z) \cos^2 \phi$, from which it follows that

$$g = g^* \left(1 - \frac{\omega^2(R + z) \cos^2 \phi}{g^*} \right) \tag{1.9}$$

By combining Eqs. (1.6) and (1.9) the mass of the earth may be calculated from a measured value of g. Numerical evaluation is left to problem 6. The numerical value of $\omega^2 R/g^*$ is 3.44 $\times$ 10^{-3}, from which it follows that the effect of the earth's rotation is to make the value of g at the equator 3.37 dyn g^{-1} less than at the poles. As pointed out earlier, the oblateness of the earth makes g^* dependent on latitude; the result is that the measured value of g is 983.2 dyn g^{-1} at the pole, 980.6 at 45° latitude, and 978.0 at the equator.

1.6   Geopotential

In order for a body to be raised from sea level, work must be done against the force of gravity. If we recall that *work* is defined by the line integral $\int \mathbf{F} \cdot d\mathbf{s}$ where $d\mathbf{s}$ represents a differential displacement, then the work done per unit mass in displacing a body in the field of gravity is expressed by

$$\Phi \equiv \int_0^z \frac{F_g}{m} \, dz = \int_0^z g \, dz \tag{1.10}$$

Equation (1.10) defines the *geopotential*, the potential energy per unit mass. The importance of the geopotential derives from the fact that at any point in the field, no matter how complicated, the geopotential is unique; that is, it is a function of position and does not depend on the path followed by the mass (m) in reaching the height, z. Consequently, the line integral of the force of gravity around any closed path must equal zero; this is the condition required for existence of a potential function. The potential function arises again in Chapter III in the form of the electrostatic potential and in Chapter VI in the form of the scalar magnetic potential. The potential and its relation to vector fields is discussed in Appendix I.G.

Because of this property of the field of gravity, the energy required to put a satellite into orbit or to travel to any specified point in the solar system may be calculated easily. And geopotential may be used in place of height to specify position in the vertical direction. The geopotential (Φ) due to the earth alone is found by substituting Eqs. (1.6) and (1.9) in Eq. (1.10) with the result

$$\Phi = \frac{GMz}{R(R + z)} - \omega^2 \cos^2 \phi \left(R + \frac{z}{2}\right) z \qquad (1.11)$$

If potential energy is calculated with respect to a nonrotating coordinate system, only the first term on the right-hand side of Eq. (1.11) is used; this is called *gravitational potential* and is useful in problems in which rotation of the earth is irrelevant.

1.7 Satellite Orbits

If Eq. (1.1) is substituted into Eq. (1.2), there emerges the differential equation which governs the motion of bodies in a gravitational field, $dv/dt = -(GM/r^2)(\mathbf{r}/r)$. For a small mass in the gravitational field of a specified large mass, the solution is fairly simple; the result is that the small mass moves along a path which is described as a conic section, that is, it may be elliptic, parabolic, or hyperbolic. The shape of the path is determined by the initial velocity and the initial position of the small mass. For certain critical conditions the path is parabolic. If the initial speed exceeds the critical value, the small body moves along a hyperbola; an example is the path of a satellite projected from the earth out into the solar system never to return to the neighborhood of the moon or earth. For an initial speed less than critical, the path is elliptic; examples are the orbits of earth satellites, the moon's orbit about the earth, and the earth's orbit about the sun. We may understand these motions intuitively if we imagine the vector sum of the gravitational force exerted by the large mass and the apparent centrifugal force. For exact balance of these forces, the orbit is circular; if the two are not balanced, acceleration occurs. The earth accelerates as it "falls" toward the sun (gravitational force exceeds centrifugal) from July to January and decelerates as it "rises" against the gravitational pull from January to July. In the case of the earth, accelerations are small because the eccentricity of the orbit is only 0.017. In the case of some satellites projected from the earth large eccentricities have been established.

If we properly can assume balance between the apparent centrifugal force and the gravitational force, certain important results are easily obtained. This balance may be expressed by

$$\frac{GM}{(R + z)^2} = \Omega^2 (R + z) \qquad (1.12)$$

where Ω represents the angular frequency of the satellite with respect to the fixed stars. From Eq. (1.12) we may observe that as the height of the satellite decreases, the angular frequency increases. An exercise is provided in problem 7. If the earth's orbit is assumed to be circular, Eq. (1.12) can be used to calculate the mass of the sun. Applied to the earth-moon system the equation permits calculation of the mass of the earth, as required in problem 8.

The total energy per unit mass needed to put the satellite into a polar orbit may be developed by adding the gravitational potential to the kinetic energy per unit mass $[\frac{1}{2}\Omega^2(R + z)^2]$. It follows that the total energy per unit mass is expressed by

$$T = \frac{GM}{R + z}\left(\frac{z}{R} + \frac{1}{2}\right) \tag{1.13}$$

The energy per unit mass required to enable a body to escape from the earth's gravitational field (that is, to follow a parabolic or hyperbolic path) is easily found from the equation for gravitational potential [Eq. (1.11) without the last term on the right]. We see immediately that the *escape energy* per unit mass is $g_0{}^*R$. The absolute velocity (velocity with respect to the solar system) which must be imparted to a free projectile (or a molecule) in order for it to escape from the gravitational field of the earth is found by equating the kinetic energy and the escape energy. The *escape velocity* is then expressed by

$$v_{es} = (2g_0{}^*R)^{1/2} \tag{1.14}$$

The escape velocity for the earth is about 11 km sec^{-1} but for the moon is only about 2.5 km sec^{-1}. The absolute velocity of a point fixed on the surface of the earth at the equator is 0.47 km sec^{-1}, so that there is economy in launching satellites from low latitude in the direction of the earth's rotation (toward the east).

1.8 Hydrostatic Equation

Although all planetary bodies (planets, moons, asteroids) possess gravitational fields, not all possess atmospheres, for there is constant escape of gas molecules. Rate of escape depends on strength of the gravitational and magnetic fields and on the velocities of the molecules near the outer limit of the atmosphere. Uncharged molecules which move upward with speeds in excess of the escape velocity and which fail to collide with other molecules or to become ionized leave the planet's gravitational field and are lost to the atmosphere. Ionized molecules moving in the magnetic field are also strongly influenced by the induced electromagnetic force discussed in Chapter VI. In the case of the moon virtually the whole atmosphere has escaped, but in the case of the earth it is not known whether the total mass of the atmosphere is increasing through release of gas from the solid earth and by collection of solar and interstellar matter or decreasing through escape of gas from the upper atmosphere.

The presence of the atmosphere as a shell surrounding the earth can now be recognized as a direct consequence of the earth's gravitational field. Each molecule of air is attracted toward the center of mass of the earth by the force of gravity and is restrained from falling to the earth by the upward force exerted by collision with a molecule below it. This collision produces a downward force on the lower molecule, and this force is balanced, in the mean, by collision with a still lower molecule. Therefore, molecules above a horizontal reference surface exert a downward force on the molecules below the surface, a force which is called the *weight* of the gas above the surface. Because weight is proportional to the force of gravity, the weight of the atmosphere includes the effect of the earth's rotation but does not include inertial effects which arise from accelerations measured with respect to the rotating earth. Under static conditions (accelerations negligible) the weight of a vertical column of unit base cross section extending from the earth to the top of the atmosphere is equal to the atmospheric pressure, and the weight therefore may be measured with a barometer. The vertical column is illustrated in Fig. 1.3. Since g varies only slightly within the layer

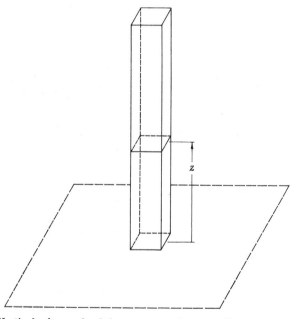

FIG. 1.3. Vertical column of unit base cross section extending from the earth's surface to the top of the atmosphere.

which contains virtually the entire atmosphere, the mass of air contained in the vertical column under static conditions is very nearly proportional to the barometric pressure. At sea level there is about one kilogram above each square centimeter of the earth's surface, and this mass rarely changes by more than ±3%.

The force of gravity exerted on a unit volume of air at any point in the column shown in Fig. 1.3 is expressed by ρg where ρ represents the mass per unit volume (density). Consequently, the pressure at any height, z, is expressed by

$$p = \int_z^\infty \rho g \, dz \qquad (1.15)$$

One of the most important equations in atmospheric physics, the *hydrostatic equation*, is now derived by differentiating Eq. (1.15) with respect to height holding x, y, and t constant. This yields

$$-\frac{\partial p}{\partial z} = \rho g \qquad (1.16)$$

The hydrostatic equation states that pressure decreases upward at a rate equal to the product of density and force of gravity per unit mass; this result is the same for incompressible and for compressible fluids. It represents a generalization of Archimedes' Principle which states that a body immersed in a fluid experiences an upward force equal to the weight of the displaced fluid. Problem 9 provides an illustrative exercise. The density of a compressible fluid like air depends on the pressure, so that we cannot integrate Eqs. (1.15) or (1.16) without knowing density as a function of pressure. Integration of the hydrostatic equation will be discussed in the following chapter; here, we shall only point out that density of air decreases with decreasing pressure, so that the rate of pressure decrease itself decreases with height entirely apart from the decrease of g with height. We may understand now how it is possible that the compressible atmosphere extends to very great distances from the earth.

The vertical walls shown in Fig. 1.3 are, of course, only mental aids to understanding; they do not confine the air in any way. It is possible to use this mental aid because under static conditions each vertical column sustains each adjacent vertical column by horizontal pressure. If mass is added to one column, say by an airplane entering the column, the increased pressure within the lower part of the column leads to unbalanced horizontal forces and to accelerations. These lead, in turn, to increased pressure in adjacent columns, and the net result is that the weight of the airplane is distributed over a very large area of ground surface. This is, of course, a most fortunate property of the atmosphere.

The hydrostatic equation may be written in other forms which are particularly revealing for certain applications to be discussed later. First, if Eq. (1.16) is written in the form

$$-\frac{1}{\rho}\frac{\partial p}{\partial z} = g \qquad (1.17)$$

the right-hand side represents the force of gravity per unit mass. We have required in the derivation of the hydrostatic equation that accelerations be negligible, so that Newton's Second Law requires that forces be balanced. The force which balances the force of gravity in Eq. (1.17) is called the *pressure force* per unit mass. It is easy to show that the horizontal components of the pressure force per unit mass may be expressed by $-(1/\rho)(\partial p/\partial x)$ and $-(1/\rho)(\partial p/\partial y)$, respectively.

A second form of the hydrostatic equation may be written if we recognize from Eq. (1.10) that $d\Phi = g\,dz$; therefore, Eq. (1.16) may be written

$$-\frac{\partial p}{\partial \Phi} = \rho \qquad (1.18)$$

In this transformation the geopotential has replaced the height as the dependent variable; in so doing the number of variables has been reduced by one. Geopotential is used as the vertical coordinate in most atmospheric applications in which energy plays an important role, particularly in the field of large scale motions.

We may develop general *equations of motion* by substituting the sum of the forces acting on an air parcel into Newton's Second Law [Eq. (1.2)]. We shall do this for the vertical component of the general vector equation. If the frictional force per unit mass is represented by f_z, then the sum of the pressure force, gravity force and frictional force must equal the acceleration. This is stated in the form*

$$\frac{dv_z}{dt} = -\frac{1}{\rho}\frac{\partial p}{\partial z} - g - f_z \qquad (1.19)$$

It is now evident that the hydrostatic equation properly may be regarded as the special form of the equation of motion which applies if the acceleration and frictional force are negligible. A change of vertical component of velocity of 10 cm sec^{-1} in one second makes dv_z/dt amount to about 1% of the pressure and gravity forces; because this is a large rate of change of velocity, we may conclude that the hydrostatic equation is accurate except in exceptional circumstances, viz., thunderstorms, tornadoes, extreme turbulence.

Analogous equations of motion may be derived for the two horizontal

---

* A very small "Coriolis acceleration" given by $2\omega v_x \cos\phi$, where v_x represents the velocity component toward the east, has been neglected here.

coordinates; the force of gravity does not appear, and as a result the magnitude of the pressure force is much smaller.

1.9 Distribution of Sea Level Pressure

With occasional exceptions sea level pressure varies with time and place between about 980 and 1040 millibars (1 mb = 10^3 dyn cm^{-2}), and the mean value for the world is about 1013 mb. Therefore, from Eq. (1.15) we find that the total mass of the atmosphere is 5×10^{18} kg. The mean distribution of pressure at sea level for the month of January is shown in Fig. 1.4. The outstanding features are the semi-permanent systems of high and low pressure. In general, high pressure predominates at about 30° north and south latitude, low pressure predominates at high latitudes and in the tropics, and the lowest pressures are found in the regions of Iceland and the Aleutian Islands. We may recognize from the discussion of Section 1.8 that the horizontal pressure force per unit mass is directed from high to low pressure and is represented by

$$-\frac{1}{\rho} \nabla_{\mathrm{H}} p$$

Therefore, where pressure varies by one millibar per degree of latitude, the pressure force per unit mass amounts to 10^{-1} dyn g^{-1}. The pressure force plays a crucial role in generation of atmospheric motions, and this will constitute an important part of a succeeding volume. Here, we shall point out only that Fig. 1.4 suggests that pressure differences are not quickly equalized by the pressure force. Fig. 1.4 also reveals that in the winter hemisphere pressure is higher over land than over sea, whereas in the summer hemisphere pressure is higher over the sea than over land.

1.10 Atmospheric Tides

Newton's Third Law requires that the gravitational forces which are exerted by the moon and sun on the earth and the atmosphere are equal and opposite to those exerted by the earth on the moon and sun. For clarity we shall consider only the earth-moon system. The earth and moon form a system of two coupled masses in rotation about their common center of mass. Because the mass of the earth is about eighty times that of the moon, the center of mass is at a point about one-eightieth of the distance between the centers of the earth and moon or about sixteen hundred kilometers below the earth's surface. The earth and moon rotate about this point once in 27.3 days as illustrated in Fig. 1.5a. The diurnal rotation of the earth is ignored here. By virtue of the 27.3-day rotation each unit mass

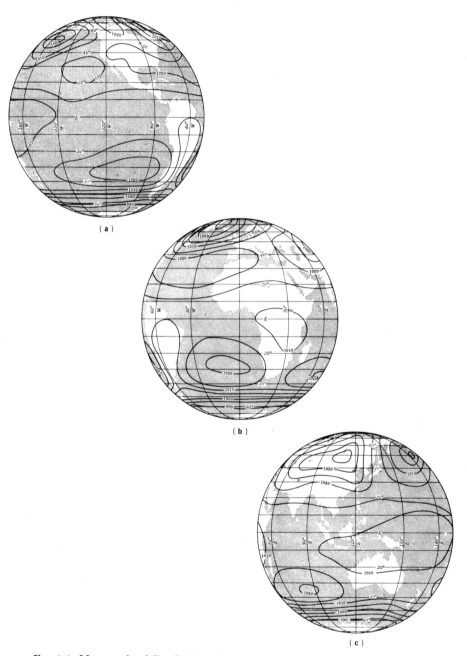

FIG. 1.4. Mean sea level distribution of pressure in millibars for the month of January [after Y. Mintz and G. Dean, *Geophys. Research Paper* **17**, Geophys. Research Dir., Cambridge, Massachusetts, p. 35 (1952)].

within the earth and atmosphere experiences equal and parallel centrifugal forces, and the total of these forces must be equal and opposite to the total gravitational force exerted by the moon. However, gravitational force is inversely proportional to the square of the distance from the moon, so that

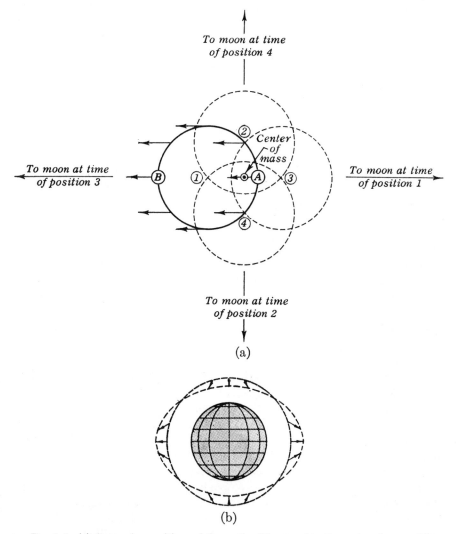

FIG. 1.5. (a) Successive positions of the earth with respect to the center of mass of the earth and moon. Parallel arrows represent centrifugal force when the earth is in position 1.

(b) The resultants of the centrifugal and gravitational forces (the tide generating force) and the consequent distortion of the atmospheric shell.

gravitational force is greater on the side facing the moon than it is on the side away from the moon. The resultant of the two forces, called the tide generating force, is illustrated in Fig. 1.5b.

To formulate the tide generating force we may replace the centrifugal force per unit mass by the gravitational force acting at the distance (D) of the center of the earth from the moon, that is by GM/D^2 where M represents the mass of the moon. The gravitational force acting at point A is $GM/(D - R)^2$ so that the difference is given by

$$GM\left(\frac{1}{D^2} - \frac{1}{(D-R)^2}\right) \approx -\frac{2GMR}{D^3} \tag{1.20}$$

so long as R is much less than D. At point B on the opposite side of the earth the result is the same except for the sign. The numerical value is about 10^{-4} dyn g^{-1}. These forces at A and B act vertically. At any other point the tide generating force and its horizontal component may be calculated in a similar manner; they must be everywhere less than the vertical forces calculated at A and B. This calculation is required in problem 10.

The effect of the tide generating force is to accelerate air toward the sublunar point (A) and toward the point on the earth opposite the sub-lunar point (B). The result is that bulges in the atmosphere tend to form on the sides toward the moon and away from the moon while a depression tends to form between the bulges. Relative to a point on the rotating earth, these bulges or waves must be expected to travel around the earth each day and to produce a semi-diurnal lunar tide. An analogous semi-diurnal solar tide is also to be expected. The two components are superimposed at times of full or new moon and are out of phase at the intervening lunar phases. Amplitudes of both should be largest in low latitudes and least in high latitudes.

Continuous records of sea level pressure indeed reveal both the semi-diurnal lunar and solar pressure waves. But, although the tide generating force exerted by the moon is two and a half times that exerted by the sun, the amplitude of the solar wave is about 1.5×10^3 dyn g^{-1} (1.5 mb) in low latitudes, while the amplitude of the lunar wave is less than 10^2 dyn g^{-1}. This anomaly was formerly thought to result from the fact that the atmosphere possesses a natural resonant period very close to the 12-hr solar wave period; therefore, the 12-hr solar component is amplified while the 12.5-hr (approximate) lunar component is not amplified. But resonance is so critically dependent on the vertical temperature and wind distribution, which is variable in space and time, that it appears unlikely that the large semi-diurnal tide should be attributed to this cause. It remains an interesting question whether the circumstance of resonant tuning is fortuitous or is a natural consequence of planetary history.

In addition to the semi-diurnal tides which result from gravitation, a diurnal solar tide results from the expansion of air which accompanies heating by the sun. Because solar heating varies greatly from place to place, the amplitude of the diurnal tide is highly variable. In general, however, it is comparable to the semi-diurnal amplitude at low latitudes and decreases toward the poles. The dominant component of the semi-diurnal tide probably should be considered as the first harmonic of the diurnal force due to solar heating.

The amplitude of the semi-diurnal pressure wave of course decreases upward from the surface of the earth, but the amplitude of the velocity wave increases markedly above a height of 30 km. In fact, the velocity is so large that, in spite of the low density of the region between 50 and 100 km, more than half the kinetic energy of the tide is contained in this layer. There is evidence of tidal velocities of tens or hundreds of meters per second at these heights, and these have important consequences to the ionospheric electrical current discussed in Section 6.8. At about 30 km the phase of the wave shifts abruptly with height by 180°. Above 100 km the diurnal wave produced by solar heating probably dominates the semi-diurnal wave as the result of viscous damping, but conclusions in this area must be drawn *cum grano salis.*

List of Symbols

		First used in Section
D	Separation of centers of earth and moon	1.10
f_z	Vertical component of frictional force per unit mass	1.8
$\mathbf{F}$	Vector representing force	1.2
g	Scalar force of gravity per unit mass	1.5
$\mathbf{g}$	Vector force of gravity per unit mass	1.5
g^*	Scalar gravitational force per unit mass	1.4
$\mathbf{g}^*$	Vector gravitational force per unit mass	1.4
g_0^*	Scalar gravitational force per unit mass at sea level	1.4
G	Universal gravitational constant	1.2
M, m	Masses	1.2
p	Pressure	1.8
r	Scalar separation of two points	1.2
$\mathbf{r}$	Vector separation of two points	1.2
R	Radius of earth	1.4
t	Time	1.3
T	Total kinetic plus potential energy per unit mass	1.7
v_x	Velocity component toward the east	1.8
v_z	Vertical component of velocity	1.8
v_{es}	Scalar escape velocity	1.7
$\mathbf{v}$	Vector velocity	1.3
x, y	Horizontal coordinates	1.8
z	Height above sea level	1.4

List of Symbols (Continued)

		First used in Section
ρ	Density	1.8
ϕ	Latitude	1.5
Φ	Geopotential	1.6
ω	Scalar angular frequency of the earth	1.5
Ω	Scalar angular frequency of a satellite	1.7

Problems

1. Show that Eq. (1.2) may be transformed to $\tau = d\mathbf{p}/dt$ where τ represents the torque (defined by $\mathbf{r} \times \mathbf{F}$) and $\mathbf{p}$ represents the angular momentum (defined by $m\mathbf{r} \times \mathbf{v}$).

2. Show by integration over the volume of a sphere of uniform density ρ that the gravitational force between the sphere of mass M and an external point mass m is given by Eq. (1.1) where r represents the distance of mass m from the center of the sphere.

3. Find the gravitational force between the sphere of problem 2 and the point mass m if the point mass is located within the sphere of radius R.

4. Find the height above the earth at which the vertical component of the gravitational force due to the earth and the moon vanishes

(a) along a line joining the centers of the earth and moon,

(b) along a line at 30° to the line joining the centers of the earth and moon.

5. Find the angle between $\mathbf{g}^*$ and $\mathbf{g}$ at the surface of the earth as a function of latitude. What is its maximum numerical value, and at what latitude does it occur?

6. Calculate the mass of the earth if the measured value of the force of gravity per unit mass at sea level is 980 dyn g^{-1} at latitude 30°.

7. Find the periods of revolution for satellites in circular orbits at heights of 200 and 600 km above the earth. What are the apparent periods as seen from a point on the earth if the satellites move from west to east around the equator?

8. Calculate the mass of the earth assuming that the moon moves in a circular path about the earth once in 27.3 days at a distance of 384,400 km from the earth. Compare the result with problem 6.

9. Find the net upward force exerted by the atmosphere on a spherical balloon of 1-meter radius if air density is 1.25×10^{-3} g cm^{-3}, force of gravity per unit mass is 980 dyn g^{-1} and the mass of the balloon and its gas is 1.00 kg.

10. Calculate the lunar tide generating force at a point in the atmosphere where the moon appears at an elevation angle of 45°.

General References*

Sears, *Principles of Physics*, Volume 1: *Mechanics Heat and Sound*, gives an excellent introduction to mechanical concepts and to Newton's laws. There are other recent introductory physics texts which are equally good.

---

* General references which provide necessary background or which substantially extend the material presented are listed at the end of each chapter. References are listed in the order of their relevance to the subject matter. Complete references are listed in the Bibliography.

Haltiner and Martin, *Dynamical and Physical Meteorology*, and other texts in meteorological theory treat the hydrostatic equation somewhat more generally than is attempted here.

Joos, *Theoretical Physics*, is one of many classical physics texts which provide a good concise account of the theory of planetary (satellite) motion. The serious student of any branch of physics can hardly be introduced too early to this splendid book.

Siebert, "Atmospheric Tides" (in *Advances in Geophysics*, Volume 7) reviews observational and theoretical evidence relating to the thermal and gravitational sources of atmospheric tides. The account is notably coherent and illuminating and yet calls attention to the important uncertainties which remain.

Wilkes, *Oscillations of the Earth's Atmosphere*, discusses both observations and theory. Although the subject is treated on an advanced level, the gaps which remain in understanding are easily recognized in this brief monograph.

CHAPTER II

Properties of
Atmospheric Gases

"Work expands so as to fill the time (and space)
available for its completion." CYRIL NORTHCOTE PARKINSON

IN THE PREVIOUS CHAPTER we have found it useful to look upon the earth and its atmosphere from the outside; the external point of view has helped to focus attention on certain important macroscopic properties of the atmosphere. In this chapter we shall investigate other properties for which an internal view is more useful. We can imagine the internal view as made by a particularly versatile and powerful microscope, although we realize that our understanding comes from many observations of varied sorts, both direct and indirect.

First, we recognize that the gaseous ocean surrounding the earth consists of countless particles, or molecules, which move about in random motion. The chaos of these motions may be reduced to tractable order by considering the statistical properties of a very large number of molecules. We then may replace the concept of innumerable minute molecules in random motion by the concept of a continuous fluid medium with continuous properties; however, wherever it is useful to do so, these continuous properties can be interpreted in statistical terms. Statistical treatment of molecular behavior falls under the branch of physics known as *statistical mechanics*, and the study of the related continuous macroscopic properties falls under the branch known as *thermodynamics*.

2.1 Molecular Behavior of Gases

The atomic structure of matter was introduced as a philosophical speculation by the Greek philosophers Empedocles and Democritus, and Dalton in 1802 introduced this concept as a scientific hypothesis.

The characteristic behavior of gases involved in chemical reactions inspired Avogadro to suggest in 1811 that equal volumes of different gases at the same pressure and temperature contain equal numbers of molecules. It has been possible to verify this hypothesis and to use it for determining the relative masses of the various atoms. The mass of the carbon twelve nucleus has been assigned the value 12 exactly, and this is used as a standard with which the masses of all atoms are compared.* A quantity of any substance whose mass in grams is equal to its molecular mass is called a *mole*. The volume occupied by a mole of gas at standard atmospheric pressure and 0°C is 22.4 liters [(22413.6 ± 0.6) cm³ mol⁻¹] and is the same for all gases. The number of molecules contained by a mole is therefore a constant and is called *Avogadro's number* (N_0).

Avogadro's number has been determined by several methods. The method which is easiest to understand and which gives a very accurate determination is based upon observations of the electrolysis of solutions. When an electric field is created between electrodes placed in a solution, the positive ions in the solution migrate toward the cathode and the negative ions migrate toward the anode. If each ion carries just one fundamental quantity of charge, the charge of the electron, the number of ions which transfer their charge to the electrodes is equal to the total transferred charge divided by the charge of the electron. By measuring the mass deposited on the electrode, the number of ions deposited per mole (Avogadro's number) may be calculated. The most accurate determination of this number is at present

$$N_0 = (6.02295 \pm 0.00016) \times 10^{23} \text{ mol}^{-1}$$

The empirical evidence that any substance consists of a very large number of molecules has led to the development of several statistical theories designed to explain the macroscopic properties of matter. These theories utilize idealized models of the molecules. A particularly simple model will be discussed in Section 2.3.

2.2 Composition of Air

The atmosphere is composed of a group of nearly "permanent" gases, a group of gases of variable concentration, and various solid and liquid particles. If water vapor, carbon dioxide, and ozone are removed, the remaining gases have virtually constant proportions up to a height of about ninety kilometers. Table 2.1 shows the concentrations of these permanent constituents of air.

From Table 2.1 it is apparent that nitrogen, oxygen, and argon account for 99.997% of the permanent gases in the air. Although all these gases are considered invariant, very small changes in space or time may be observed. The uniformity of the proportions is produced by mixing associated with atmospheric motions. Above a height of 90 km the proportion of the lighter

---

* Adopted by the General Assembly of the International Union of Physics and Applied Physics, September 1960, Ottawa, Canada.

TABLE 2.1[a]

PERMANENT CONSTITUENTS OF AIR

Constituent	Formula	Molecular mass	% by volume
Nitrogen	N_2	28.016	78.110 ± 0.004
Oxygen	O_2	31.9986	20.953 ± 0.001
Argon	Ar	39.942	0.934 ± 0.001
Neon	Ne	20.182	$(18.18 \pm 0.04) \times 10^{-4}$
Helium	He	4.003	$(5.24 \pm 0.004) \times 10^{-4}$
Krypton	Kr	83.80	$(1.14 \pm 0.01) \times 10^{-4}$
Xenon	Xe	131.3	$(0.087 \pm 0.001) \times 10^{-4}$
Hydrogen	H_2	2.016	0.5×10^{-4}
Methane	CH_4	16.043	2×10^{-4}
Nitrous oxide	N_2O	44.015	$(0.5 \pm 0.1) \times 10^{-4}$

[a] Computed using the C^{12} nucleus as the base for molecular mass. From E. Glueckauf, The composition of atmospheric air, *Compendium Meteorol.*, pp. 3–10 (1951).

gases increases with height as diffusion becomes more important relative to mixing. Diffusive equilibrium and the process of diffusion are discussed in Sections 2.6 and 2.16.

The major variable gases in air, taken judiciously from a variety of sources, are listed in Table 2.2. Variations in CO_2 are caused by combustion,

TABLE 2.2

VARIABLE CONSTITUENTS OF AIR

Constituent	Formula	Molecular mass	% by volume
Water	H_2O	18.005	0 to 7
Carbon dioxide	CO_2	44.009	0.01 to 0.1 (near the ground) average 0.032
Ozone	O_3	47.998	0 to 0.01
Sulfur dioxide	SO_2	64.064	0 to 0.0001
Nitrogen dioxide	NO_2	46.007	0 to 0.000002

absorption and release by the oceans, and by photosynthesis; these variations have a significant effect on atmospheric absorption and emission of infrared radiation. Variations in water vapor and ozone are important in processes which are discussed in later sections. Other gases not included in Table 2.2 are chiefly products of combustion and are found in the atmosphere in variable concentration.

The solid and liquid particles which are suspended in the atmosphere (the *aerosol*) play an important role in the physics of clouds. The distribution and properties of these particles are discussed in Sections 3.4 and 3.8.

2.3 Elementary Kinetic Theory

A simple theory of dilute gases based on classical mechanics may be developed by assuming that the gas consists of molecules which have their mass concentrated in very small spherical volumes and which collide with each other and with the walls surrounding them without loss of kinetic energy. The molecules are assumed to exert no forces on each other except when they collide. With this model it is possible quite accurately to account for the behavior of the permanent gases under atmospheric pressure and temperature. The great power of kinetic theory, which goes far beyond the elementary theory developed here, may be attributed to the fact that, whereas thermodynamics is concerned with equilibrium states only, kinetic theory recognizes no such limitation.

Consider N molecules in a rectangular volume V. The molecules are separated by distances large compared with their own diameters. Between collisions they move in straight lines with constant speed. Since the walls are considered perfectly smooth, there is no change of tangential velocity in a collision with the walls. If it is assumed that the molecules are distributed uniformly, the number of molecules in any volume element is

$$dN = n \, dV \qquad (2.1)$$

where n represents the number of molecules per unit volume. It is also assumed that the distribution of molecular velocities is the same in all directions. It will be shown in Section 2.15 that these last two assumptions are valid for the condition of maximum probability. Problem 1 at the end of the chapter concerns the estimation of a smallest volume increment for which the assumption of a uniform distribution still is true.

To find an expression for the number of molecules striking a wall per unit area and per unit time, consider the small element dA of the wall sketched in Fig. 2.1 and construct the normal to the element and a reference plane through the normal. We now ask how many molecules traveling in the particular direction θ, ϕ and with specified speed v strike the surface in time dt. This means that we consider all cases between θ and $\theta + d\theta$, ϕ and $\phi + d\phi$, v and $v + dv$. Now construct the cylinder of length $v \, dt$ whose axis lies in the direction θ, ϕ as shown in the figure. All the molecules with the specified speed and direction in the so-constructed cylinder hit the wall in time dt, and also all the molecules with this speed and direction that hit the wall are contained by the cylinder.

The next question is: How many of these molecules are there in the

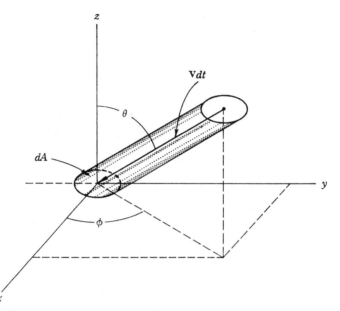

Fig. 2.1. Volume element containing all molecules that hit the area dA in the time interval dt coming from the direction ϕ, θ with speed v.

cylinder? Equation (2.1) shows that the total number is $nv \, dt \, dA \cos \theta$. If dn_v represents the total number of molecules with speeds between v and $v + dv$ per unit volume, the number of molecules with the required speed in the cylinder is $dn_v \, v \, dt \, dA \cos \theta$. These molecules are distributed uniformly over all directions, and, with the aid of Appendix I.H, the number with speed v in the direction ϕ, θ may be expressed by

$$v \, dn_v \, dt \, dA \, \cos \theta \, \frac{\sin \theta \, d\theta \, d\phi}{4\pi}$$

This number is the same for each cylinder with an angle θ, so integration over ϕ yields the number of molecules hitting the wall in area dA and in time dt from all directions defined by the ring between θ and $\theta + d\theta$. The result of this integration is

$$\tfrac{1}{2}v \, dn_v \sin \theta \cos \theta \, d\theta \, dt \, dA \qquad (2.2)$$

The number of molecules hitting the area dA in time dt is found by integrating from $\theta = 0$ to $\theta = \pi/2$ to be

$$\tfrac{1}{4}v \, dn_v \, dt \, dA$$

Consequently, the molecules with speeds between v and $v + dv$ experience

$$\tfrac{1}{4}v \, dn_v$$

collisions with the wall per unit area per unit time. The total number of collisions per unit area and per unit time is obtained by integrating over the entire range of speeds. If the average speed is defined by

$$\bar{v} \equiv \frac{1}{n} \int_0^n v \, dn_v \tag{2.3}$$

then the total number of collisions per unit area per unit time may be written in the form

$$\tfrac{1}{4} n \bar{v}$$

In problem 2 an expression is developed for the total number of collisions per unit area per unit time per unit solid angle for molecules of all velocities.

2.4 Equation of State of an Ideal Gas

A molecule which collides with the wall experiences a change in momentum. If m represents the mass of the molecule and θ the angle of its direction with the vertical, the momentum change illustrated in Fig. 2.2 is given by

$$mv \cos \theta - (-mv \cos \theta) = 2mv \cos \theta$$

Therefore, it follows from Eq. (2.2) that the change in momentum due to all collisions taking place per unit area per unit time coming from molecules with directions defined by the ring between θ and $\theta + d\theta$ is expressed by

$$mv^2 \, dn_v \sin \theta \cos^2 \theta \, d\theta$$

and, upon integrating between 0 and $\pi/2$, the change in momentum from collisions coming from all directions is given by

$$\tfrac{1}{3} mv^2 \, dn_v$$

Finally, the change of momentum per unit time may be equated to the force exerted on the surface dA. The force is then expressed by

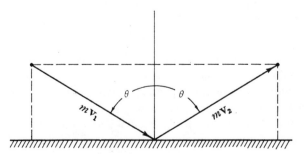

FIG. 2.2. The momentum vectors of a moving molecule before and after making a perfectly elastic collision with a plane wall.

$$dF = \tfrac{1}{3}m \left(\int_0^n v^2 \, dn_v \right) dA$$

and the pressure exerted on the surface is

$$p \equiv \frac{dF}{dA} = \tfrac{1}{3}m \int_0^n v^2 \, dn_v$$

This may be written in the form

$$p = \tfrac{1}{3}mn\overline{v^2} \tag{2.4}$$

where the average value of the square of the speed is defined by

$$\overline{v^2} \equiv \frac{1}{n} \int_0^n v^2 \, dn_v \tag{2.5}$$

Upon eliminating n from Eq. (2.4)

$$pV = \tfrac{1}{3}mN\overline{v^2} \tag{2.6}$$

Equation (2.6) may be written in the form

$$p\alpha_m = \tfrac{1}{3}mN_0\overline{v^2} \tag{2.7}$$

where α_m represents the volume containing one mole of the gas, the *molar specific volume*.

We now define the *temperature* as proportional to the translational kinetic energy of the molecules, and write

$$\tfrac{1}{2}m\overline{v^2} \equiv \tfrac{3}{2}kT \tag{2.8}$$

where the factor $\tfrac{3}{2}k$ is the constant of proportionality, and k is known as the *Boltzmann constant*. Upon substituting Eq. (2.8) into (2.7)

$$p\alpha_m = kN_0T$$

which is the *equation of state* for an *ideal gas*. This equation is usually written in the form

$$p\alpha_m = RT \tag{2.9}$$

where R, the *gas constant*, replaces kN_0. The Boltzmann constant (k) may be interpreted as the gas constant for a single molecule. Avogadro's number (N_0), the Boltzmann constant (k), and the gas constant (R) are universal, that is, they are the same for any ideal gas.

It is also useful to express the equation of state in terms of *specific volume* (α) rather than the molar specific volume (α_m). By definition

$$\alpha \equiv \frac{\alpha_m}{M}$$

where M represents molecular mass. Therefore, Eq. (2.9) may be written

$$p\alpha = \frac{R}{M}\,T \equiv R_m T \tag{2.10}$$

where R_m is the specific gas constant for a gas with molecular mass M.

The numerical value of R may be calculated using the results of experiment. A specific mass of a gas brought into contact with melting ice at standard atmospheric pressure always reaches the same volume after a long exposure. Similar behavior is observed if a volume of gas is brought into contact with boiling water at standard atmospheric pressure. In each case the average kinetic energy of the gas molecules approaches the average kinetic energy of the contacting molecules, and it is recognized that the systems in contact have uniform temperature and are in equilibrium. It further has been observed that whenever two different masses of gas (or *systems*) are brought into equilibrium with melting ice, they are also in equilibrium with each other. This is a general thermodynamic principle called the *zeroth law of thermodynamics*.

Now, Eq. (2.9) predicts that at constant pressure a volume of an ideal gas varies linearly with T and that the constant of proportionality is the same for all ideal gases. By comparing volumes of various gases at the steam point of water (s) with their volumes at the ice point (i), it is found that the ratio V_s/V_i is constant for all the permanent gases. If we arbitrarily assign to the difference in temperature between the ice and steam points the value 100 degrees, the equation of state gives

$$\frac{V_s}{V_i} = \frac{T_s}{T_0} = \frac{T_0 + 100}{T_0}$$

Solving for T_0 it is found that the temperature of the ice point on this scale, referred to as the absolute or Kelvin scale, is

$$T_0 = (273.155 \pm 0.015)\ {}^\circ\text{K}$$

By subtracting $273.155{}^\circ\text{K}$ from the absolute temperature, the Celsius temperature is defined. Thus, on the Celsius scale the ice and steam points are at $0{}^\circ\text{C}$ and $100{}^\circ\text{C}$, respectively. It is now possible to determine from Eq. (2.9) that R is given by

$$R = (8.31432 \pm 0.00034) \times 10^7\ \text{erg mol}^{-1}{}^\circ\text{K}^{-1}$$

It follows that $k = (1.38044 \pm 0.00007) \times 10^{-16}\ \text{erg }{}^\circ\text{K}^{-1}$. A definition of temperature which is independent of kinetic theory has been given by Kelvin using the Second Law of Thermodynamics.

The root mean square (rms) speed of the molecules may now be calculated from Eq. (2.8) for any temperature if the mass of the molecules and the Boltzmann constant are known. In this way the rms speed for nitrogen molecules in air at 280°K is found to be 500 m sec⁻¹. This speed is proportional to the square root of the temperature and inversely proportional to the square root of the mass. In problem 3 it is required to relate this information to the escape velocity.

The kinetic theory developed so far only gives an adequate equation of state for the special group of permanent gases. All other substances apparently obey more complicated equations of state. This is not surprising because the molecular model used as the basis of this theory is particularly simple. In Section 2.17 a more complicated model will be described in relation to real gases.

The equation of state is the corner stone of thermodynamics. The development just completed emphasizes that the equation of state refers to a system in *equilibrium*, that is, to a system in which every statistical sample of molecules exhibits the same average condition. The pressure, specific volume and temperature are called the *state variables*. It is clear that two state variables define the state of a system, and the endeavor of thermodynamics is to describe the characteristic properties of matter during various processes entirely in terms of the state variables. It is evident that only a special class of processes can be described, processes that are so gradual that at each instant the state of the system can be defined. In other words, the process evolves from one equilibrium state to the next. This type of process has the characteristic that it is *reversible*. Processes that involve nonequilibrium conditions of the system are *irreversible* processes and, in general, cannot be described by thermodynamics. Although the limitations of thermodynamics in describing the properties of matter are severe, the number of atmospheric processes that can be described by thermodynamics should not be underestimated.

2.5   The Velocity Distribution of Molecules

In the previous section we have found it possible to derive the equation of state utilizing only the average kinetic energy of the molecules. Certain properties of a gas in equilibrium depend not only on the mean molecular energy but also on the distribution of energies (the proportion of molecules with each particular energy) and the associated velocity distribution.

The distribution of a single velocity component, say v_x, can be represented in normalized form by

$$\phi(v_x) \equiv \frac{1}{n}\frac{dn}{dv_x} \tag{2.11}$$

Integration yields

$$\int_{-\infty}^{\infty} \phi(v_x)\, dv_x = 1 \tag{2.12}$$

Similarly, velocity distributions can be defined in the y and z directions. It is intuitively clear that for a gas in equilibrium the velocity distribution must be independent of the direction in which the components are chosen, therefore

$$\phi(v_x) = \phi(v_y) = \phi(v_z) \quad \text{for} \quad v_x = v_y = v_z$$

Furthermore, ϕ should be an even function of the velocity components, so that

$$\phi(-v_x) = \phi(v_x), \text{ etc.}$$

For simplicity only a two-dimensional velocity distribution is considered; later, the result will be generalized to three dimensions.

It follows from the point just made that all the molecules with an x velocity component lying between v_x and v_{x+dx} have the same distribution in the y direction as all the molecules in the system. Using Eq. (2.11) the number of molecules with velocity components between v_x and v_{x+dx} and between v_y and v_{y+dy}, as shown in Fig. 2.3, may be expressed by

$$d^2n = n \ \phi(v_x) \ \phi(v_y) \ dv_x \ dv_y$$

But the distribution is independent of the coordinate system, so that in the system defined by the axis connecting O and P in Fig. 2.3 the same number may be expressed by

$$d^2n = n \ \phi(v_2) \ \phi(0) \ dv_x \ dv_y$$

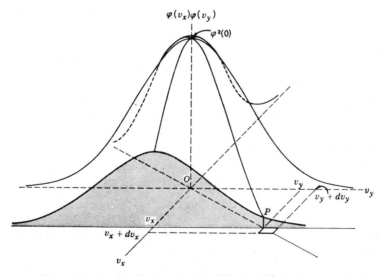

FIG. 2.3. The molecular distribution function [Eq. (2.13)] for the two velocity components, v_x and v_y.

where $v_2{}^2 \equiv v_x{}^2 + v_y{}^2$. Because the intersections of the function with vertical planes, as shown in Fig. 2.3, are geometrically similar, it follows that

$$\phi(v_2)\phi(0) = \phi(v_x)\phi(v_y) \tag{2.13}$$

This functional equation can be solved by first differentiating with respect to v_x and next differentiating with respect to v_y. From the two relations so obtained there results

$$\frac{1}{v_x}\frac{\phi'(v_x)}{\phi(v_x)} = \frac{1}{v_y}\frac{\phi'(v_y)}{\phi(v_y)}$$

where the prime denotes the derivative. The left-hand side of this equation is independent of v_y and the right-hand side is independent of v_x. This is only possible when each side is equal to the same constant, say $-\lambda$. Therefore

$$\frac{1}{v_x}\frac{\phi'(v_x)}{\phi(v_x)} = -\lambda$$

Integration of this equation yields

$$\phi(v_x) = A \exp\left[-(\lambda/2)v_x{}^2\right] \tag{2.14a}$$

Similarly

$$\phi(v_y) = A \exp\left[-(\lambda/2)v_y{}^2\right] \tag{2.14b}$$

and

$$\phi(v_x{}^2 + v_y{}^2)^{1/2} = A \exp\left[-(\lambda/2)(v_x{}^2 + v_y{}^2)\right] \tag{2.14c}$$

The constant of integration is found by substituting Eqs. (2.14) into (2.13), which yields

$$A = \phi(0)$$

The value of $\phi(0)$ is found by substituting one of Eqs. (2.14) into (2.12) and carrying out the integration. Therefore

$$\phi(0) \int_{-\infty}^{\infty} \exp\left(-\frac{\lambda}{2}v_x{}^2\right) dv_x = \phi(0) \int_{-\infty}^{\infty} \exp\left(-\frac{\lambda}{2}x^2\right) dx = 1$$

and because $\int_{-\infty}^{\infty} \exp\left[-(\lambda x^2/2)\right] dx = (2\pi/\lambda)^{1/2}$

$$\phi(0) = \left(\frac{\lambda}{2\pi}\right)^{1/2} \tag{2.15}$$

It remains to determine λ; this may be accomplished by relating λ to $\overline{v^2}$. To do this Eqs. (2.14) must be extended to the three-dimensional case by adding an equation expressing the distribution for $\phi(v_z)$. Upon carrying out once again the steps that led to Eq. (2.13), it is found that

$$\phi(v)\phi^2(0) = \phi(v_x)\phi(v_y)\phi(v_z)$$

where $v^2 = v_x^2 + v_y^2 + v_z^2$. The corresponding number of molecules in a volume increment in three dimensional momentum space is

$$d^3n = n \ \phi(v_x) \ \phi(v_y) \ \phi(v_z) \ dv_x \, dv_y \, dv_z$$
$$= n \ \phi(v) \ \phi^2(0) \ dv_x \, dv_y \, dv_z$$

or, upon transforming to spherical coordinates and integrating over a spherical shell, the number of molecules in this shell is

$$dn = 4\pi n v^2 \phi^2(0)\phi(v) \ dv \tag{2.16}$$

From definition (2.5) the average of the square of the velocities is given by

$$\overline{v^2} = 4\pi\phi^2(0) \int_0^\infty v^4\phi(v) \ dv$$

and, upon substituting Eqs. (2.14) and (2.15) into this equation

$$\overline{v^2} = 4\pi \left(\frac{\lambda}{2\pi}\right)^{3/2} \int_0^\infty v^4 \exp\left(-\frac{\lambda}{2} v^2\right) dv = \frac{3}{\lambda}$$

This result may be combined with Eq. (2.8) applied to molecules of identical mass with the result

$$\lambda = \frac{m}{kT}$$

If this is substituted back into the distribution function (2.14), the one-dimensional distribution is found to be

$$\phi(v_x) = \left(\frac{m}{2\pi kT}\right)^{1/2} \exp\left(-\frac{mv_x^2}{2kT}\right)$$

or, from Eq. (2.11)

$$\frac{dn}{dv_x} = n\left(\frac{m}{2\pi kT}\right)^{1/2} \exp\left(-\frac{mv_x^2}{2kT}\right) \tag{2.17}$$

The three-dimensional velocity distribution is obtained by substituting Eq. (2.17) into (2.16) with the result

$$\frac{dn}{dv} = 4\pi n v^2 \left(\frac{m}{2\pi kT}\right)^{3/2} \exp\left(-\frac{mv^2}{2kT}\right) \tag{2.18}$$

Equations (2.17) and (2.18) are represented graphically in Fig. 2.4. An application of the velocity distribution is given in problem 4. Figure 2.4 shows that substantial proportions of the molecules have kinetic energies very different from the mean value and that the mean energy is greater than the most probable energy.

2.6 The Atmosphere in Equilibrium

We are now in a position to investigate the effect of the earth's field of gravity on the energy or velocity distribution of the molecules and on the

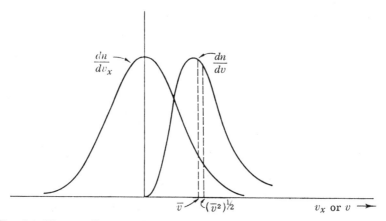

Fɪɢ. 2.4. The one-dimensional velocity distribution (v_x) and the three-dimensional velocity distribution (v) represented by Eqs. (2.17) and (2.18), respectively.

height dependence of density or pressure. For this purpose substitute the equation of state (2.10) into the hydrostatic equation (1.18), yielding

$$\frac{\partial p}{\partial \Phi} = -\frac{p}{R_m T} \tag{2.19}$$

For the case of temperature independent of height (isothermal case) Eq. (2.19) can be integrated immediately yielding the vertical pressure distribution in the form known as the *barometric equation*

$$p = p(0) \exp \left(-\Phi/R_m T \right)$$

and, by using again the equation of state, the vertical density distribution is found in the form

$$\rho = \rho(0) \exp \left(-\Phi/R_m T \right) \tag{2.20}$$

The question arises now whether an isothermal atmosphere is also an atmosphere in equilibrium. The molecules traveling upward in the field of gravity lose kinetic energy and gain potential energy and vice versa, and this suggests that the conditions of isothermal temperature distribution and velocity distribution independent of height may not be compatible.

Because the density is proportional to the number of molecules per unit volume, Eq. (2.20) requires that

$$n = n(0) \exp \left(-\Phi/R_m T \right) = n(0) \exp \left(-m\Phi/kT \right) \tag{2.21}$$

The form of the exponent is similar to the exponent in the velocity distribution function [Eq. (2.17)]. The difference is that in Eq. (2.21) the potential energy takes the place of the kinetic energy, $\frac{1}{2}mv^2$. Consider now two levels of constant geopotential, Φ_1 and Φ_2, so close together that collisions occurring between the planes may be neglected. The number of molecules passing through level Φ_1 per unit area per unit time with velocities between v_z and $v_z + dv_z$ is expressed by Eq. (2.17) as

$$v_z(\Phi_1) \, dn_1 = n_1 v_z(\Phi_1) \left(\frac{m}{2\pi k T_1}\right)^{1/2} \exp\left(-\frac{m v_z^2(\phi_1)}{2 k T_1}\right) dv_{z1} \qquad (2.22)$$

The same number of molecules pass through level Φ_2. Because the molecules form part of the distribution at the new level, the number is expressed by

$$v_z(\Phi_1) \, dn_1 = v_z(\Phi_2) \, dn_2 = n_2 v_z(\Phi_2) \left(\frac{m}{2\pi k T_2}\right)^{1/2} \exp\left(-\frac{m v_z^2(\Phi_2)}{2 k T_2}\right) dv_{z2} \qquad (2.23)$$

where the index 2 refers to the level Φ_2. The z component of the velocity has been affected by the field of gravity in going from level Φ_1 to Φ_2. Therefore

$$\tfrac{1}{2} v_z^2(\Phi_2) = \tfrac{1}{2} v_z^2(\Phi_1) - \Delta\Phi \qquad (2.24)$$

where $\Delta\Phi = \Phi_2 - \Phi_1$. Differentiation yields

$$v_z(\Phi_2) \, dv_{z2} = v_z(\Phi_1) \, dv_{z1} \qquad (2.25)$$

If Eqs. (2.24) and (2.25) are substituted into Eq. (2.23)

$$v_z(\Phi_1) \, dn_1 = n_2 \exp\left(\frac{m\Delta\Phi}{k T_2}\right) v_z(\Phi_1) \left(\frac{m}{2\pi k T_2}\right)^{1/2} \exp\left(-\frac{m v_z^2(\Phi_1)}{2 k T_2}\right) dv_{z1}$$

Now, provided $T_2 = T_1$ and $n_2 = n_1 \exp(-m\Delta\Phi/kT_1)$, this equation reduces to Eq. (2.22). Therefore, compression of the gas with decreasing height occurs at the expense of potential energy only, and an atmosphere in equilibrium is also an atmosphere with uniform temperature. This conclusion holds for a gas in equilibrium in any potential field.

As a consequence of Dalton's law, which states that under equilibrium conditions the partial pressure of each component of a mixture of gases is the same as the pressure exerted by that gas if it occupied the volume alone, the hydrostatic equation may be integrated independently for each constituent of a mixture of gases. The partial densities of two constituents are then represented by

$$\rho_1 = \rho_1(0) \exp(-\Phi/R_{m1}T) \qquad \rho_2 = \rho_2(0) \exp(-\Phi/R_{m2}T)$$

where the indices 1, 2 refer to the two constituents. Additional similar equations may be written for other constituents. From a series of these equations the height distribution of each constituent may be calculated easily. The heavy gases have highest concentrations near the ground, and the light gases are most abundant at higher levels. An exercise concerning *diffusive equilibrium* is given in problem 5.

Diffusive equilibrium is not found in the lowest 90 kilometers of the atmosphere because turbulent mixing maintains constant proportions of the permanent gases. However, the tendency toward diffusive equilibrium becomes increasingly important with increasing height above 90 kilometers.

2.7 Conservation of Energy

The Principle of Conservation of Energy states that *it is not possible to create or destroy energy*. This is an independent fundamental principle which is, like the other fundamental principles already introduced, based upon intuitive insight into nature. It cannot be derived from other principles or laws.

The Principle of Conservation of Energy requires, however, that we think in particular ways about transformations of energy from one form to another. We must recognize that two systems can exchange energy through work done by one on the other and by heat transfer, and that the energy transferred from one by these transfer processes (work and heat) must appear as energy in the other. This insight developed gradually during the first half of the nineteenth century as the result of observations from widely different fields. Rumford in 1798 drew attention to the concepts later to be known as work, heat, and internal energy as a result of observations of cannon boring, R. J. Mayer in 1842 from physiological observations made the first clear statement of the principle, and Joule in 1847 determined accurately from laboratory measurements the mechanical equivalent of heat, thereby verifying the principle.

Energy may take many different forms: kinetic, internal, potential (of various kinds), chemical, nuclear, etc. We shall consider only kinetic energy of the mean motion, kinetic energy of the molecular motions called internal energy, and the potential energy associated with the field of gravity, and we shall assume that other forms of energy do not change during atmospheric processes. The total energy is the sum of these three energies, and change in the specific total energy is expressed by

$$de_t = du + d\Phi + \tfrac{1}{2}dv^2 \qquad (2.26)$$

where u represents internal energy per unit mass.

The Principle of Conservation of Energy requires that the change in total energy equals the sum of heat transferred to the system and work done on the system, and it may be written in the form

$$\boxed{de_t = dh + dw} \qquad (2.27)$$

where h represents the heat added per unit mass and w work done per unit mass. There are about as many books in which w is defined as positive for work done *by* the system as for work done *on* the system. The resulting confusion is not necessarily serious.

To evaluate the work done on the system (dW) it is most convenient to consider the rate at which work is done on a parcel enclosed at a particular instant within the boundaries shown in Fig. 2.5. The rate of work done at each face may be represented by the product of the normal velocity and the

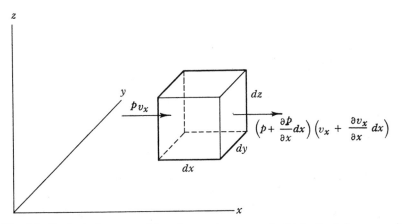

FIG. 2.5. Product of pressure (p) and normal component of velocity (v_x) on opposite faces of a differential fluid element.

pressure at the face. Therefore, upon expanding the pressure and velocity on the right y, z face in a Taylor series, the rate of work done at the two y, z faces is

$$\frac{dW_x}{dt} = p(x, t)v_x(x, t)\, dy\, dz - \left[\left(p + \frac{\partial p}{\partial x}\, dx\right)\left(v_x + \frac{\partial v_x}{\partial x}\, dx\right)\right] dy\, dz$$

or

$$\frac{dW_x}{dt} = -\left(p \frac{\partial v_x}{\partial x} + v_x \frac{\partial p}{\partial x}\right) dx\, dy\, dz$$

This equation may be extended to three dimensions by writing similar expressions for the other faces and adding. This yields for the total rate of work done on the air parcel

$$\frac{dW}{dt} = -(p\nabla \cdot \mathbf{v} + \mathbf{v} \cdot \nabla p)\, dx\, dy\, dz$$

and, upon dividing by the mass of the parcel, the work rate per unit mass or the *specific work* rate is

$$\frac{dw}{dt} = -\frac{1}{\rho}(p\nabla \cdot \mathbf{v} + \mathbf{v} \cdot \nabla p) \tag{2.28}$$

The use of the vector operator ∇ is discussed in Appendix I.B. The first term on the right represents rate of work done by compression (external work), and the second term represents rate of work done by the pressure field (internal work). Equation (2.28) is a complete statement of the rate of work done on the parcel. However, it proves to be useful to express (2.28) in an alternate form which is reached in Section 2.9 after an intermezzo in Section 2.8.

2.8   Conservation of Mass

The last of the great conservation principles states that *the mass of a closed system remains constant,* and this applies to all systems for which mass-energy transformations are unimportant. The same principle applied to an open system may be recognized as stating that *the net mass flux into the system equals the rate of increase of mass of the system.* This is expressed by

$$\oint \rho \mathbf{v} \cdot \mathbf{n} \, dA = - \iiint \frac{\partial \rho}{\partial t} \, dx \, dy \, dz \qquad (2.29)$$

where $\mathbf{n}$ represents the normal unit vector outwardly directed from the surface. In Appendix I.G the surface and volume integrals are related by Gauss' theorem in the form

$$\oint \rho \mathbf{v} \cdot \mathbf{n} \, dA = \iiint \nabla \cdot (\rho \mathbf{v}) \, dx \, dy \, dz \qquad (2.30)$$

Equations (2.29) and (2.30) may be combined for any volume; therefore

$$\boxed{\nabla \cdot (\rho \mathbf{v}) = -\frac{\partial \rho}{\partial t}} \qquad (2.31)$$

Equation (2.31) is known as the Equation of Continuity. It may be applied to any *conservative* scalar quantity. The left-hand side may be expanded to $\rho \nabla \cdot \mathbf{v} + \mathbf{v} \cdot \nabla \rho$; and, upon recognizing the expansion of the total derivative, Eq. (2.31) may also be written

$$\nabla \cdot \mathbf{v} = -\frac{1}{\rho} \frac{d\rho}{dt} \qquad (2.31a)$$

By comparing Eqs. (2.28) and (2.31a) it is evident that the expansion work is proportional to the individual rate of density change.

2.9   First Law of Thermodynamics

The rate of work done on the parcel may now be expressed by introducing Eq. (2.31a) into (2.28) in the form

$$\frac{dw}{dt} = \frac{p}{\rho^2}\frac{d\rho}{dt} - \frac{1}{\rho}\mathbf{v}\cdot\nabla p$$

or

$$\frac{dw}{dt} = -p\frac{d\alpha}{dt} - \alpha\mathbf{v}\cdot\nabla p \qquad (2.32)$$

which is, like Eq. (2.28), a general expression for the rate of work done. The second term is negligible when the pressure is uniform (in most laboratory work, for example) but, in general, work done by the atmospheric pressure field is not negligible.

If Eqs. (2.26) and (2.32) are introduced into (2.27), the statement of conservation of energy becomes

$$dh = du + d\Phi + \tfrac{1}{2}dv^2 + p\,d\alpha + \alpha\mathbf{v}\cdot\nabla p\,dt \qquad (2.33)$$

This equation suggests that there are many possible energy transformations. However, the number of possible transformations is limited by Newton's Second Law and the Second Law of Thermodynamics. If Newton's Second Law in the form of Eq. (1.19) is multiplied by v_z and if friction is neglected, the rates of change of kinetic and potential energy are related to the rate at which work is done by the pressure field according to

$$\frac{1}{2}\frac{dv_z^2}{dt} + \frac{d\Phi}{dt} = -\alpha v_z \frac{\partial p}{\partial z}$$

This equation may be generalized for the three-dimensional case to the form

$$\tfrac{1}{2}dv^2 + d\Phi = -\alpha\mathbf{v}\cdot\nabla p\,dt$$

This equation states that the sum of kinetic and potential energies is changed only by virtue of the work done by the pressure field. Each term in this equation appears in Eq. (2.33), and combination therefore yields

$$dh = du + p\,d\alpha \qquad (2.34)$$

This result is called the First Law of Thermodynamics. It may be used as the definition of heat added to the system per unit mass.

The internal energy is by definition the kinetic energy of molecular motions of the molecules, and therefore is proportional to the temperature. If the volume is held constant, $d\alpha = 0$, and all the heat added to the system is used for the increase of internal energy. For the ideal gas discussed earlier, Eq. (2.8) shows that

$$dh = du = \tfrac{1}{2}mn^*\,d(\overline{v^2}) = \tfrac{3}{2}n^*k\,dT \qquad (2.35)$$

where n^* represents the number of molecules per unit mass. The relation between internal energy and temperature will be further discussed in Sections 2.10 and 2.11.

The last term in Eq. (2.34), the expansion work, can be represented in a graph of p versus α as illustrated in Fig. 2.6. If the state is changed from a to b along curve 1, then the area under the curve represents the expansion work done by the system

$$\int_a^b p \, d\alpha$$

If the state now is changed from b to a along curve 2, then an amount of work is done on the system which corresponds to the area under this new

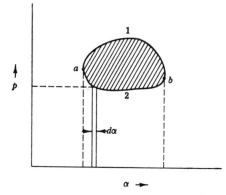

FIG. 2.6. Work represented on the pressure-specific volume (p–α) diagram.

curve. The area enclosed by the two paths, 1 from a to b and 2 from b to a, represents the work per unit mass performed by the system in a complete cycle. The complete integration described is called the *line integral* around a closed path and is written

$$\oint p \, d\alpha$$

An example is provided by problem 6.

2.10   Equipartition of Energy

The idealized point masses utilized in Sections 2.3 and 2.4 exhibit energy which may be specified by the three degrees of freedom corresponding to the three velocity components. If Eq. (2.8) is written in the form

$$\tfrac{1}{2}m(v_x^2 + v_y^2 + v_z^2) = \tfrac{3}{2}kT$$

it is clear that the average energy per degree of freedom is $\tfrac{1}{2}kT$. From the discussion of distribution of molecular velocities in Section 2.5 it is also clear that the total energy must be equally divided among the three degrees of freedom. This is a special case of *equipartition of energy.*

In the case of a more complex molecule whose energy is shared by rotational as well as translational motion, the molecule is said to possess more than three degrees of freedom. The same intuitive statement made in Section 2.5 in establishing the uniform distributions of velocity components may be extended to include the rotational degrees of freedom. It follows that the *total energy is equally divided among each degree of freedom, and each component of energy is equal to $\frac{1}{2}kT$.* This is a statement of the law of equipartition of energy. The law holds accurately for molecules which obey classical mechanics; it *does not* apply to degrees of freedom corresponding to internal vibrations, in which case quantum mechanical interactions are important. It now may be recognized that the general form of the dependence of internal energy change on temperature change is, in place of Eq. (2.35),

$$du = \tfrac{1}{2}jn^*k \, dT \tag{2.36}$$

where j represents the number of degrees of freedom. Whereas a monatomic molecule has three translational degrees of freedom, a diatomic molecule has, in addition, two degrees of freedom corresponding to rotation about the two axes perpendicular to the line connecting the atoms. Similarly, a triatomic molecule possesses three rotational degrees of freedom. The vibrational energies of air molecules are usually negligible.

2.11 Specific Heat

The heat required to raise the temperature of a unit mass of a substance by one degree is called the *specific heat* of that substance; it is defined by

$$c \equiv \frac{dh}{dT}$$

This definition is not sufficient to determine a unique value for c, because the circumstances under which the heat is supplied to the system are not yet specified. It may be that the system expands or contracts or maintains a constant volume. In each of an infinite number of possible cases a different value for the specific heat would be found. In the case of an ideal gas Eqs. (2.34) and (2.36) lead to the general form

$$c = \frac{\tfrac{1}{2}jn^*k \, dT + p \, d\alpha}{dT} \tag{2.37}$$

It is convenient to distinguish between the specific heat at constant volume (c_v) and the specific heat at constant pressure (c_p). It follows from Eq. (2.37) that the *specific heat at constant volume* is defined by

$$c_v \equiv \tfrac{1}{2}jn^*k$$

Upon combining this with Eq. (2.36) the differential of internal energy is

$$du = c_v \, dT \qquad (2.38)$$

and because $n^*k = R_m$

$$c_v = \tfrac{1}{2}jR_m \qquad (2.39)$$

The *specific heat at constant pressure* can be determined if $p \, d\alpha$ is eliminated from Eq. (2.37) by use of the equation of state and if pressure is then held constant. This yields

$$c_p = (\tfrac{1}{2}j + 1)R_m \qquad (2.40)$$

Two useful relations can be derived from Eqs. (2.39) and (2.40). Subtraction gives

$$c_p - c_v = R_m \qquad (2.41)$$

and division gives

$$\frac{c_p}{c_v} = \frac{j+2}{j}$$

The physical properties of molecules discussed so far are summarized for the major constituents of the atmosphere in Table 2.3. From this table it is clear that the specific heat as computed from Eqs. (2.39) and (2.40) corresponds closely to the observed values.

TABLE 2.3

MOLECULAR PROPERTIES OF ATMOSPHERIC GASES

Gas	Degrees of Freedom			J g^{-1} °K^{-1}			c_p/c_v (obs., 18–25°C)[a]
	trans.	rot.	vib.	R_m	c_p	c_v	
Monatomic gases							
Helium (He)	3			2.076	5.226	3.148	1.63
Argon (Ar)	3			0.2081	0.531	0.318	1.648
Diatomic gases							
Hydrogen (H$_2$)	3	2		4.124	14.058	9.970	1.407
Nitrogen (N$_2$)	3	2		0.2967	1.004	0.7460	1.401
Oxygen (O$_2$)	3	2		0.2598	1.05	0.750	1.396
Triatomic gases							
Water vapor (H$_2$O)	3	3	0.3	0.4617	1.912	1.450	[1.32]
Carbon dioxide (CO$_2$)	3	3	0.3	0.1889	0.822	0.632	1.293
Ozone (O$_3$)	3	3		0.1732			
Dry air				0.2870			1.402

[a] J. D'Ans and E. Lax, "Taschenbuch fur Chemiker und Physiker." Springer, Berlin, p. 1052, 1943.

2.12 Entropy

From the discussion of Section 2.9 it follows that work done by a system depends on the process it goes through and not alone on the initial and final states of the system; therefore dw is not expressible as an exact differential by expansion in terms of the state variables. The same consideration holds for heat. In Appendix I.E properties of the exact differential are discussed.

In order to transform the First Law of Thermodynamics [Eq. (2.34)] to a form containing only exact differentials, divide this equation by the temperature and eliminate the pressure by use of the equation of state. This yields

$$\frac{dh}{T} = c_v \frac{dT}{T} + R_m \frac{d\alpha}{\alpha} \qquad (2.42)$$

The right-hand side of Eq. (2.42) is the sum of two exact differentials, so it follows that the left-hand side is also an exact differential. It is convenient, therefore, to define the differential of the *specific entropy* by

$$\frac{dh}{T} \equiv ds \qquad (2.43)$$

For any cyclic process

$$\oint ds = 0$$

Entropy will prove to be very useful in discussing the Second Law of Thermodynamics in Section 2.15 and in deriving the Clausius-Clapeyron equation in Section 2.16.

2.13 Isentropic Processes and Potential Temperature

A process in which no heat is added to the system is called an *adiabatic* process; if the adiabatic process is also reversible the entropy remains constant and the process is called *isentropic*. In this case

$$dh = ds = 0$$

An adiabatic process is not isentropic if it is irreversible, as follows from the discussion in Section 2.15. Atmospheric processes are seldom strictly adiabatic or strictly reversible, but many cases occur in which the isentropic approximation is useful and sufficiently accurate. For this reason the terms adiabatic and isentropic are often used interchangeably.

If Eq. (2.38) is introduced into Eq. (2.34), the isentropic form of the First Law is written

$$c_v \, dT + p \, d\alpha = 0 \qquad (2.44)$$

Upon combining the equation of state with Eq. (2.42), the temperature, the pressure, or the specific volume can be eliminated. After some simple algebraic manipulations the following isentropic equations are obtained:

$$\frac{c_v}{R_m}\frac{dT}{T} + \frac{d\alpha}{\alpha} = 0 \tag{2.45a}$$

$$\frac{c_p}{R_m}\frac{dT}{T} - \frac{dp}{p} = 0 \tag{2.45b}$$

and

$$\frac{c_p}{c_v}\frac{d\alpha}{\alpha} + \frac{dp}{p} = 0 \tag{2.45c}$$

These equations can be integrated giving the following results, first obtained by Poisson in 1823:

$$T\alpha^{R_m/c_v} = \text{const} \tag{2.46a}$$

$$Tp^{-R_m/c_p} = \text{const} \tag{2.46b}$$

$$p\alpha^{c_p/c_v} = \text{const} \tag{2.46c}$$

These equations are widely used in thermodynamics. In problem 7 an example of a possible application is given. In atmospheric studies Eqs. (2.46b) and (2.46c) are used most frequently.

Potential Temperature

Because the pressure varies with height according to the hydrostatic equation, an air parcel moving vertically with constant entropy experiences a change in temperature specified by Eq. (2.46b). The constant of integration can be evaluated from a particular state of the system. If a pressure surface is chosen as a reference, then the characteristic constant is determined by the temperature at that level. The *potential temperature* (θ) is defined as that temperature which would result if the air were brought isentropically to 1000 mb. Equation (2.46b) then can be written

$$Tp^{-\kappa} = \theta(1000 \text{ mb})^{-\kappa} = \text{const}$$

where κ represents R_m/c_p, or

$$\theta = T\left(\frac{1000 \text{ mb}}{p}\right)^{\kappa} \tag{2.47}$$

Poisson's equation is most frequently used in this form. The close connection between potential temperature and entropy is illustrated by problem 8.

The "Adiabatic" Chart

Since θ and 1000 are constants, Eq. (2.47) may be treated as a linear relation between T and p^{κ} and expressed by

$$T = \theta(1000 \text{ mb})^{-\kappa}p^{\kappa}$$

Now, if T and p^κ are used as coordinates, lines of constant potential temperature may be drawn having a slope given by

$$\theta(1000 \text{ mb})^{-\kappa}$$

To determine the potential temperature corresponding to any combination of p and T, it is necessary only to find the intersection of the appropriate p and T lines and to read the value of the potential temperature line passing through that point. In this way potential temperature may be determined graphically on the *"adiabatic"* chart as illustrated by Fig. 2.7. Graphical determination of the change in temperature experienced by a parcel in moving isentropically through a certain pressure interval also is illustrated.

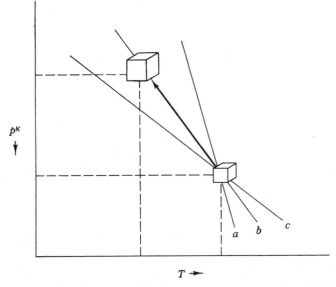

FIG. 2.7. Path followed on the *adiabatic chart* by a parcel displaced isentropically upward. Lines a, b, and c represent three possible vertical temperature distributions (the stable, the adiabatic, and the unstable, respectively). The coordinates are T (temperature) and p^κ (pressure raised to the R_m/c_p power).

2.14 Static Stability

We propose to examine the problem of vertical displacement in an atmosphere in which vertical accelerations are negligible, that is, in an atmosphere in which the hydrostatic equation holds. If the force acting on the displaced parcel in Fig. 2.7 tends to bring it back to its original position, the atmosphere is called statically *stable;* if the parcel is again in balance with its surroundings, the atmosphere is called *adiabatic,* and if the parcel tends to move away from its original position the atmosphere is statically *unstable*.

The net force on a parcel of air displaced from its equilibrium position may be evaluated by use of Archimedes' Principle or, more generally, from the hydrostatic equation. The result of problem 7 of Chapter I shows that the net upward force on a submerged body is proportional to the difference in densities of the fluid environment and the submerged body. The net upward force per unit volume on an air parcel of density ρ' in an environment of density ρ is then given by

$$F = (\rho - \rho')g$$

By Newton's Second Law this force is proportional to an upward acceleration given by

$$a = \frac{F}{\rho'} = \left(\frac{\rho - \rho'}{\rho'}\right)g$$

Upon eliminating the density by introducing Eq. (2.10) and recalling that $\alpha = 1/\rho$, the acceleration is given by

$$a = g\,\frac{(p/T) - (p'/T')}{p'/T'}$$

But it has been pointed out earlier that fluid in equilibrium must assume the pressure of its surroundings. Although acceleration is involved in this problem, in many cases this departure from equilibrium conditions does not affect significantly the pressure of the moving fluid. Therefore, $p \approx p'$, and

$$a = g\left(\frac{T' - T}{T}\right) \tag{2.48}$$

If a parcel having the same initial temperature (T_0) as its surroundings is displaced vertically from height z_0, and if heat transfer from the parcel is negligible, the temperature changes according to Eq. (2.45b). If the pressure is replaced as the vertical coordinate by use of the hydrostatic equation, the *adiabatic lapse rate* is given by

$$\Gamma \equiv -\left(\frac{dT}{dz}\right)_{\text{ad}} = \frac{g}{c_p} \tag{2.49}$$

Integration of Eq. (2.49) gives for the temperature of the displaced parcel

$$T' = T_0 - \Gamma\,\Delta z$$

where Δz represents the vertical displacement. The surrounding temperature changes with height at a rate defined by $\gamma \equiv -\partial T/\partial z$. In this case integration gives for the temperature of the environment at the height of the displaced parcel

$$T = T_0 - \gamma\,\Delta z$$

If these two temperatures are introduced into Eq. (2.48), the acceleration of the displaced parcel is expressed by

$$a = -\frac{g\,\Delta z}{T}\,(\Gamma - \gamma)$$

or, upon defining the *static stability* by

$$s_z \equiv -\frac{a}{g\,\Delta z} = -\frac{a}{\Delta\Phi}$$

it follows that

$$s_z = \frac{1}{T}\,(\Gamma - \gamma) \tag{2.50}$$

The static stability is the downward acceleration experienced by the displaced parcel per unit of geopotential. A refinement of Eq. (2.50) is discussed in problem 9. Although the static stability has been derived using the adiabatic assumption, Eq. (2.50) holds also in the more general case if Γ is interpreted as the rate of change of temperature with height following the parcel as it is displaced. In the case of condensation (upward displacement within clouds) this rate of change of temperature is discussed in Section 2.20.

It is often useful to express the stability in terms of the potential temperature. Differentiation of Eq. (2.47) with respect to z gives

$$\frac{1}{T}\frac{\partial T}{\partial z} = \frac{1}{\theta}\frac{\partial\theta}{\partial z} + \frac{R_m}{c_p}\frac{\partial p}{p\partial z}$$

Upon introducing the hydrostatic equation, the static stability is

$$s_z = \frac{1}{T}\left(\frac{g}{c_p} + \frac{\partial T}{\partial z}\right) = \frac{1}{\theta}\frac{\partial\theta}{\partial z} = \frac{\partial(\ln\theta)}{\partial z} \tag{2.51}$$

Equation (2.51) shows that where the potential temperature increases with height the atmosphere is statically stable, and vice versa.

2.15 Thermodynamic Probability and Entropy*

Although the development so far has concerned systems in equilibrium, processes in which the state of the system changes must introduce departures from the equilibrium condition. Atmospheric processes seldom involve such great departures from equilibrium that the equations developed in preceding sections become invalid, but there is something to be learned from a detailed look at molecular behavior and its relation to equilibrium. Only the case of a monatomic gas will be considered, and it will be assumed that no heat is added or work done on the gas.

---

* This section contains advanced material which is used in Section 4.3, but is otherwise not essential for understanding the following sections and chapters.

The molecules which constitute the system move with various velocities in a random fashion and make elastic collisions with other molecules. In order to completely describe the state, the positions and the momenta of all the molecules must be specified. The state of each molecule can be specified by six numbers indicating the three coordinates in space and the three momentum components. We introduce now a six-dimensional space, the *phase space*, whose coordinates are those of ordinary space and those of *momentum space*. The state of each molecule is now specified by a point in phase space (x, y, z, p_x, p_y, p_z) where p_x, p_y, and p_z are the three momentum components, mv_x, mv_y, mv_z, respectively. Now divide the phase space into many small cells

$$\Delta x\, \Delta y\, \Delta z\, \Delta p_x\, \Delta p_y\, \Delta p_z = H$$

and count the number of molecules belonging to each cell. This number is called the *occupation number*.

The state of the system could be described by giving the complete set of occupation numbers corresponding to all the cells in the system. This describes a *microstate* of the system. However, observational techniques are insufficiently precise to give a complete account of the microstate; only the number of particles in large groups of cells can be estimated. This gross specification is called the *macrostate*.

Obviously there are many possible microstates corresponding to a particular macrostate. In order to assign a probability to a macrostate, it is necessary to know the relative probabilities of individual microstates. It is assumed that each microstate is equally probable. The probability of a macrostate is then proportional to the number of microstates that belong to it. The *thermodynamic probability* (W) of the macrostate is now defined as the number of microstates which belong to it. Because a system in equilibrium will most of the time be in or close to the macrostate with the highest probability, the immediate problem is to find an expression for the number of microstates belonging to a macrostate, and then to find the macrostate which makes this number a maximum. If two systems, A and B, are combined into a single system, $A + B$, each microstate of component A of the system can combine with all the microstates of component B to give a new microstate for the new system. The number of microstates of the components therefore multiply together to give that of the combined system. In other words the thermodynamic probability W_{A+B} of a macrostate of system $A + B$ is equal to the product of the thermodynamic probabilities of system A and system B, so

$$W_{A+B} = W_A \cdot W_B \tag{2.52}$$

To find the number of microstates contained by a macrostate it is necessary to divide the large number of cells into groups in such a way that the macrostates of all the groups together define the macrostate of the system. Suppose that each group contains g cells and the total number of cells is G, then the number of groups is G/g. Consider now the ith group; all the cells of this group are numbered from i_1 to i_g. The cells are arranged in a series in such a way that after each cell symbol the number of identical particles contained by that cell is indicated by a number of zeros. For example, this series might look as follows:

$$i_1 i_2 i_3 0 i_4 0 i_5 0 0 i_6 i_7 \ldots i_g 0 \qquad (2.53)$$

We see, for instance, that cell i_5 contains 2 particles, i_1 zero, etc. The order of the symbols and zeros may be changed in all possible ways without change in the macrostate, except that the series cannot begin with a zero. Therefore, there are $g(g + n_i - 1)!$ sequences of series (2.53) belonging to the same macrostate, where n_i represents the total number of particles in group i. Each sequence represents a microstate, but there are many repetitions. Permuting the i's is not necessary, as all the microstates can be represented by sequences in which the cells are in the order of their numbers. Furthermore, permuting the zeros makes no difference, because the particles are identical. It follows that the thermodynamic probability of the ith group is represented by

$$W_i = \frac{g(g + n_i - 1)!}{g! \, n_i!} = \frac{(g + n_i - 1)!}{(g - 1)! \, n_i!} \qquad (2.54)$$

When the ith group is combined with the other groups forming the system, their probabilities multiply and the probability of the system is given by

$$W = \prod_{i=1}^{G/g} W_i = \prod_{i=1}^{G/g} \frac{(g + n_i - 1)!}{(g - 1)! \, n_i!} \qquad (2.55)$$

where the symbol $\prod_{i=1}^{G/g}$ means that the product of all W_i's must be taken between the limit 1 and G/g.

Now, recall Stirling's approximation in the form

$$\ln (P!) = \sum_{j=1}^{P} \ln j \approx \int_1^P \ln j \, dj = P \ln P - P$$

and apply to Eq. (2.54) with the result

$$\ln W_i = (g + n_i - 1) \ln (g + n_i - 1) - (g - 1) \ln (g - 1) - n_i \ln n_i$$

By assuming $g \gg 1$ this may be written

$$\ln W_i = g[(1 + N_i) \ln (1 + N_i) - N_i \ln N_i] \qquad (2.56)$$

where $N_i = n_i/g$ represents the average occupation number. Substituting Eq. (2.56) into (2.55) after taking the logarithm, the logarithm of the probability becomes

$$\ln W = \sum_{i=1}^{G/g} g[(1 + N_i) \ln (1 + N_i) - N_i \ln N_i]$$

$$= \sum_{1}^{G} [(1 + N_i) \ln (1 + N_i) - N_i \ln N_i] \qquad (2.57)$$

because the quantity summed is identical over each cell group.
In order that Eq. (2.57) express a maximum

$$\frac{\delta \ln W}{\delta N_i} = 0$$

This condition applied to the right-hand side of Eq. (2.57) yields

$$\delta \ln W = \sum [\delta (1 + N_i) \ln (1 + N_i) + (1 + N_i) \delta \ln (1 + N_i)$$

$$- \delta N_i \ln N_i - N_i \delta \ln N_i] = \sum \ln \left(1 + \frac{1}{N_i}\right) \delta N_i = 0 \qquad (2.58)$$

Because the total number of molecules (N) does not change

$$\sum \delta N_i = 0 \qquad (2.59)$$

Also, the internal energy of the system does not change because it is assumed that no heat is supplied to and no work is done on the system. The internal energy can be given by adding the energies of the individual molecules (u_i) associated with each cell and then adding all the energies of the cells. Therefore

$$U = \sum u_i N_i$$

and because $\delta(U) = 0$

$$\sum u_i \delta N_i = 0 \qquad (2.60)$$

In order to determine the distribution N_i Lagrange's method of undetermined multipliers may be used. Equations (2.59) and (2.60) are multiplied, respectively, by the undetermined constants α' and β. Addition of these equations to (2.58) then gives

$$\sum \left\{\ln \left(1 + \frac{1}{N_i}\right) - \alpha' - \beta U_i\right\} \delta N_i = 0$$

The choice of the δN_i is independent of the expression in brackets, so for any value of i the bracket must vanish, and

$$N_i = \frac{1}{\exp (\alpha' + \beta u_i) - 1} \qquad (2.61)$$

For an ideal gas the cells in phase space may be made so small that $N_i \ll 1$, which means that $1/N_i \gg 1$, so that unity in Eq. (2.58) can be neglected. Therefore, Eq. (2.61) may be written

$$N_i = \exp\left(-\alpha' - \beta u_i\right)$$

And, because $u_i = \frac{1}{2}mv_i^2$ for a monatomic gas

$$N_i = \exp\left[-\left(\alpha' + \frac{1}{2}\beta mv_i^2\right)\right] \tag{2.62}$$

It remains to evaluate α' and β.

If all cells occupying the spherical shell in phase space defined by the velocity interval between v and $v + dv$ are considered, the number of cells may be expressed by

$$dG = \frac{4\pi V m^3 v^2}{H}\, dv$$

The number of molecules in the velocity interval is therefore

$$N_i dG = \frac{4\pi V m^3 N_i v^2}{H}\, dv = \frac{4\pi V m^3 v^2}{H} \exp\left[-\left(\alpha' + \frac{\beta mv_i^2}{2}\right)\right] dv$$

This number is easily obtained from Eq. (2.18) if it is assumed that the velocity distribution of the gas in equilibrium corresponds to the most probable distribution. Therefore

$$\frac{4\pi m^3 N_i v^2}{H} = 4\pi nv^2 \left(\frac{m}{2\pi kT}\right)^{3/2} \exp\left(-\frac{mv^2}{2kT}\right)$$

or, because v and v_i are identical

$$N_i = nH(2\pi mkT)^{-3/2} \exp\left(-mv_i^2/2kT\right) \tag{2.63}$$

Equation (2.63) is identical to Eq. (2.62) if α' is chosen to be given by

$$\alpha' = -\ln\{nH(2\pi mkT)^{-3/2}\}$$

and

$$\beta = \frac{1}{kT} \tag{2.64}$$

The thermodynamic probability now may be computed from Eqs. (2.57) and (2.63). If it is again assumed that the cells are so small that $N_i \ll 1$, Eq. (2.57) may be simplified to

$$\ln W = N - \sum N_i \ln N_i$$

Upon substituting Eq. (2.63) into the logarithm on the right-hand side and recognizing that $\sum N_i(mv_i^2/2kT) = \frac{3}{2}N$, the summation yields

$$\ln W = N\{\tfrac{5}{2} - \ln(nH) + \tfrac{3}{2}\ln(2\pi mk) + \tfrac{3}{2}\ln T\} \tag{2.65}$$

Now, if Eq. (2.65) is differentiated and it is recognized that $dN = 0$ and that $dn/n = -d\alpha/\alpha$ where α represents specific volume

$$\frac{k}{Nm}\, d(\ln W) = kn^*\left(\frac{3}{2}\frac{dT}{T} + \frac{d\alpha}{\alpha}\right)$$

The right-hand side looks familiar, and upon referring to the definitions of Section 2.11 and Eqs. (2.42) and (2.43)

$$\frac{k}{Nm} d(\ln W) = ds \tag{2.66}$$

Equation (2.66) shows that a system having a maximum probable state also must have maximum entropy. Consequently, a system in equilibrium has maximum entropy, and if the thermodynamic probability and the entropy have not reached their maxima, the system is not in equilibrium. The state of the system must be expected to change until this maximum is reached. This insight is the foundation for the Second Law of Thermodynamics.

The theory developed here does not give a clue to how fast equilibrium will be reached from a specified nonequilibrium state. In Section 2.16 a simple model of transport mechanism is introduced which permits quantitative prediction of the rate at which equilibrium is approached.

2.16 Second Law of Thermodynamics and Transfer Processes

It is common experience that heat flows from a warm to a cold system. We wait before drinking a hot cup of coffee because we know that the hot coffee will lose heat to its surroundings and therefore will become cooler. The change takes place only in one direction and the change continues until equilibrium is reached; that is, until the coffee has the same temperature as its surroundings.

Observations like this form the basis for the *Second Law of Thermodynamics*. In the middle of the nineteenth century this law was formulated in a number of equivalent statements by Carnot, Clausius, and Lord Kelvin. One form of the Second Law states that *it is impossible to construct a device whose sole effect is the transfer of heat from a cooler to a hotter body.*

No exceptions to the Second Law are known. However, there is a subtle difference between the status of the Second Law and the status of the other fundamental principles stated earlier. For example, Conservation of Energy is presumed to hold for all systems no matter how small and for all time increments. On the other hand, the Second Law of Thermodynamics (and also the Zeroth Law) is a statement of probability which has meaning only if applied to a statistically significant number of molecules. The certainty and precision of the Second Law increase with the number of molecules contained in the system and with the time interval to which the law is applied.

If the Second Law is applied to the entire universe it must be concluded that the universe is striving towards an equilibrium characterized by uniform temperature and maximum entropy. In such a state no further thermodynamic processes are possible, and the universe has reached a "heat death." The Second Law states only the direction in which a process takes place, but not how fast the changes occur or how much time must elapse before equilibrium is established. Although it may appear that we have been following a trail from *pontus to pillatus*, we are now in sight of a most important generalization.

Changes in the state of a gas come about through transfer by random molecular motions. The general theory of the transfer process is very complex, but elementary kinetic theory again provides useful and instructive results.

Consider a system in which the temperature and the concentration of molecules varies linearly in the x direction as illustrated in Fig. 2.8. Each

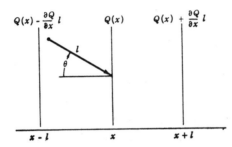

FIG. 2.8. Distribution of the property Q and the contribution to transport of Q made by a molecule moving through the distance l at the angle θ to the x axis.

molecule carries with it a characteristic property (for example, its mass or momentum or kinetic energy) which may be designated by Q. The rate of transfer of Q by a single molecule is expressed by $\mathbf{v}Q$. At any instant half of the molecules have a component in the positive x direction. Of these molecules the "average" molecule which passes through a unit area may be visualized as having traveled a distance l, called the *mean free path*, since its last collision. It therefore passes the unit area with properties characteristic of the region a distance l to the left of the unit area. Therefore, the first two terms of a Taylor expansion give for the rate of transport of the property Q from the left through the unit cross section

$$\frac{1}{2} \int_0^{\pi/2} \left(n\bar{v}Q - \frac{\partial(n\bar{v}Q)}{\partial x} l \cos\theta \right) \cos\theta \sin\theta \, d\theta = \frac{1}{4}n\bar{v}Q - \frac{1}{6}l \frac{\partial(n\bar{v}Q)}{\partial x}$$

The transport of Q by those molecules moving in the negative x direction is

$$-\tfrac{1}{4}n\bar{v}Q - \tfrac{1}{6}l\frac{\partial}{\partial x}(n\bar{v}Q)$$

and the net rate of transport per unit area in the x direction is

$$F_Q = -\tfrac{1}{3}l\frac{\partial(n\bar{v}Q)}{\partial x} \qquad (2.67)$$

For three dimensions the flux density of Q is given by

$$\mathbf{F}_Q = -\tfrac{1}{3}l\nabla(n\bar{v}Q) \qquad (2.68)$$

The mean free path may be calculated by considering a molecule to move through a field of stationary identical molecules as illustrated in Fig. 2.9.

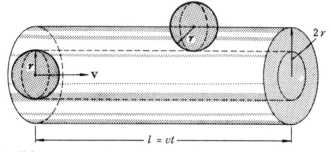

Fig. 2.9. Volume swept out by a molecule of radius r moving through a field of similar molecules.

Collisions occur with all molecules whose centers are within a distance $2r$ of the center line of the moving molecule. Therefore, in time t there occur $4\pi r^2 n v t$ collisions; and the average distance between successive collisions is given by

$$l = \frac{vt}{4\pi r^2 n v t} = \frac{1}{4\pi r^2 n} \qquad (2.69)$$

Equation (2.69) is correct only for a gas with all but one molecule at rest, that is to say, for no gas at all. In a real gas the molecules on the average move less than the "average" distance between collisions because, in effect, they are struck from the sides and rear by faster moving molecules. Computation for the Maxwellian distribution of velocities yields

$$l = \frac{1}{4\sqrt{2}\pi r^2 n}$$

Because this differs from Eq. (2.69) by only the constant numerical factor $\sqrt{2}$, Eq. (2.69) will be used in the following discussion.

Upon returning to Eq. (2.68) and replacing Q successively by the molecular mass, momentum, and kinetic energy, the respective flux densities may be obtained for ideal gases in the form

$$\mathbf{F}_m = -\tfrac{1}{3}l\nabla(nm\bar{v}) = -\tfrac{1}{3}l\bar{v}[\nabla\rho + \tfrac{1}{2}\rho\nabla(\ln T)] \qquad (2.70)$$

$$\mathbf{F}_{mv} = -\tfrac{1}{3}l\nabla(nm\bar{v}^2) = -l\,\nabla p \qquad (2.71)$$

$$\mathbf{F}_{\frac{1}{2}jmv^2} \equiv \mathbf{F}_h = -\tfrac{1}{18}\,l\nabla(njm\bar{v}\bar{v}^2)$$

$$= -\frac{j}{12}\,l\bar{v}R_m\rho[\nabla T + 2T\nabla(\ln p)] \qquad (2.72)$$

where the identity $nm \equiv \rho$, and the kinetic definition of temperature [Eq. (2.8)] and Eq. (2.36) have been used. Equation (2.70) shows that mass flux results from gradients of density and temperature. Equation (2.71) shows that momentum flux is proportional to pressure gradient, and Eq. (2.72) relates the heat flux to the gradients of pressure and temperature.

Momentum flux may result from a temperature gradient which in turn leads to the pressure gradient shown in Eq. (2.71). For example, rotation of the "radiometer" vanes in the "wrong" direction may be explained by the force resulting from the temperature gradient established at the surface of the vanes. Light shining on the black vanes creates a temperature and pressure gradient which exerts a force normal to the vane, whereas little pressure gradient is developed at the reflecting surface. In vacuum the instrument rotates the other way due to radiation pressure.

Heat Conduction

For the special but important case of no mass flux the heat flux may be expressed as a function of temperature gradient alone. To do this set $\mathbf{F}_m$ equal to zero with the result that

$$\nabla(\ln p) = \tfrac{1}{2}\,\nabla(\ln T)$$

Upon using this to eliminate $\nabla(\ln p)$ from Eq. (2.72), the heat flux density is expressed by

$$\mathbf{F}_h = -\tfrac{1}{6}\,jl\bar{v}R_m\rho\,\nabla T$$

and Eq. (2.39) permits writing this in the form

$$\mathbf{F}_h = -\tfrac{1}{3}\,l\bar{v}\rho c_v\,\nabla T \qquad (2.73)$$

Because the ratio of heat flux density to temperature gradient defines the *thermal conductivity* (k), Eq. (2.73) shows that

$$k = \frac{l\bar{v}\rho c_v}{3}$$

Upon replacing l and ρ by their equivalents

$$k = \frac{mc_v\bar{v}}{12\pi r^2} \qquad (2.74)$$

Equation (2.74) relates the thermal conductivity entirely to the molecular properties of the gas. Other things being equal conductivity is inversely related to molecular size, and this is consistent with observations. In fact, a more exact form of Eq. (2.74) has been applied to measurements of heat conductivity in order to estimate molecular size.

Shear Stress

Consider a macroscopic velocity distribution in which the x-component varies linearly with distance from the surface as shown in Fig. 2.10. The

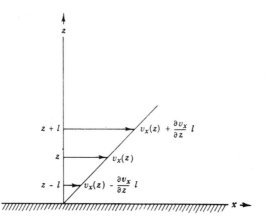

FIG. 2.10. Velocity profile $(v_x(z))$ for the case of linear shear.

discussion is limited to a single plane because evaluation of the three-dimensional momentum flux requires use of second-order tensors which have been omitted from this book. For the important atmospheric case of vertical shear of the x-component, Eq. (2.68) takes the form

$$F_{(mv_x)_z} = -\tfrac{1}{3}nm\bar{v}l\,\frac{\partial v_x}{\partial z} = -\tfrac{1}{3}\rho\bar{v}l\,\frac{\partial v_x}{\partial z} \tag{2.75}$$

This flux density of momentum may be visualized with the aid of Fig. 2.10 as follows. Downward moving molecules pass through a horizontal plane with more x momentum than those which move upward through the plane. Subsequent collisions then impart a positive x momentum to the region below the plane and a negative x momentum to the region above the plane. Newton's Second Law shows that the time rate of change of momentum must equal a "viscous" force, and Eq. (2.75) shows that this force must be proportional to the mean velocity shear. The viscous force therefore may be expressed by

$$F_x' = \eta\,\frac{\partial v_x}{\partial z} \tag{2.76}$$

where η is called the *viscosity* of the fluid and $F_x{}' = -F_{(mv_x)_x}$. Equations (2.75) and (2.76) together show that the viscosity is given by

$$\eta = \tfrac{1}{3}\rho\bar{v}l = \frac{m\bar{v}}{4\pi r^2}$$

Diffusion

Equation (2.70) expresses the mass flux density resulting from combined density and temperature gradients. The most important atmospheric applications are those in which a gas of low concentration varies in concentration and therefore diffuses through air whose density is essentially uniform. For example, Eq. (2.70) may be written for the diffusion of water vapor in the form

$$\mathbf{F}_w = -\tfrac{1}{3}l_w\bar{v}_w[\nabla\rho_w + \tfrac{1}{2}\rho_w\nabla(\ln T)]$$

$$= -\frac{1}{3}\frac{l_w\bar{v}_w}{R_{mw}T}\,[\nabla e - \tfrac{1}{2}e\nabla(\ln T)] \tag{2.77}$$

where the "w" refers to water vapor. If temperature and vapor pressure gradients corresponding to saturation in the atmosphere are substituted into Eq. (2.77) [see Eq. (2.88)], the second term is about 2% of the first. Other terms which have not been mentioned here may be comparable to this magnitude, so that the last term is usually omitted in atmospheric applications.

It is convenient to define the *diffusion coefficient* (D) by the equation

$$\mathbf{F}_w \equiv -D\nabla\rho_w \tag{2.78}$$

from which it follows that the molecular equivalent of the diffusion coefficient is

$$D = \tfrac{1}{3}l_w\bar{v}_w$$

The Heat Conduction and Diffusion Equations

A general differential equation governing the time and space variations of any conservative quantity may be derived by writing the continuity equation for an open system [Eq. (2.30)] for the general conservative molecular property Q in the form

$$\frac{\partial(nQ)}{\partial t} = -\nabla \cdot \mathbf{F}_Q \tag{2.79}$$

For a closed system the number of molecules contained in the system is constant, and Q changes only by flux through the boundaries. Therefore, the equation corresponding to (2.79) may be written

$$n\frac{\partial Q}{\partial t} = -\nabla \cdot \mathbf{F}_Q \tag{2.80}$$

For the special case in which Q represents the mass of the water vapor molecule, nQ represents the water vapor density (ρ_w); and Eq. (2.78) may be substituted into Eq. (2.79) with the result

$$\frac{\partial \rho_w}{\partial t} = \nabla \cdot (D \nabla \rho_w)$$

For uniform diffusivity

$$\frac{\partial \rho_w}{\partial t} = D \nabla^2 \rho_w \tag{2.81}$$

which is the *diffusion equation* for water vapor. It is clear that it holds only in the absence of a temperature gradient.

For the important though special case of heat flux in air which is free to expand at constant pressure, the work done in expanding must be considered. The well-known box, this time closed, is considered to expand in response to heat flux through the walls as shown in Fig. 2.11. For constant

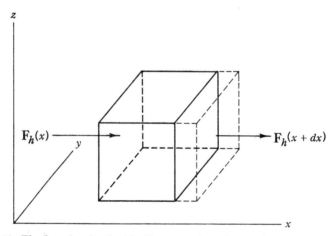

FIG. 2.11. The flux of molecular kinetic energy through opposite faces of a differential closed system. The dashed outline depicts expansion of the box at constant pressure as a result of increased temperature.

pressure the heat added to the system per unit mass may be expressed by

$$dh = d(u + p\alpha)$$

The term *specific enthalpy* is used to designate $u + p\alpha$. For an ideal gas

$$dh = c_p \, dT$$

so that $c_p T$ represents the specific enthalpy for an ideal gas. Now, it is clear that in the example under consideration, although the heat flux density through the walls is expressed by Eq. (2.73), it is the enthalpy which is conserved within the box. Therefore, the enthalpy per molecule may be substituted for Q in Eq. (2.80) with the result

$$\rho c_p \frac{\partial T}{\partial t} = \nabla \cdot k \nabla T$$

where n has been replaced by $n^*\rho$. For uniform conductivity and this takes the form

$$\frac{\partial T}{\partial t} = \frac{k}{\rho c_p} \nabla^2 T \qquad (2.82)$$

This is referred to as the *heat conduction equation*, although it is clear that it is a special case. The combination $k/\rho c_p$ is called the *thermal diffusivity* for air. It plays the same role in the heat conduction equation as D plays in the diffusion equation.

Examples of heat conduction are provided in problems 10 and 11.

Heat Conduction and Entropy

The Second Law of Thermodynamics implies that heat flows from the warmer to the cooler regions of a system and that entropy flows in the same direction. There is, however, an interesting difference between the two fluxes, which is most easily illustrated for the case of uniform heat flux in the x direction. For this case temperature decreases linearly in the x direction. Because an increment of entropy is inversely proportional to temperature, the entropy flux must continually increase in the x direction. This case illustrates that entropy is created when heat transfer takes place. The rate of entropy production can be evaluated by deriving the equation analogous to Eq. (2.79) for a nonconservative property and substituting the definition of entropy and the heat flux equation. The time rate of production of entropy following the heat flux is then expressed by

$$\rho \frac{d(\Delta s)}{dt} = \frac{k}{T^2} \left(\frac{\partial T}{\partial x} \right)^2$$

In this case the rate of production of entropy is proportional to the square of the temperature gradient and to the conductivity. When the temperature gradient vanishes, the entropy production stops, and the system approaches equilibrium.

2.17 Real Gases and Changes of Phase

The thermodynamic properties of an ideal gas apply with good accuracy to unsaturated air. However, when air is saturated and change of phase occurs there are important deviations from the ideal gas behavior. In this case a more realistic model is needed than the elementary one presented in

Sections 2.3 and 2.4. The elementary theory and the derived equation of state for an ideal gas imply that the volume may be made as small as desired by increasing the pressure or by decreasing the temperature. The gas may be imagined as composed of a group of mass points which individually and in the sum occupy no volume. These points can collide with the walls and exert forces on it; but they cannot collide with each other. The first modification to the molecular model, therefore, is to introduce a finite volume for the molecules.

The assumption that the molecules exert no forces upon each other except when they collide is an oversimplification. Experiment shows that ·attractive forces exist between molecules, and that the attractive force increases with decreasing separation. However, when the separation is reduced beyond approximately one "diameter," the molecules strongly repel each other. Figure 2.12 illustrates the intermolecular force as a

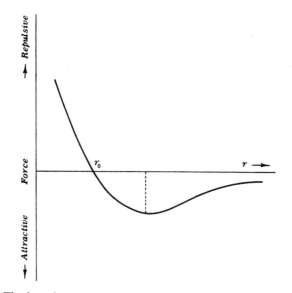

FIG. 2.12. The force between two molecules as a function of the separation of their centers.

function of the distance, and this subject is discussed again in Section 3.1. The separation (r_0) at which the force vanishes represents a position of minimum potential energy and therefore represents the mean distance between molecular centers in the liquid state. Molecules which are displaced from the equilibrium separation oscillate about r_0 just as two weights attached to a spring oscillate. The oscillations about r_0 are referred to as *thermal motion* of the molecules. If the temperature increases, the oscillations become more and more violent until the kinetic energy of the molecule at r_0 is larger than the potential energy required to overcome the attracting force of the other molecule, and the two molecules fly apart. Therefore, the second modification to the molecular model is to assume that the molecules exert an attractive force on each other.

Van der Waals' Equation

An equation of state taking into account the finite volume of the molecules and the attractive force between molecules was first deduced by van der Waals in 1873. He reasoned that as pressure increases or temperature decreases the volume must approach a small but finite value b, and that therefore α should be replaced in the equation of state by $\alpha - b$. Also, he represented the pressure at a point within the fluid as the sum of the pressure exerted by the walls (p) and the pressure exerted by the intermolecular attraction (p_i). To evaluate p_i consider the molecules near the walls. Each molecule is subject to an internal force which is proportional to the number of attracting molecules and therefore to the density. But the number of molecules near the boundary subject to the attractive force also is proportional to the density. It follows that the internal pressure due to the intermolecular force field is proportional to the square of the density. Consequently, van der Waals' equation of state may be written

$$\left(p + \frac{a}{\alpha^2}\right)(\alpha - b) = R_m T \qquad \text{or} \qquad p = \frac{R_m T}{\alpha - b} - \frac{a}{\alpha^2} \qquad (2.83)$$

where a and b are constants which depend upon the gas. It should be pointed out that van der Waals' equation represents a second approximation but not an exact statement of the equation of state for real gases, as indeed should be obvious from the largely intuitive derivation of the equation.

Van der Waals' equation is a cubic equation in α and consequently has a maximum and a minimum. At high temperatures the curves approach those computed from the equation of state for an ideal gas, whereas for low temperatures considerable difference exists. In Fig. 2.13 the 10°C isotherm for water vapor may be compared with the 10°C van der Waals' isotherm using values of a and b computed from problem 2.13.

Experimental Behavior of Real Gases

If a real gas under low pressure is compressed isothermally at temperature T, the volume decreases to the point α_2 as illustrated in Fig. 2.13. At this time drops of liquid begin to appear, and further decrease in volume occurs with no increase in pressure. When the volume indicated by α_1 is reached, all the gas has condensed into liquid; a *change of phase* is said to have occurred between α_2 and α_1. Appreciable reduction in volume beyond α_1 is possible only under great pressure. Conversion from the liquid phase to the gaseous phase occurs if the pressure is released isothermally. Van der Waals' equation is obeyed closely in the liquid and gaseous phases; however, between α_2 and α_1 the pressure is constant, so the van der Waals' equation is not valid during change of phase.*

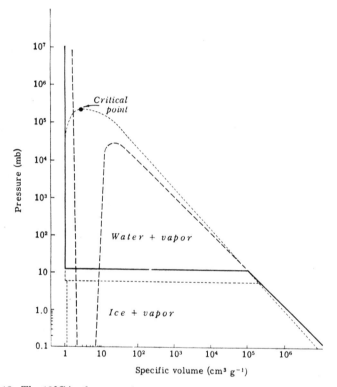

FIG. 2.13. The 10°C isotherm on the pressure-specific volume (p–α) diagram observed for water (solid line) and computed from van der Waals' equation (dashed line). The regions of mixed phases are bounded by dotted lines.

If the change of phase occurs at a temperature T' which is higher than T, it is found that the change in volume ($\alpha_2' - \alpha_1'$) is less than $\alpha_2 - \alpha_1$. It follows that the work expended or the heat absorbed in vaporization (latent heat) varies inversely with the temperature. At a certain critical temperature T_c, $(dp/d\alpha)_T$ and $(\partial^2 p/\partial\alpha^2)_T$ both vanish at a single point called the *critical point*. At this point it is not possible to distinguish sharply between the liquid and gaseous phases. At higher temperatures the gas

---

* The important role played by condensation nuclei in the condensation process is discussed in Chapter III.

cannot be liquefied by increase of pressure. The critical point is a characteristic property of all real gases. For water this point is specified by

$$T_c = 647°K \qquad p_c = 226 \times 10^6 \text{ dyn cm}^{-2} \qquad \alpha_c = 3.1 \text{ cm}^3 \text{ g}^{-1}*$$

Problem 12 requires calculation of the van der Waals' constant for water
vapor from these values.

So far only the transition from gaseous to liquid phase or vice versa has
been considered, but other phase changes are also important. In general,
the transition between liquid and solid occurs at a lower temperature than
does the transition from vapor to liquid. However, at the *triple point* all
three phases may exist in equilibrium. For water, the triple point occurs at
a pressure of 6.11 mb (1 mb = 10^3 dyn cm^{-2}) and a temperature of
0.0098°C. At pressures below the triple point water vapor may be converted
directly into ice without passing through the liquid phase. The boundaries
between the three phases are shown in Fig. 2.14. The dashed line represents
the equilibrium vapor pressure over liquid water at temperatures below
273°K (*supercooled water*). The fact that the equilibrium vapor pressure
over supercooled water is higher than the equilibrium vapor pressure
over ice at the same temperature has an important consequence which is
discussed in Section 3.9.

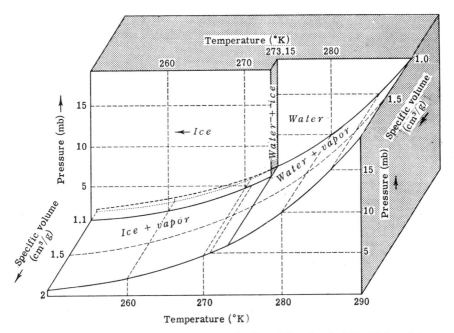

FIG. 2.14. Phase diagram for water vapor in the neighborhood of the triple point.

* Osborne, N. S., Stimson, H. F., Ginnings, D. C., *J. Res. Nat. Bur. Stand.*, **18**,
389–448 (RP 983) 1937.

Latent Heat of Vaporization

Suppose a system is in the liquid state indicated by the point α_1 in Fig. 2.13. In order to change the phase from liquid to vapor at constant temperature and constant pressure, a characteristic quantity of heat, called the *latent heat of vaporization* (L), must be added. The First Law of Thermodynamics gives

$$L = \int_{h_1}^{h_2} dh = \int_{u_1}^{u_2} du + p_s \int_{\alpha_1}^{\alpha_2} d\alpha = u_2 - u_1 + p_s(\alpha_2 - \alpha_1) \quad (2.84)$$

where index 1 refers to the liquid phase and index 2 to the vapor phase. The latent heat of vaporization consists of two terms: a change in internal energy of the system and expansion work done by the system.

2.18   Clausius-Clapeyron Equation

The vapor pressure at which change of phase occurs at constant temperature is called the *saturation* or *equilibrium vapor prsssure*. In order to determine the dependence of saturation vapor pressure on temperature, Eq. (2.84) may be combined with definition (2.43) with the result

$$L = T(s_2 - s_1) = u_2 - u_1 + p_s(\alpha_2 - \alpha_1) \quad (2.85)$$

Therefore

$$u_1 + p_s\alpha_1 - Ts_1 = u_2 + p_s\alpha_2 - Ts_2 \quad (2.86)$$

where p_s represents saturation vapor pressure. The combination $u + p_s\alpha - Ts$ is called the *Gibbs function* and is represented by G. The Gibbs function is a function of the state only, and Eq. (2.86) shows that it is constant during an isothermal change of phase.

Now consider the isothermal change of phase at the temperature, $T + dT$, and the corresponding pressure, $p_s + dp_s$. The Gibbs function now has another constant value, $G + dG$. From the definition of G

$$dG = du + p_s\,d\alpha - T\,ds + \alpha\,dp_s - s\,dT$$

Upon combining this with Eqs. (2.34) and (2.43)

$$dG = \alpha\,dp_s - s\,dT$$

But $G_1 + dG_1 = G_2 + dG_2$, so also $dG_1 = dG_2$ and

$$\alpha_1\,dp_s - s_1\,dT = \alpha_2\,dp_s - s_2\,dT$$

This combined with Eq. (2.85) yields the Clausius-Clapeyron equation in the form

$$\frac{dp_s}{dT} = \frac{L}{T(\alpha_2 - \alpha_1)} \quad (2.87)$$

This equation can be applied successfully to water vapor. Under normal atmospheric conditions $\alpha_2 \gg \alpha_1$, so α_1 may be neglected. If it is assumed that water vapor behaves closely as an ideal gas, then Eq. (2.87) can be combined with (2.10) yielding

$$\frac{de_s}{e_s} = \frac{L}{R_{mw}} \frac{dT}{T^2} \tag{2.88}$$

where p_s is replaced by e_s, the saturation water vapor pressure and R_{mw} is the specific gas constant for water vapor. If L is assumed constant, Eq. (2.88) can readily be integrated. Using $L = 2.500 \times 10^3$ J g^{-1}, $R_{mw} = 0.4617$ J g^{-1} °K^{-1} and also the fact that at a temperature of 0°C the saturation vapor pressure is 6.11 mb, Eq. (2.88) yields

$$\log_{10} e_s = 9.4051 - \frac{2353}{T} \tag{2.89}$$

The accuracy of this equation can be found by comparing the vapor pressures computed from Eq. (2.89) with observed saturation vapor pressures. In problem 13 the error made by Eq. (2.89) is considered.

An equation similar to Eq. (2.89) can be derived for the saturation vapor pressure over ice; this is examined in problem 14.

2.19 The Moist Atmosphere

The proportions of dry air, water vapor, and liquid or solid water vary within rather wide limits in the atmosphere as results of evaporation, condensation, and precipitation. The amount of moisture contained by a certain parcel of air may be considered one of the specific properties of the parcel which, like temperature and pressure, specify its state.

Humidity Variables

The amount of water vapor the air contains can be expressed in the following ways:

The *specific humidity* (q) is the mass of water vapor per unit mass of air.

The *mixing ratio* (w) is the mass of water vapor contained by a unit mass of dry air.

The mixing ratio and specific humidity are related by

$$w = \frac{q}{1 - q} \quad \text{and} \quad q = \frac{w}{1 + w}$$

which is easily verified. Both quantities are usually expressed in the units grams per kilogram (g/kg).

The *vapor pressure* (e) is the partial pressure exerted by water vapor. Assuming that the equation of state for an ideal gas can be applied to water vapor, it is possible to express the vapor pressure in terms of the mixing ratio or the specific humidity. The error introduced in this way is examined in problem 15. The equation of state for an ideal gas applied to water vapor is

$$e\alpha_w = R_{mw}T \tag{2.90}$$

where α_w is the specific volume of water vapor and R_{mw} is the specific gas constant for water vapor. If Eq. (2.10) is divided by Eq. (2.90)

$$w \equiv \frac{\rho_w}{\rho_a} = \frac{R_{ma}}{R_{mw}}\frac{e}{p - e} \equiv \epsilon\frac{e}{p - e} \tag{2.91}$$

The molecular masses of water vapor and dry air are, respectively, 18.015 and 28.97, from which it follows that $\epsilon = 0.622$. Because $e \ll p$, $w \ll 1$, and $q \ll 1$

$$w \approx q \approx \epsilon\frac{e}{p} \quad \text{and} \quad \frac{dw}{w} \approx \frac{de}{e}$$

The *relative humidity* (r) is the ratio of the actual vapor pressure to the saturation vapor pressure, so*

$$r \equiv \frac{e}{e_s} \approx \frac{w}{w_s} \approx \frac{q}{q_s} \tag{2.92}$$

Among the several humidity variables, most frequent use is made of vapor pressure and mixing ratio, but the other variables are useful in certain problems. For example, relative humidity is an important parameter in the study of the environment of plants and animals because it is a rough indicator of the rate of evaporation from animal or vegetable tissue. For this reason, we are more immediately aware of changes in relative humidity than of changes in specific humidity or vapor pressure.

The Gas Constant and Specific Heats of Unsaturated Air

Because the proportion of water vapor in the atmosphere is variable, the gas constant for moist (that is, natural) air is variable. This is evident from the equation of state for moist unsaturated air written in the form

$$p\alpha_{wa} = R_{mwa}T \tag{2.93}$$

where the symbol R_{mwa} is slightly shorter than the five words it represents and is equivalent to

$$R_{ma}(1 - q) + R_{mw}q = \left[1 + \left(\frac{1}{\epsilon} - 1\right)q\right]R_{ma}$$
$$= [1 + 0.61q]R_{ma} \approx (1 + 0.61w)R_{ma}$$

In order that the equation of state contain only three variables rather than

---

* Although several efforts have been made in recent years to topple this definition of relative humidity from its Victorian pedestal, no unanimity exists; and we can see no compelling reason for change.

the four in Eq. (2.93), the variation of temperature and of humidity may be combined by defining the *virtual temperature* according to

$$T^* \equiv (1 + 0.61q)T \tag{2.94}$$

The equation of state may then be written

$$p\alpha_{wa} = R_{ma}T^*$$

The virtual temperature is the temperature of dry air having the same pressure and specific volume as the moist air. For a specific humidity of 10 g/kg the virtual temperature exceeds the temperature by about 2°C. The difference between temperature and virtual temperature may be critical in determination of static stability and important in calculation of the vertical pressure distribution by integration of the hydrostatic equation.

The influence of humidity on the specific heats is expressed by the equations

$$c_p = (1 + 0.90q)c_{pa} \approx (1 + 0.90w)c_{pa}$$

and

$$c_v = (1 + 1.02q)c_{va} \approx (1 + 1.02w)c_{va}$$

where the subscript "*a*" refers to dry air.

2.20   Saturation Adiabatic Processes

In a *saturation adiabatic process* condensation (or evaporation) and consequent release of latent heat occurs within the system, but no heat is added to or taken from the system. For example, an air parcel lifted adiabatically experiences decrease in temperature at the adiabatic rate until saturated. Further lifting is accompanied by release of latent heat within the parcel; consequently, the rate of decrease in temperature is less than the adiabatic rate. The First Law of Thermodynamics for this process may be written

$$-L\,dw_s = c_v\,dT + p\,d\alpha$$

where w_s represents the mass of water condensed per unit mass of air. This expression combined with the equation of state in differential form yields

$$-L\,dw_s = c_p\,dT - \alpha\,dp$$

and upon introducing the hydrostatic equation

$$\frac{dT}{dz} = -\frac{g}{c_p} - \frac{L}{c_p}\frac{dw_s}{dz}$$

But

$$\frac{dw_s}{dz} = \frac{dw_s}{dT}\frac{dT}{dz}$$

so that the *saturation adiabatic lapse rate* is given by

$$\Gamma_s \equiv -\left(\frac{dT}{dz}\right)_s = \frac{g}{c_p + L(dw_s/dT)} = \frac{\Gamma}{1 + (L/c_p)(dw_s/dT)} \quad (2.95)$$

It follows from Eqs. (2.88) and (2.91) that dw_s/dT is positive, therefore the saturation adiabatic lapse rate is less than the adiabatic lapse

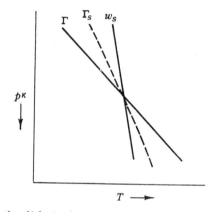

FIG. 2.15. The *pseudo-adiabatic chart*. The coordinates are T (temperature) and p^κ (pressure raised to the R_m/c_p power), Γ represents an adiabatic line, Γ_s represents a pseudo-adiabatic line, and w_s represents a line of constant saturation mixing ratio.

rate. Since supersaturation rarely exceeds 1%, saturated air which is rising or sinking follows rather closely the rate of temperature change given in Eq. (2.95). When water in liquid or solid form is removed from the system by precipitation, the change of state is referred to as *pseudo-adiabatic*. The equations for the pseudo-adiabatic process differ only slightly from the corresponding equations for the saturation adiabatic process, and this difference usually is ignored.

Stability of Saturated Air

Equation (2.50) expresses the stability of a layer of air as a function of the difference between the vertical rate of change of temperature experienced by rising air and the vertical rate of temperature change within the layer. If the rising air is saturated, the stability may be written

$$s_z = \frac{1}{T}(\Gamma_s - \gamma)$$

A lapse rate greater than the pseudo-adiabatic but less than the adiabatic is statically stable for unsaturated air but is statically unstable for saturated air.

The Pseudo-Adiabatic Chart

Lines which describe the temperature change experienced by saturated air in rising or sinking in the atmosphere may be added to the adiabatic chart by introducing appropriate numerical values in (2.95) or other equations derived from it. The resulting *pseudo-adiabatic chart* may be used to determine temperature change experienced by either saturated or unsaturated air. The chart usually includes lines which indicate w_s as a function of temperature and pressure (height). The w_s lines may be constructed by using Eq. (2.90) combined with the Clausius-Clapeyron equation in the form of Eq. (2.89). Upon eliminating e_s, w_s is expressed as a function of p and T, and then may be plotted on the pseudo-adiabatic chart as illustrated in Fig. 2.15. By using the three sets of lines it is possible to determine graphically the point at which unsaturated air becomes saturated during ascent and to determine the temperature change during ascent.

The heat released in condensation within clouds represents an important source of energy for the atmosphere. In the following simple example the pseudo-adiabatic chart provides an estimate of the heating brought about by condensation. Air which is forced upward by mountains cools adiabatically until the vapor pressure equals the equilibrium vapor pressure. This point is determined by the intersection of the appropriate adiabatic lines with the w_s line equal to the mixing ratio of the rising air. From this point to the top of the mountain the rising air cools pseudo-adiabatically, and a cloud is formed. If precipitation falls from the cloud, the total water per unit mass is reduced; therefore, if the air descends on the leeward side of the mountain, the air warms pseudo-adiabatically only until the new value of w_s is reached. From this point the air warms adiabatically. Consequently, the temperature on the leeward side of a mountain may be considerably higher than on the windward side. An example of such a sequence is examined in problem 16. Although the strong dry wind in the lee of a mountain, called *foehn* or *chinook*, is often attributed to this mechanism, in the most marked foehns air descends from far above the top of the mountain and is warmed adiabatically on the entire descent.

2.21 Distribution of Temperature and Water Vapor

Equation (2.19) and the pressure at the surface together specify the vertical pressure distribution as a function of the temperature distribution.

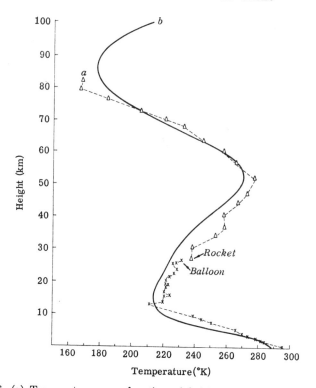

FIG. 2.16. (a) Temperature as a function of height observed at Fort Churchill, Canada (59°N) at 2330 CST on July 23, 1957 [after W. G. Stroud, W. Nordberg, W. R. Bandeen, F. L. Batman, and P. Titus, *J. Geophys. Research* **65**, 2307 (1960)].

(b) Average temperature of the International Reference Atmosphere (see Appendix II.E).

Because, as is illustrated in Fig. 2.16, the temperature distribution is usually a complicated function of height, Eq. (2.19) must be integrated numerically. An example of such an integration is given in problem 17.

The Average Vertical Temperature Distribution

The atmosphere consists of a series of nearly spherical layers each characterized by a distinctive vertical temperature distribution. These layers can usually be identified in soundings of the temperature, like the one labeled "a" in Fig. 2.16. The characteristic features are even more clearly defined in curve b of Fig. 2.16, which represents the international reference atmosphere. Although many observed temperature profiles resemble closely the international reference atmosphere, deviations of 20°K or more are observed frequently.

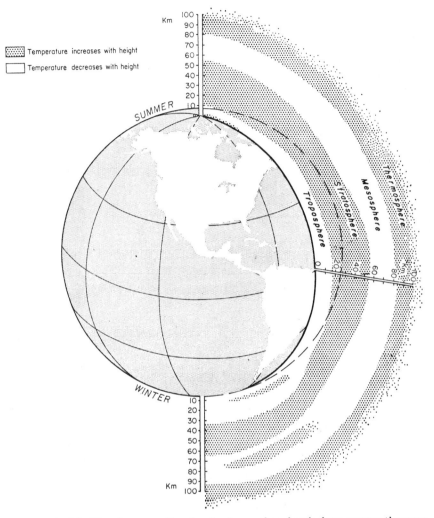

Fig. 2.17. Meridional cross section showing for northern hemisphere summer the gross features of the temperature distribution.

Figure 2.17 indicates the gross features of the temperature distribution to 100 kilometers. The lowest layer, characterized for the most part by decreasing temperature with height, is called the *troposphere*, which means the turning or changing sphere. This layer contains about 80% of the total atmospheric mass. The troposphere is most closely influenced by the energy transfer that takes place at the earth's surface through evaporation and heat conduction. These processes often create horizontal and vertical temperature gradients which may lead to the development of atmospheric motions and the upward transport of heat and water vapor. But rising air tends to cool adiabatically or pseudo-adiabatically; as a result temperature decreases with height and clouds form in the troposphere. A warning is in order; the motions which are referred to here are extremely complex and should not be imagined simply as closed circuits in a vertical plane. A discussion and a summary of the energy transfer processes in the troposphere is given in Chapter V, and consideration of atmospheric motions is planned for a subsequent volume.

The vertical extent of the troposphere varies with season and latitude. In tropical regions it is usually 16 to 18 kilometers. Over the poles the extent in summer is about 8 to 10 kilometers, but in the winter the troposphere may be entirely absent.

The troposphere is usually bounded at the top by a remarkably abrupt increase of static stability with height. The surface formed by this virtual discontinuity of lapse rate is called the *tropopause;* it is particularly clearly defined in tropical regions. In middle and high latitudes the tropopause often is not clearly defined, particularly in the region of severe storms.

The statically stable layer above the troposphere is called the *stratosphere* (stratified sphere). It extends upward to a height of about 50 kilometers where the temperature is comparable to earth surface temperature. Above the tropopause the temperature increases first slowly with height to about 20 kilometers and then much more rapidly. This temperature distribution is associated with absorption of ultraviolet solar radiation by the ozone which is present between heights of 20 and 50 kilometers. Many details of the stratosphere have not yet been thoroughly observed, and the interaction between troposphere and stratosphere is not thoroughly understood. The top of the stratosphere is a surface of maximum temperature, which is called the *stratopause.*

Above 50 kilometers the temperature decreases with height to a minimum of about 200°K at about 80 kilometers. This layer of decreasing temperature with height is called the *mesosphere* (middle sphere) and has been less fully observed than the regions above or below it. Information gathered from meteor trails and rockets indicates that wind speeds up to 150 m sec^{-1} occur in this region. The similarity between the temperature distributions in troposphere and mesosphere suggests the existence of analogous processes. Absorption of solar radiation in the region of the stratopause provides the energy source, and resulting motions with accompanying adiabatic expansion and cooling may account for the temperature distribution in this layer. Mysterious noctilucent clouds whose origin and composition are unknown, are sometimes observed near the region of lowest temperature, the *mesopause.*

Above a height of 80 or 90 kilometers the temperature seems to increase continuously with height, probably until a temperature of 1500°K or more is reached at a height of about 500 kilometers. This extensive region is referred to as the *thermosphere*. The character of the thermosphere differs from the lower layers because

(a) ionization of air molecules and atoms occurs,
(b) dissociation of molecular oxygen and other constituents takes place,
(c) diffusion is more important than mixing, resulting in relatively high concentration of light gases in the upper layers.

Molecular oxygen is increasingly dissociated above about 80 kilometers, and above 110 kilometers practically all oxygen is in atomic form. At these heights because of the large mean free path diffusion becomes more important than mixing, and in accordance with Dalton's law the heavier gases tend to be less concentrated with increasing height.

Above about 110 kilometers atomic oxygen increases in relative abundance with height while molecular nitrogen decreases. Above 500 kilometers helium probably replaces atomic oxygen as the dominant constituent, and above about 3000 kilometers atomic hydrogen replaces helium. These estimates are based on diffusive equilibrium in an isothermal temperature distribution of about 1500°K above 500 kilometers. These conditions permit the existence of a population of neutral hydrogen atoms which are formed by dissociation of water vapor and methane in the lower thermosphere. The neutral atoms diffuse upward to the base of the *exosphere* (perhaps 500–1000 kilometers), and the more energetic may then escape from the earth's gravitational field. Helium escapes more slowly, and the rate of escape of the heavier gases is probably negligible.

Chapman has suggested a contrary temperature distribution. He postulates increasing temperature with height from the base of the thermosphere to the solar corona, which he visualizes as surrounding the atmosphere. According to this view the temperature probably reaches ionizing level for hydrogen ($3 \times 10^{4°}$K) at a height between 10,000 and 100,000 kilometers. Above this height the atmosphere is composed of protons and electrons, and the motions of the particles are strongly influenced by the earth's magnetic field.

In this region the high temperatures of the solar corona may give rise to thermal conduction downward. The magnitude of this heat flux is extremely uncertain because neither the temperature gradient nor the thermal conductivity in the upper thermosphere is known. Nevertheless, rough estimates of the heat flux density indicate a magnitude between 0.2 and 5 erg cm^{-2} sec^{-1}. The mean solar radiant flux density in this region is

1.382×10^{6} erg cm^{-2} sec^{-1}, so that the coronal heat flux is of little direct consequence to the energy transfer in lower layers of the atmosphere.

Although Chapman's picture of the upper atmosphere is speculative and not generally accepted, all investigators agree that the region above about 500 kilometers contains an uncertain and varying population of ionized particles. It is appropriate, therefore, to think of the two populations of neutral and ionized particles as existing together more or less independently of one another. Because these ionized particles are bound by the earth's magnetic field, it is appropriate to refer to this region as the *magnetosphere*. The terms exosphere and magnetosphere refer to the same region in space but to different populations and to different phenomena.*

Because collisions are infrequent in the exosphere or magnetosphere, the concept of temperature needs to be reexamined. Neutral particles may have relatively low energies corresponding to the kinetic temperature at the base of the exosphere, whereas ionized particles may have a very much higher energy. Because the definition of temperature is based on a Maxwellian distribution of particle energies, it is not possible to assign a unique temperature in this region. It is conceivable that the "temperature" of the magnetosphere increases with height as suggested by Chapman, while the temperature of the coexisting exosphere is nearly constant with height.

We may still be in for surprises as new data from the exosphere and magnetosphere become available. Perhaps the most interesting and important aspects are related to the effects of the earth's magnetic field on charged particles. These will be discussed in Chapter VI.

Average Latitudinal Temperature and Water Vapor Distribution

The average temperature distribution plotted on an atmospheric cross section on a plane through the axis of the earth is illustrated in Fig. 2.18. The troposphere extends to a maximum height near the equator and to minimum height over the poles. We see also that in summer the troposphere is deeper than in winter.

The average latitudinal water vapor distribution is illustrated in Fig. 2.19. Highest concentration of water vapor is found at low latitudes near the earth's surface where mixing ratio as high as 40 g kg^{-1} or more may be observed. Water vapor concentration normally decreases upward and toward the poles. The distribution is, like the temperature, highly variable in time and space. Decrease of concentration by one or two orders of magnitude occurs frequently in a vertical distance of 10 kilometers or a horizontal distance of 10,000 kilometers.

---

* The regions below the exosphere, in which the statistical properties of the gas permit the definition of pressure, may be collectively referred to as the *barosphere*.

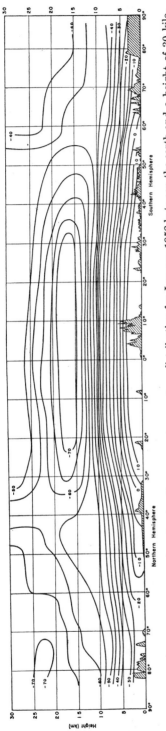

FIG. 2.18. Meridional cross section showing the average temperature distribution for January 1958 between the earth and a height of 30 kilometers at 75°W longitude (after U. S. Weather Bureau, "Monthly Mean Aerological Cross Sections." U. S. Govt. Printing Office, Washington, D.C., 1961).

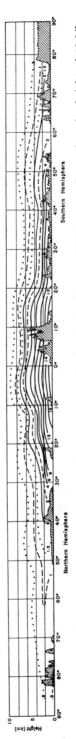

FIG. 2.19. Meridional cross section showing the average mixing ratio distribution for January 1958 between the earth and a height of 10 kilometers at 75°W longitude (after U. S. Weather Bureau, "Monthly Mean Aerological Cross Sections." U. S. Govt. Printing Office, Washington, D.C.,1961).

List of Symbols

First used
in Section

a	Acceleration	2.14
A	Area	2.3
a, b	Constants in van der Waals' equation	2.17
c	Specific heat	2.11·
c_p	Specific heat at constant pressure	2.11
c_v	Specific heat at constant volume	2.11
D	Diffusivity	2.16
e	Vapor pressure	2.18
e_t	Total energy per unit mass	2.7
F	Force	2.4
F_Q	Flux density of property Q	2.16
g	Force of gravity per unit mass (magnitude)	2.14
G	Gibbs function, number of cells	2.15, 2.18
h	Heat added per unit mass	2.7
H	Volume of cell in phase space	2.13
j	Number of degrees of freedom	2.10
k	Boltzmann's constant, thermal conductivity	2.4, 2.16
l	Mean free path length	2.16
L	Latent heat of vaporization	2.17
m	Mass of a molecule	2.3
M	Molecular mass, based on the C_{12} nucleus	2.1, 2.4
n	Number of molecules per unit volume	2.3
n^*	Number of molecules per unit mass	2.7
N	Number of molecules in a system	2.3
N_0	Avogadro's number	2.1
p	Pressure	2.4
p_x, p_y, p_z	Components of momentum in momentum space	2.15
q	Specific humidity	2.19
Q	Conservative molecular property	2.16
r	Radius of molecule, relative humidity	2.16, 2.19
R	Universal gas constant	2.4
R_m	Specific gas constant for a gas with molecular mass M	2.4
s	Specific entropy	2.12
S	Entropy	2.12
s_z	Static stability	2.14
$\mathbf{s}$	Vector denoting distance of displacement	2.7
t	Time	2.3
T	Temperature	2.4
T^*	Virtual temperature	2.19
u	Specific internal energy	2.7
v	Speed of a molecule	2.3
$\bar{v}$	Mean speed of molecules in a system	2.3
$\mathbf{v}$	Velocity vector	2.16
V	Volume of a system	2.3
w	Work per unit mass, mixing ratio	2.7, 2.19
W	Work, thermodynamic probability of a system	2.7, 2.15
x, y, z	Cartesian coordinates	2.14

		First used in Section
α	Specific volume	2.4
α_m	Molar specific volume	2.4
γ	Lapse rate	2.14
Γ	Adiabatic lapse rate	2.14
Γ_s	Pseudo-adiabatic lapse rate	2.20
ϵ	Ratio of specific gas constants of air and water vapor	2.19
η	Coefficient of viscosity	2.16
κ	R_m/c_p	2.13
θ	Angle between normal to area and direction in which molecule moves, potential temperature	2.3, 2.13
ρ	Density	2.14
ϕ	Azimuth angle, distribution function	2.3, 2.5
Φ	Geopotential	2.7

Subscripts

a	Dry air
c	Critical
s	Saturation
w	Water vapor
wa	Moist air
x, y, z	Components corresponding to axes
1	Liquid phase
2	Vapor phase

Problems

1. If 10^6 molecules are required in order to ensure a statistically uniform distribution of velocities in all directions, what is the minimum volume in which the state can be defined at standard atmospheric conditions (1013 mb and 0°C)?

2. Show that the number of collisions per unit wall area per unit time and per unit solid angle can be expressed by

$$\frac{1}{4\pi} n\bar{v} \cos \theta$$

3. At what height would a hydrogen molecule reach the escape velocity if the temperature of the molecule is 5000°K?

4. If the temperature is 1000°K at a height of 200 km, what fraction of the molecules with molecular mass 2 will have at least the escape velocity?

5. Suppose that the atmosphere is completely at rest and in diffusive equilibrium. Compute at what height 10% O_2 and 90% N_2 is found when at the surface the temperature is 300°K, the pressure is 1000 mb, and the composition is 20% O_2 and 80% N_2.

6. What is the expansion work done by a system in the cyclic process shown in the adjoining figure? Expansion occurs at constant pressure from α_1 to α_2, compression occurs at constant temperature from α_2 to α_1, and cooling occurs at constant volume from p_2 to p_1. Assume that $\alpha_1 = 10^3$ cm³ g⁻¹, $\alpha_2 = 2 \times 10^3$ cm³ g⁻¹, and T at α_2 and p_1 is 500°K. The system contains air with molecular mass 29. See figure page 76.

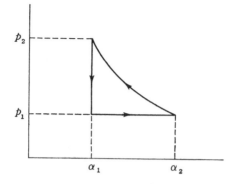

7. A cyclic reversible process following two isotherms and two adiabatics is called a Carnot cycle. Compute the work done per unit mass by a system containing dry air when it completes a Carnot cycle defined by (a) an isothermal expansion at 300°K from 1000 mb to 500 mb followed by (b) an adiabatic expansion until the temperature is 200°K, (c) an isothermal compression and adiabatic compression in such a manner that the original point of 300°K and 1000 mb is reached.

How much heat has been supplied to the system during the two isothermal processes?

8. Show that a change in potential temperature is related to a change in entropy by

$$ds = c_p \, d \ln \theta$$

9. Show that when a parcel with temperature T' moves isentropically in air of temperature T the lapse rate following the parcel is given by

$$-\frac{dT'}{dz} = \frac{T'}{T} \, \Gamma$$

10. Solve Eq. (2.82) for steady state conduction of heat through a wall of thickness D and conductivity k if the temperature is maintained at one side at T_0 and at the other side at T_1. Choose the x axis perpendicular to the planes of the wall.

11. Assume that the earth is a homogeneous sphere with thermal conductivity $k = 8 \times 10^4$ erg cm^{-1} °K^{-1} sec^{-1} and constant surface temperature $T = 285$°K. If the temperature near the surface increases downward by 1°C (100 m)$^{-1}$, what is the temperature 1000 km from the center? Assume steady state conditions and also that all the heat that flows out is generated within the core of 1000-km radius.

12. Show that the van der Waals' constants may be expressed by

$$a = \frac{27}{64} \frac{(R_m T_c)^2}{p_c} \qquad b = \frac{R_m T_c}{8 p_c}$$

and evaluate a and b for water vapor. Remember that at the critical point $\partial^2 p / \partial \alpha^2 = 0$ and $\partial p / \partial \alpha = 0$.

13. The observed saturation vapor pressure for water vapor at 20°C and at 30°C is 23.38 mb and 42.43 mb, respectively (1 mb = 10³ dyn cm^{-2}). Compute the error made by using Eq. (2.89) at these temperatures.

14. Derive an equation similar to Eq. (2.89) for the saturation vapor pressure over ice. Use for the latent heat of vaporization and the specific gas constant for water vapor

$$L = 2.834 \times 10^3 \text{ J g}^{-1} \quad \text{and} \quad R_{mw} = 0.4617 \text{ J g}^{-1}\,°\text{K}^{-1}.$$

15. Compute the error that is made in the specific volume of water vapor when the equation of state for an ideal gas is used instead of van der Waals' equation for temperature of 300°K and vapor pressure of 15 mb.

$$T_c = 647°\text{K} \qquad p_c = 226 \times 10^6 \text{ dyn cm}^{-2}$$
$$\alpha_c = 3.1 \text{ cm}^3 \text{ g}^{-1} \qquad R_{mw} = 0.4617 \text{ J g}^{-1}\,°\text{K}^{-1}$$

16. Air at a temperature of 20°C and mixing ratio $w = 10 \text{ g kg}^{-1}$ is lifted from 1000 mb to 700 mb by moving over a mountain barrier. Determine from a pseudo-adiabatic chart its temperature after descending to 900 mb on the other side of the mountain. Assume that 10% of the water vapor is removed by precipitation during the ascent. Describe the thermodynamic state of the air at each significant point.

17. Compute the height of the 700, 500, 300, 100, 50, and 25 mb surfaces from the following radiosonde sounding, taken at Ft. Churchill, 00 GMT, 24 July, 1957. The height of the ground surface is 30 meters above sea level and the surface pressure is 1007 mb.

Pressure (mb)	Temperature (°C)	Pressure (mb)	Temperature (°C)
1007	27.5	214	−58.2
1000	27.8	200	−57.5
988	28.1	176	−59.6
850	15.9	164	−52.1
755	8.0	150	−54.1
700	5.1	141	−55.6
648	2.5	100	−50.4
500	−11.5	50	−46.9
468	−14.2	40	−47.0
400	−23.9	25	−43.4
300	−40.0	12	−33.7
250	−50.1		

General References

Allis and Herlin, *Thermodynamic and Statistical Mechanics*, gives an excellent introduction to statistical mechanics and thermodynamic probability.

Sears, *An Introduction to Thermodynamics, the Kinetic Theory of Gases and Statistical Mechanics*, covers about the same material as Allis and Herlin and gives a clear account of the kinetic theory of gases.

Sommerfeld, *Thermodynamics and Statistical Mechanics*, is a beautiful book giving a more advanced account of the principles used in this chapter.

Holmboe, Forsythe, and Gustin, *Dynamic Meteorology*, gives a useful account of the thermodynamic effects of water vapor.

Ratcliffe, *Physics of the Upper Atmosphere*, contains a number of valuable articles by various authors on physical problems of the upper atmosphere. In this chapter special use has been made of the first chapter by S. Chapman.

Kennard, *Kinetic Theory of Gases*, is a solid book on a more advanced level. It contains a wealth of information and is clearly written.

CHAPTER III

Properties and Behavior
of Cloud Particles

PART I: GROWTH

3.1 Intermolecular Force and Surface Tension

"The parts of all homogeneal hard Bodies which fully touch one another, stick together very strongly. . . . their Particles attract one another by some Force, which in immediate Contact is exceeding strong, at small distances performs the Chymical Operations above mentioned, and reaches not far from the Particles with any sensible Effect."* In these words Sir Isaac Newton described the intermolecular force which he recognized as underlying chemical reactions, wetting or resistance to wetting, capillary phenomena, solubility, crystallization, and evaporation. It has required the full development of electrodynamics and quantum mechanics to elucidate these forces to the level that predictions are possible for a few relatively simple systems. Here, we shall be satisfied to give a partial, intuitive and brief account of these forces. Figure 2.11 indicates the dependence of the total intermolecular force on separation.

At very small separations, so small that the electron clouds of the molecules overlap, strong repulsive *valence* forces dominate. The theory of the valence force, which is developed from quantum mechanics, predicts that this force is approximately an exponential function of distance. Beyond the effective range of the valence force several forces combine to give a complex dependence on separation. *Electrostatic* contributions may be calculated from Coulomb's Law if the geometrical arrangement and electrical nature of the molecules are known. *Induction* contributions may arise through the dipoles induced in nonpolar molecules by nearby charges or polar molecules. As illustrated in Fig. 3.1a, the induction force is attractive regardless of the orientation of the dipole. If the dipole in Fig. 3.1a were replaced by a nonpolar molecule as illustrated in Fig. 3.1b, there would still arise an attractive force between the molecules. This force may be understood from the fact that the electrons in a nonpolar molecule are constantly in oscillation, so that at a certain instant the molecule assumes a dipole configuration. An induced dipole is created from the adjacent mole-

* Sir Isaac Newton, "Opticks" (4th ed., London, 1730), 406 pp. Dover, New York 1952.

cule, and an attractive force arises between them. In both examples of induction the force is inversely proportional to the seventh power of the distance, so at small separations large attractive forces exist. There are other contributions to the intermolecular force, but the induction forces are probably of greatest importance in understanding surface properties of liquids.

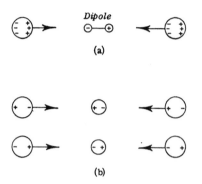

FIG. 3.1. (a) Induced charges and induced forces in two nonpolar molecules in the neighborhood of a neutral dipole.

(b) Induced charges and induced forces produced by the instantaneous dipole configurations of a nonpolar molecule.

The water molecules which at a certain instant form the free surface of a body of water are subjected to intermolecular attractive forces exerted by the nearby molecules just below the surface. The opposing forces exerted on the surface molecules by the air molecules outside are very much less. If, now, the surface area is increased by changing the shape of the container or by other means, molecules must be moved from the interior into the surface layer, and work must be done against the intermolecular force. The work required to increase the surface area by one unit stores potential energy in the surface; it is called *surface energy* or *surface tension* and has units of energy per unit area or force per unit length.

Because of surface tension a volume of liquid tends to assume a shape with minimum area-to-volume ratio. Therefore, small masses tend strongly to assume spherical shapes; in the case of larger masses forces which are mass dependent may distort or destroy the spherical shape. Spherical drops experience an internal pressure due to surface tension which may be calculated in the following way.

Imagine a small spherical drop of radius r which is divided in half by a hypothetical plane as shown in Fig. 3.2. Surface tension acting across the plane holds the edges of the sphere together; the total force may be expressed by $2\pi r\sigma$ where σ represents surface tension or surface energy per unit area. The two halves of the sphere are held apart by the equal pressure forces exerted normal to the plane surface and given by $\pi r^2 p_i$ where p_i represents the internal pressure due to surface tension. For equilibrium between the two forces, the internal pressure due to surface tension is given by

$$p_i = \frac{2\sigma}{r}$$

The surface tension of pure water at 0°C is about 75 dyn cm⁻¹ and is very slightly dependent on radius. This equation shows that a droplet of 1 μ radius experiences an internal pressure of about 1.5 atmospheres.

3.2 Equilibrium Vapor Pressure over a Curved Surface

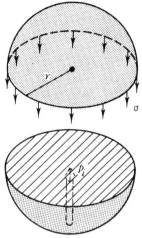

FIG. 3.2. The surface tension force per unit length (σ) and the internal pressure force per unit area (p_i) for a dissected drop of radius r.

The surface energy associated with a curved surface has an important effect on the equilibrium vapor pressure and the rate of evaporation from droplets. The difference between the equilibrium vapor pressure over a flat surface (e_s) and over a spherical (or curved) surface (e_c) may be calculated from consideration of the change in surface energy which accompanies decrease of surface area.

The surface energy released by a droplet during a change in radius (dr) is given by the surface tension multiplied by the change in area and therefore is equal to

$$8\pi\sigma r \, dr$$

And the surface energy released per unit mass of evaporated water is consequently given by $(2\sigma/r)\alpha_1$ where α_1 represents the specific volume of the liquid water in the drop. The heat required to evaporate one gram of liquid water is now, instead of Eq. (2.85), expressed by

$$L = T(s_2 - s_1) = u_2 - u_1 + e_c(\alpha_2 - \alpha_1) - (2\sigma/r)\alpha_1$$

or, in a form analogous to Eq. (2.86)

$$u_1 + \left(e_c + \frac{2\sigma}{r}\right)\alpha_1 - Ts_1 = u_2 + e_c\alpha_2 - Ts_2 = G$$

The term, $e_c + (2\sigma/r)$, represents the pressure inside a droplet of radius r. Evidently, the Gibbs function is constant during an isothermal change of phase.

We now consider an isothermal change of phase at the same temperature but for a droplet of radius $r + dr$. The Gibbs function now has another constant value $G + dG$. Following a similar argument as is given in Section 2.18, the difference in Gibbs function between the two cases may be expressed by

$$dG = \alpha_1 \, de_c - 2\sigma\alpha_1 \frac{dr}{r^2} = \alpha_2 \, de_c$$

and, because $\alpha_2 - \alpha_1 \approx \alpha_2 \approx R_w T/e_c$

$$R_w T \frac{de_c}{e_c} = -2\sigma\alpha_1 \frac{dr}{r^2}$$

where R_w represents the specific gas constant for water vapor. The equilibrium vapor pressure over the droplet can be determined by integrating from a flat surface ($r = \infty$) to droplet radius r, because over a flat surface the equilibrium vapor pressure is identical to the saturation vapor pressure. This yields

$$R_w T \ln \frac{e_c}{e_s} = \frac{2\sigma\alpha_1}{r_0}$$

or

$$e_c = e_s \exp \left(\frac{2\sigma}{\rho_w R_w T r} \right) \tag{3.1}$$

where ρ_w represents the water density. Equation (3.1) shows that the vapor pressure required for condensation on very small droplets may be very large. Surface tension itself is dependent on curvature, but this does not invalidate the conclusion that the effect of curvature on equilibrium vapor pressure would make droplet growth impossible at minute radii if it were not for other very powerful effects. Problem 1 requires a calculation using Eq. (3.1).

The original formulation of Eq. (3.1) is due to Lord Kelvin, who deduced it from the experimental fact of capillary rise in small tubes. The reader is invited to reproduce Lord Kelvin's deduction in problem 2.

3.3   Equilibrium Vapor Pressure over Solutions

In the light of the previous section, how is it possible that condensation of water in air is a common occurrence?

There are in the atmosphere many small dust particles on some of which condensation can proceed without the powerful inhibiting effect of curved surface. Some of these dust particles are soluble in water (*hygroscopic*), others are wettable but insoluble, and others are water resistant (*hydrophobic*). The hygroscopic particles or nuclei are the most favorable for condensation; for this reason their effect will be discussed.

Imagine a substance having essentially zero vapor pressure to be dissolved in water. The molecules of the solute are distributed uniformly through the water and some of them occupy positions in the surface layer. Now compare a unit surface area before and after solute is added to the water. The proportion of the surface area occupied by water molecules will be reduced in the approximate ratio $n/(n + n')$ where n represents the number of molecules of solute and n' the number of molecules of water.

This assumes that the molecules retain their identity in solution (are undissociated), that the cross-sectional areas of individual particles of solute and water are the same, and that both types of particles are distributed uniformly over the surface. It follows that the number of escaping molecules and the equilibrium vapor pressure should be reduced in the same ratio. These considerations are expressed as Raoult's law for ideal solutions in the form

$$\frac{e_s - e_h}{e_s} = \frac{n}{n + n'}$$

where e_h represents the equilibrium vapor pressure over the solution. For dilute solutions Raoult's law may be rewritten

$$\frac{e_h}{e_s} = 1 - \frac{n}{n'} \tag{3.2}$$

For dilute solutions in which the dissolved molecules are dissociated Eq. (3.2) must be modified by multiplying n by i, the number of dissociated ions per molecule. In dilute solution the NaCl molecule dissociates into two ions and the $(NH_4)_2SO_4$ molecule dissociates into three ions. The number of dissociated ions of solute of mass M may now be expressed by iN_0M/m_s where N_0 represents Avogadro's number and m_s the molecular mass of solute. The number of water molecules (n') in mass m may be expressed by N_0m/m_w where m_w represents the molecular mass of water. For a spherical water drop Eq. (3.2) may now be written

$$\frac{e_h}{e_s} = 1 - \frac{3im_wM}{4\pi r^3 \rho_w m_s} \tag{3.3}$$

Equation (3.3) shows that for a specific mass of dissolved material the equilibrium vapor pressure over a solution decreases rapidly with decreasing radius of the drop. As a result, it is possible that large drops evaporate while small drops grow by condensation.

The effects of surface tension and of hygroscopic substance are of opposite sign and both increase in importance with decreasing size of the droplet. A general equation which represents both effects may be written by combining Eqs. (3.1) and (3.3) and expanding the exponential function in a Taylor series. Where e_{hc} represents the equilibrium vapor pressure over a droplet of solution, the leading terms of this expansion are

$$\frac{e_{hc}}{e_s} = 1 + \frac{2\sigma}{r\rho_w R_w T} - \frac{3im_wM}{4\pi r^3 \rho_w m_s} \tag{3.4}$$

Equation (3.4) is illustrated in Fig. 3.3. The maximum of each curve represents a critical radius which can be calculated from Eq. (3.4). Problem 3 provides an exercise using Eq. (3.4).

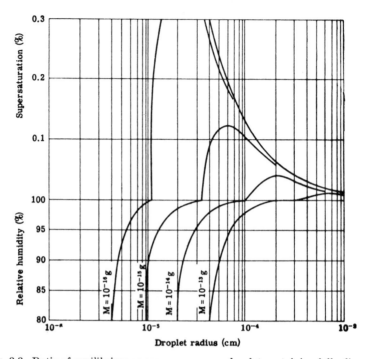

Fig. 3.3. Ratio of equilibrium vapor pressure over droplet containing fully dissociated NaCl of mass M to that over flat water surface (expressed in per cent) plotted as a function of radius as computed from Eq. (3.4). Surface tension of 75 erg cm^{-2} and temperature of 273°K have been used.

Several conclusions may be drawn from Fig. 3.3. (a) In an isolated population of droplets containing equal masses of solute, the very small droplets must grow at the expense of the larger droplets. (b) Beyond a critical size the larger droplets may grow at the expense of the smaller. (c) If supersaturation of the vapor with respect to a flat surface is limited, say to about 0.1%, droplets must contain more than 10^{-15} gram of NaCl (or other similar solute) in order to reach critical size. (d) Small droplets may exist in equilibrium with moist air of relative humidity far below 100%.

3.4   Distribution and Properties of Aerosols

Suspended solid and liquid particles (aerosols) are present in the atmosphere in enormous numbers, and their concentration varies by several orders of magnitude with time and in space. They vary from about 5×10^{-7} to 20×10^{-4} centimeters (5×10^{-3} to 20 μ) in "effective" radius. These particles play crucial roles both in condensation and in the formation of ice crystals. In addition, the aerosols participate in chemical processes, they influence the electrical properties of the atmosphere, and in large concentrations they may be annoying, dangerous and even lethal. Radioactive aerosols may be used as tracers of the air motion but also have their uniquely hazardous aspects.

Representative distributions by number and volume are shown in Figs. 3.4 and 3.5, respectively. As we realize from casual observation, the con-

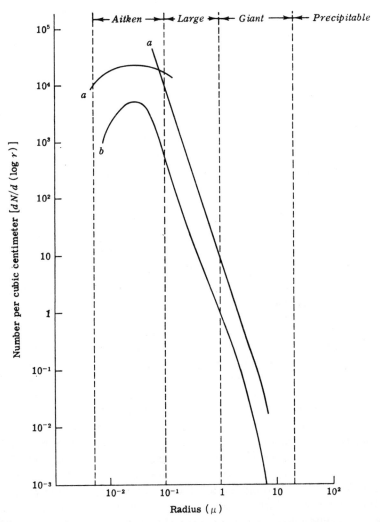

FIG. 3.4. Size distribution of aerosol particles for (a) Frankfurt, Germany, and (b) Zugspitze, Germany [after C. E. Junge, *Advances in Geophys.* **4**, 8 (1959)].

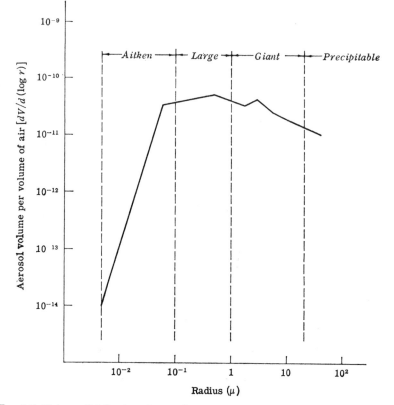

FIG. 3.5. Volume distribution of aerosol particles for Frankfurt, Germany [after C. E. Junge, *Advances in Geophys.* **4**, 8 (1958)].

centration in the troposphere decreases with height and with distance from populated areas. Except for areas near industries or other sources, the distributions exhibit remarkable similarity, which may be explained by the physical processes of *coagulation* and *fallout*. If one imagines an initial distribution with equal numbers of particles of every size, the particles of radius less than 10^{-2} μ quickly become attached to larger particles due to Brownian motion, while the particles of radius greater than 20 μ are sufficiently heavy to precipitate out. Both processes are rather sharply dependent on size, so that they produce fixed limits to the distribution as shown in Fig. 3.4. Between the upper and lower limits the distribution of particles by mass is roughly uniform as shown in Fig. 3.5. Division into the three categories, Aitken, large, and giant, is arbitrary but corresponds roughly to differences in technique of observation. *Aitken* nuclei are so named for the physicist who first studied the small particles using a counter based on the rapid expansion of a chamber containing saturated air. Expansion produces adiabatic cooling until the *Aitken* particles act as condensation nuclei; the minute droplets so formed grow almost instantaneously to visible size. The Aitken nucleus counter has remained the standard instrument for detection of particles between 5×10^{-3} and 10^{-1} μ. Particles in the range from 0.1 to 1 μ, called *large* nuclei, have been collected by thermal and electrostatic precipitation, by filters, and by collection on coated slides. Particles larger than one micron in radius are called *giant* nuclei; they have been collected on coated slides, by impactor (air jets which project the particles at high speed onto a slide), and they have been detected by burning in a hydrogen flame. After collection, the large and giant nuclei are analyzed using an optical or electron microscope.

The composition of Aitken nuclei remains unknown, although it is presumed that they are formed from many different materials, industrial and natural. Due to their very small size, they usually do not provide an important source of condensation nuclei for cloud particles.

Among the large nuclei particles of ammonium sulfate appear to be most numerous, at least in industrial area. Hygroscopic nuclei are sufficiently numerous and contain sufficient mass to account for development of most of the droplets present in clouds. Junge and Manson* have called attention to a permanent aerosol layer at about 20-kilometer height in which large nuclei are found in concentrations of 10^{-1} cm^{-3}. They are probably formed by oxidation of gaseous sulfur compounds.

Away from industrial areas giant nuclei appear to be predominantly sodium chloride; they are ejected into the air from the sea surface through breaking of air bubbles in sea foam and spray. When cooling of air brings about saturation, the giant nuclei are, of course, the first to act as condensation nuclei; and they are responsible for the development of the largest droplets. Although relatively few in number, the giant nuclei play an important role in creation of the broad drop size spectrum which, as is shown later in Section 3.12, is necessary for efficient coalescence of droplets and precipitation from water clouds.

Crystalline particles play an important part in the processes of freezing and of growth of ice crystals. These processes are discussed in Section 3.7.

Very little is known about differences between aerosol particles of the same size which might make one more effective in condensation than another. Nor is it known to what extent interaction of particles is important, for example, by condensation of gases on nonhygroscopic surfaces or by coating of nonhygroscopic surfaces with hygroscopic Aitken nuclei.

---

* C. E. Junge and J. E. Manson, Stratospheric aerosol studies, *J. Geophys. Research* **66**, 2163 (1961).

In air from which all nuclei have been removed small droplets form only at vapor pressures which exceed about four times the equilibrium vapor pressure. These droplets are formed on negative ions, a few of which are always present in natural air. Condensation on positive ions requires an even greater supersaturation.

3.5 Growth of Droplets by Condensation[*]

The essential physics of equilibrium vapor pressure depends on the rather simple considerations of surface tension and hygroscopic effects discussed in Sections 3.2 and 3.3. However, under nonequilibrium conditions droplets may grow or may become smaller; the rate of growth depends on, in addition to surface tension and hygroscopic effects, the rate of diffusion of water vapor to the droplet from the surrounding air and the rate of conduction of latent heat away from the droplet.

An exact treatment of these processes is not possible because effects of air motion on the environment of the droplets and other complex interactions cannot be incorporated, but if approximations are made judiciously, satisfactory accuracy (perhaps to within 10 to 20%) is readily obtained as follows. Rate of growth of the droplet by diffusion may be expressed by neglecting the small temperature term in Eq. (2.77) and expressing the flux for a spherical surface in the form

$$\frac{dm}{dt} = 4\pi \frac{r^2 D}{R_w T} \frac{de}{dr}$$

where m represents mass of the droplet and D the diffusion coefficient. Integration may be performed by assuming that dm/dt is independent of radius. Then, upon neglecting $1/r_\infty$, the rate of growth of the droplet is given by

$$\frac{dm}{dt} = \frac{4\pi D r_0}{R_w T} (e_\infty - e_0)$$

This is equivalent to

$$r_0 \frac{dr_0}{dt} = \frac{D(e_\infty - e_0)}{\rho_w R_w T} \tag{3.5}$$

where ρ_w represents water density.

Latent heat liberated at the droplet surface when condensation occurs is diffused away from the drop also according to Eq. (2.73) in the form

$$L \frac{dm}{dt} = -4\pi r^2 k \frac{dT}{dr}$$

This equation may be integrated and the result combined with Eq. (3.5) to give

---

[*] For a fuller discussion of the role of nuclei in condensation see Fletcher's *The Physics of Rainclouds*.

$$T_0 - T_\infty = \frac{DL(e_\infty - e_0)}{kR_wT} \tag{3.6}$$

where $T \approx T_0 \approx T_\infty$.

Equations (3.4), (3.5), and (3.6) and the Clausius-Clapeyron equation (2.88) are a set of four simultaneous equations in the variables e_0, e_s, T_0, and r_0. From observations of the mass of the solute and the vapor pressure and temperature of the environment, the three unknowns can be calculated for any value of r_0. Then r_0 can be calculated as a function of time by numerical integration. Sample calculations illustrated in Fig. 3.6 show

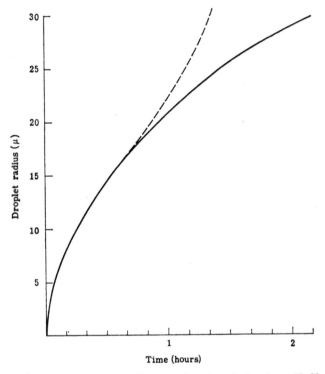

Time (hours)

FIG. 3.6. *Solid curve:* Radius of droplet as a function of time for a NaCl nucleus of 6×10^{-15} g, supersaturation of 0.05% temperature of 273°K, and pressure of 800 mb as calculated from Eqs. (2.88), (3.4), (3.5), and (3.6).

Dashed curve: Estimated radius of droplet as a function of time for growth by accretion and condensation.

that a NaCl nucleus of about 0.1 μ radius (6×10^{-15} g) in an environment of 0.05% supersaturation grows to a radius of 1 μ in about 1 second, to 8 μ in 10 minutes and to 20 μ in 1 hour. Subsequent growth by condensation is quite slow.

A good approximation for droplets larger than about 3μ can be obtained by replacing the differentials by finite differences in Eq. (2.88) and omitting Eq. (3.4). The result may be written

$$\int_{3\mu}^{r_0} r_0 dr_0 = \frac{kDR_wT_\infty^2(e_\infty - e_{\infty s})}{\rho_w(DL^2e_{\infty s} + kR_w^2T_\infty^3)} \int_0^t dt \qquad (3.7)$$

where the coefficient on the right has been taken as constant. Upon integrating Eq. (3.7) the radius is seen to increase as the square root of the time.

3.6 Growth of Droplets by Accretion

It is clear from Fig. 3.6 that growth of droplets by diffusion is not likely to account for the development of full grown raindrops. In this section the process of droplet growth by collision and coalescence with other droplets of different sizes and different fall velocities will be examined.

A falling droplet is acted on by the force of gravity and by the friction

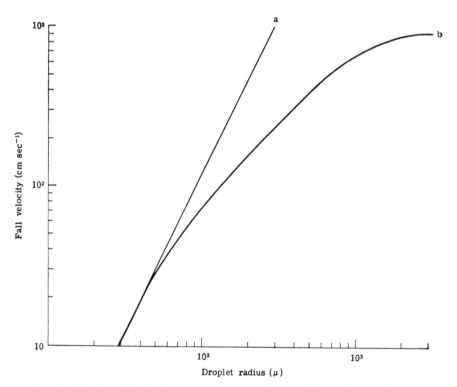

FIG. 3.7. Droplet fall velocity as a function of radius (a) calculated from Stokes' law (Eq. 3.8), (b) observed by R. Gunn and G. D. Kinzer, *J. Meteorol.* **6**, 243 (1949).

or drag force exerted by the air. At equilibrium the drag force is given by

$$F_D = \tfrac{4}{3}\pi r^3(\rho_w - \rho)g$$

where ρ represents air density. Intuition suggests strongly that the drag force for small droplets must increase with viscosity, with radius of the droplet, and with the fall speed. This is borne out by Stokes' law which holds accurately for droplets of radii less than 50 μ, and which may be written

$$F_D = 6\pi\eta r v$$

where η represents the dynamic viscosity of air and v the fall velocity of the droplet. The derivation of Stokes' law is too dependent on topics in fluid motions to be given here, but it is available in Joos, "Theoretical Physics," p. 206 (see Bibliography), and elsewhere. Because $\rho_w \gg \rho$ the fall velocity may be expressed by

$$v = \frac{2}{9}\frac{r^2\rho_w}{\eta}g \qquad (3.8)$$

For droplets with radius larger than 50 μ Eq. (3.8) overestimates the fall velocity. The computation then becomes considerably more complicated. Theoretical and experimental values are given in Fig. 3.7.

In order for a falling drop to coalesce with other smaller drops it first has to collide with these drops, and collision depends critically on the position and the radii of the two drops. Air flows around the large drop as indicated in Fig. 3.8, and the smaller droplet tends to follow the stream-

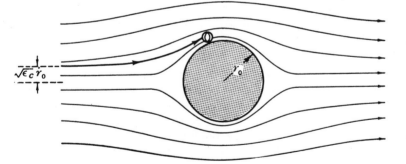

FIG. 3.8. Streamlines for air flowing slowly around a sphere and the limiting trajectory of the droplets which collide with the sphere.

lines. Because the inertia of the droplet causes it to cross the streamlines in regions where they are curved, it is easy to recognize that the trajectory of the small droplet lies between a straight line and a streamline. The precise shape of the trajectory has been calculated from the hydrodynamic equations of viscous flow by Hocking.* These computations show that it is possible to calculate the ratio of the number of droplets of a specified size that collide with the larger droplet to the total number of droplets of the same size initially lying within the volume swept out by the larger droplet, and this ratio is called the *collision efficiency*.

Langmuir† has shown that collision efficiency may be expressed as a function of the ratio of *penetration range* to radius of the large droplet, where the penetration range is the distance a small droplet would travel before coming to rest if shot into still air at the initial relative speed of the large droplet. This ratio, called the *Langmuir inertia parameter*, can be expressed by

$$K = \frac{4}{81} g \frac{\rho_w^2}{\eta^2} r^2 r_0$$

where r and r_0 are, respectively, the radii of the small and large droplets. Development of this equation is left as problem 4. Hocking has derived the equation relating K and collision efficiency, but due to its complexity it is not derived here.

From Hocking's equation the curves relating the collision efficiency to the radii of the small and large droplets can easily be generated. A significant example is given in Fig. 3.9. From this figure it is clear that a droplet should exceed 18 μ in radius if smaller droplets are to collide with it. Growth by diffusion, therefore, has to proceed until droplets with radii of at least 18 μ are formed.

Not all droplets that collide with the large droplet adhere to it. The ratio of the number of droplets which adhere to the number of droplets that collide with the larger droplet, called the *coalescence efficiency*, is highly uncertain. Laboratory experiments show that coalescence occurs readily in a strong electric field, whereas, if the field is weak or absent altogether, coalescence does not occur readily. Our knowledge of electric fields in clouds is far too fragmentary to permit accurate calculations of coalescence efficiency. It is known that the product of the coalescence and collision efficiencies (*accretion efficiency*) increases with radius of the collecting drop as is suggested by Fig. 3.9; and, of course, the relative velocity increases with radius. Therefore, coalescence increases rapidly in effectiveness as the drop increases in size. Almost immediately after the droplet reaches sufficient size to start the process of accretion, this process becomes more important than the diffusion process. Growth by accretion is illustrated

---

* L. M. Hocking, *Quart. J. Roy Meteorol. Soc.* **85**, 44 (1959).
† I. Langmuir, *J. Meteorol.* **5**, 175 (1948).

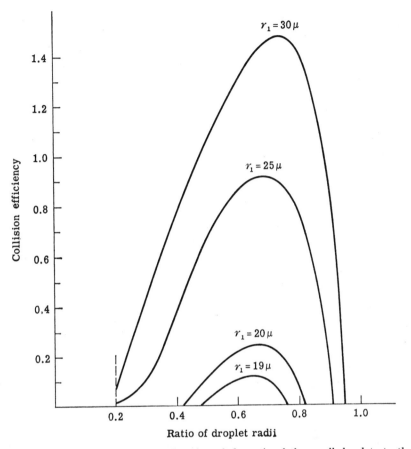

FIG. 3.9. Collision efficiency as a function of the ratio of the small droplets to the larger collecting droplet (r_1) [after L. M. Hocking, *Quart. J. Roy. Meteorol. Soc.* **85**, 45 (1959)].

by the dashed curve in Fig. 3.6 which suggests that at radii greater than 20 μ growth occurs primarily by accretion. The rate of growth by accretion can be represented by summing over the small droplets that collide with the large drop

$$\frac{dm}{dt} = \pi r_0^2 \sum_{i=1}^{n} \epsilon(r_i) v'(r_i) \, \Delta \rho_{di} \tag{3.9}$$

where ϵ represents accretion efficiency, ρ_d represents the density of liquid water of the volume swept out by the collecting drop of size r_0, v' difference in fall velocity of the collecting drop and the droplet of radius r_i, and n the number of intervals into which the small droplets are divided. Equation

(3.9) is not quite correct because growth by accretion does not occur continuously but in jumps each time a droplet coalesces with the collecting drop. If the discontinuity of accretion is taken into account the growth is somewhat faster than described by Eq. (3.9).

In considering growth of droplets by condensation and by accretion we have found that considerable uncertainty is introduced by interactions between complex processes within the cloud: processes which involve the air velocity, degree and nature of turbulence, the distributions of temperature and water vapor within the cloud, electrical conditions, and perhaps others. We must recognize that droplet growth can only be thoroughly understood as an integral part of the problem of the *macrophysical* processes of clouds. These processes have begun to be seriously investigated, and a summary of recent work is available in N. H. Fletcher's "Physics of Rainclouds" (see Bibliography).

3.7 Supercooling of Droplets

Clouds have been subjects of observation by scientists and poets and painters for hundreds of years, but for most of this time the physical properties of clouds have been largely ignored. Only since 1935 has serious attention been directed at the provocative observation that water clouds are quite common at temperatures below −30°C. In the laboratory distilled water in small droplets freezes only when cooled to below about −40°C, and there is scatter about this temperature suggesting a random process. Droplets which contain foreign insoluble material freeze at temperatures above −40°C; the freezing temperature is different for different impurities, and it increases with increasing volume.

Freezing of droplets may be explained on the basis of statistical considerations. Water molecules in thermal agitation may come into temporary alignment similar to that of an ice crystal. Such molecular aggregates may grow but also may be destroyed by random molecular motions. If an aggregate happens to grow to a size such that it is safe from the effects of thermal agitation, the whole droplet quickly freezes. The probability of growth of an aggregate to this critical size increases with decrease of temperature. The presence of a foreign particle makes the initial growth of the aggregate more probable by attracting a surface layer of water molecules on which the ice crystal lattice can form more readily than in the interior of the liquid. Freezing of a droplet requires that only one aggregate reach critical size, and therefore probability of freezing increases with volume. The probability that at least one foreign particle is present also increases with volume, so freezing temperature depends markedly on volume.

3.8 Ice Nuclei

The small solid particles which catalyse the freezing process are called *freezing nuclei*. They make up a very small part of the total atmospheric aerosol. Their concentration can be determined by observing ice crystal scintillations in a light beam in a refrigerated box or in an expansion chamber, or on a chilled metal plate. By these methods measurements have been made which show that nuclei become effective at different temperatures depending on their chemical composition. Crystalline dust particles of a particular origin may be effective at $-15°C$, volcanic ash at $-25°C$, other crystalline dusts at $-20°C$ or $-30°C$. Observations of natural air near the ground show that nuclei effective at $-5°C$ are only rarely present and in concentrations of only one per cubic meter or so. The number of effective nuclei increases with lowered temperature, and at $-20°C$ concentrations of 10^2 to 10^3 per cubic meter are often present. At temperatures below $-25°C$ more than 10^6 nuclei per cubic meter have been observed, but observations also have been made which failed to detect any ice crystals at all at temperatures above $-40°C$. Observations from aircraft also indicate great variability in nuclei concentration, both in number and in the temperature at which the nuclei may become effective. It is this natural variation in concentration of freezing nuclei which makes feasible the artificial modification of clouds. This subject is discussed in Section 3.13.

The effectiveness of a nucleus in initiating freezing seems to depend on how closely the crystal structure of the nucleus duplicates the crystal structure of ice. Therefore, ice is itself the most effective nucleating agent; ice crystals grow at temperatures just below $-0°C$. The next most effective agent known is silver iodide whose crystal form is, like ice, hexagonal with almost the same spacing of molecules within the lattice; silver iodide is effective at about $-5°C$. Then come, in order, lead iodide $(-7°C)$, ammonium iodide $(-9°C)$, and other compounds of iodine.

When an ice crystal is evaporated from a nucleus and then the same nucleus is tested in a cold chamber, crystallization sometimes occurs at temperatures just below the *freezing* point. This may result from small particles of ice which, remaining in crevices in the nucleus during evaporation, serve as freezing nuclei during the subsequent cold cycle. Such nuclei are called "trained" nuclei, and they are likely to be present among the nuclei which are left when cirrus clouds evaporate. Trained nuclei formed in this way may be very important in initiating freezing in supercooled clouds.

Freezing of supercooled droplets in the laboratory, once initiated, progresses rapidly through the cloud. Microscopic examination of the droplets during freezing helps to explain this phenomenon. When a shell of ice forms on a droplet, the water at the center is subjected to great pressure. It often is observed that water is squeezed out through a crack and forms a frozen spike which then is likely to break off. Or the internal pressure may shatter the frozen shell. In either case ice fragments are projected out into the region of the nearby supercooled droplets, and the freezing process is extended at an accelerated pace.

Ice crystals also may grow on suitable crystal structures by the transition from the vapor to the solid phase. This process is called *sublimation* or deposition, and the nuclei on which the initial sublimation occurs may be called *sublimation nuclei*. Because the same considerations of similar crystal structure apply in the cases of freezing and of sublimation, it is to be expected that the same nuclei serve as freezing and sublimation nuclei. The relative importance of the freezing and sublimation processes is uncertain. Because the probability that a cloud droplet captures a freezing nucleus is extremely small, it must be inferred that most freezing nuclei are attached to condensation nuclei before condensation occurs. This is quite possible for nuclei that have been suspended in the air for a very long time but very unlikely for artificial nuclei. Therefore, it is likely that observations of ice crystal growth following injection of nuclei are observations of the sublimation and not the freezing process. On the other hand, very little laboratory evidence has been obtained indicating that direct sublimation on nuclei occurs. Because of this fundamental uncertainty concerning the relative importance of freezing and sublimation, the nuclei responsible for ice crystal formation by either process are referred to as *ice nuclei*.

3.9   Vapor Pressure of Ice and Water Particles

From the molecular point of view, the fundamental distinction between the liquid and solid phases is a difference in energy level; that is, a molecule which makes up part of an ice crystal lattice is tightly held by intermolecular forces and is therefore in a state of low potential energy. A certain quantity of energy, called the latent heat of melting, is required to free the molecule from this low-energy state or to bring about the phase change from ice to liquid water. Still greater energy, the latent heat of vaporization, is required to bring about the phase change from liquid to water vapor. The latent heats have the following values near 0°C:

Melting: $L_{iw} = 334 \times 10^7$ erg g^{-1}
Vaporization: $L = (2500 - 2.274t\,°C) \times 10^7$ erg g^{-1}
Sublimation: $L_{iv} = (2834 - 0.149t\,°C) \times 10^7$ erg g^{-1}

At the triple point, $L_{iv} = L + L_{iw}$, as is to be expected from conservation of energy.

Now, the Clausius-Clapeyron equation shows that the equilibrium vapor pressure depends on the latent heat; it follows that the saturation vapor pressures over ice differs from the saturation vapor pressure over water at the same temperature. To calculate this difference, the Clausius-Clapeyron equation [Eq. (2.88)] may be integrated from the triple point, where the saturation vapor pressures are equal, to the temperature T. The result is

$$e_s = e_0 \exp\left(\frac{T - T_0}{R_w T T_0} L\right) \tag{3.10}$$

where e_s represents saturation vapor pressure at temperature T, e_0 and T_0 the vapor pressure and temperature at the triple point, and L the latent heat. If a similar integral is formed for the ice-vapor and liquid-vapor transition, it follows that

$$e_{sl} - e_{si} = e_0\left[\exp\left(\frac{T - T_0}{R_w T T_0} L\right) - \exp\left(\frac{T - T_0}{R_w T T_0} L_{iv}\right)\right] \tag{3.11}$$

Equation (3.11) is illustrated in Fig. 3.10. Maximum difference in vapor

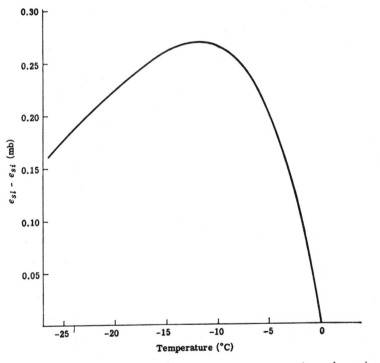

FIG. 3.10. Difference between saturation vapor pressures over water and over ice as a function of temperature calculated from Eq. (3.11).

pressure occurs at about $-12°C$. Here the difference amounts to 0.27 mb; this represents a supersaturation of 12.5% with respect to ice.

Now consider what happens in a cloud composed of supercooled water droplets when a few ice crystals are formed in it. The vapor pressure of the air in the cloud is likely to be within 0.1% of the vapor pressure of the droplets; but, as has just been calculated, may be higher than the vapor pressure of the ice crystals by as much as 0.27 millibar. Consequently, the ice crystals grow rapidly, depleting the water vapor in the cloud. The supercooled droplets now are no longer in equilibrium and evaporate in order to restore the water vapor in the cloud. In this way water substance is rapidly transferred from supercooled droplets to ice crystals.

3.10   Growth of Ice Particles

The growth of ice crystals depends on diffusion of water vapor and con-duction of heat just as does the growth of liquid droplets. However, ice crystal growth is simpler in that it does not involve change of equilibrium vapor pressure with size of the crystal; on the other hand, the complex geometry of the crystal introduces some difficulty into the calculation.

The surface of the crystal may be considered to have a uniform tempera-ture and therefore a uniform vapor pressure, and the vapor pressure at infinite distance (several crystal diameters) from the crystal is assumed uniform. The vapor pressure or vapor density in the neighborhood of the crystal may be represented by surfaces which tend to follow the contours of the crystal, but beyond the neighborhood of the crystal these surfaces approach the spherical shape as illustrated in Fig. 3.11.

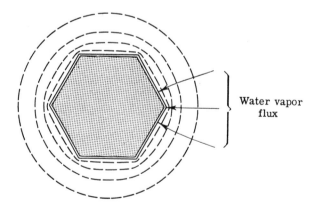

Water vapor flux

FIG. 3.11. Diffusion of water vapor toward one corner of a hexagonal ice crystal Dashed lines represent surfaces of constant vapor density and arrows represent direction of flux.

Flux of water vapor by diffusion occurs in the direction normal to the surfaces of constant vapor density and with the magnitude given by $-D\partial\rho_v/\partial n$ where n represents the coordinate measured normal to lines of constant ρ_v. Consequently, in the neighborhood of a sharp point vapor diffuses toward the point from many directions with the result that ice may accumulate more rapidly here than on flat surfaces. In spite of complexities of this sort, the diffusion occurring through a spherical surface of constant vapor density shown in Fig. 3.11 may be calculated using Eq. (2.78). The radius of the sphere is found to be represented by

$$\frac{R_wT\, dm/dt}{4\pi D(e_\infty - e_0)} = r \tag{3.12}$$

Evidently, the ratio of flux of vapor to vapor density increment is independent of the vapor density and is a characteristic of the sphere. This ratio may be called the *diffusion capacitance* of the sphere. Now, imagine that the sphere retreats slowly inward toward the crystal shape shown in Fig. 3.11 but in such a way that the original spherical vapor density surface remains undisturbed. The diffusion capacitance of the crystal depends in a complicated way on the size and shape of the crystal, but in spite of this the capacitance may be defined by the ratio of rate of water vapor flux to $4\pi D$ times the difference in vapor density between environment and the crystal surface. This concept is clarified if it is recognized that the capacitance, of course, has the dimension of length and is numerically equal to the radius of a sphere having the same surface vapor density and vapor flux as the crystal. Now, if the capacitance of the crystal is represented by C, Eq. (3.12) may be written

$$\frac{dm}{dt} = 4\pi\, \frac{DC}{R_wT}\, (e_\infty - e_0) \tag{3.13}$$

The problem of crystal growth by diffusion is an analogue of the problem of the charging of an electrical capacitor in which electrical potential replaces the vapor density increment and electrical charge replaces the ratio of vapor flux to diffusion coefficient. The electrical capacitance of a complex geometrical shape can be measured in the laboratory from the charge-to-potential ratio just as, in principle, the diffusion capacitance can be measured from the flux to vapor pressure ratio given by Eq. (3.12). Therefore, existence of the electrical analogue to the diffusion problem makes it possible to determine the diffusion capacitance of any crystal by measuring the electrical capacitance of a body of the same shape and size.

Equation (3.13) permits calculation of the rate of growth of a crystal for any ambient vapor density if its shape and size and the vapor density at the crystal surface are known. The latter, of course, depends on the temperature of the surface which is influenced by the latent heat of fusion released at the surface. Inward diffusion of water vapor is therefore limited by the rate at which latent heat can be dissipated.

If it is assumed that dissipation of the latent heat occurs solely by conduction into the air surrounding the crystal and that the crystal surface is at uniform temperature, the heat conduction problem becomes an analogue of the diffusion problem. Thus

$$L_{iv} \frac{dm}{dt} = 4\pi Ck(T_0 - T_\infty) \tag{3.14}$$

Using Eqs. (2.88), (3.13), and (3.14) the rate of crystal growth may be derived by eliminating T_0 and e_0 in the same way that they were eliminated in the case of droplet growth. The result may be written in the form

$$\frac{dm}{dt} = \frac{4\pi CDkR_w T_\infty^2 (e_\infty - e_{si\,\infty})}{L_{iv}^2 e_{si\,\infty} D + R_w^2 T_\infty^3 k} \tag{3.15}$$

where e_∞ represents ambient vapor pressure and $e_{si\,\infty}$ saturation vapor pressure over ice at the ambient temperature. Equation (3.15) shows that the rate of growth is directly proportional to the supersaturation of water vapor with respect to ice. It follows that Eq. (3.15) also describes the evaporation of ice crystals in unsaturated air. An example is given in problem 5. The rate of growth calculated from Eq. (3.15) is very much greater than the rate of growth of water droplets, primarily due to the large supersaturation with respect to ice. A sample comparison shows that a 20 μ water drop in air supersaturated by 0.05% requires about five hundred times as long to grow to a droplet of 50 μ radius as does an ice crystal of equivalent mass at $-12°C$.

3.11   Structure of Ice Crystals

The molecular structure of ice crystals has been determined by X-ray and neutron diffraction experiments. These results show that at atmospheric temperatures the basic crystal structure consists of six oxygen atoms arranged in a hexagon. Each oxygen is bonded to two hydrogens, and each hydrogen is bonded to two oxygens. Crystals may grow in various ways, however; for example, the hexagonal base may grow in two dimensions to form a hexagonal plate, or it may grow in the direction normal to the base to form a hexagonal prism. Mode of development is influenced by the external diffusion field, but it also probably depends on accidental arrangements of molecules and on impurities in the crystal lattice. The chief distinctive forms of ice crystals found abundantly in clouds are illustrated in Fig. 3.12.

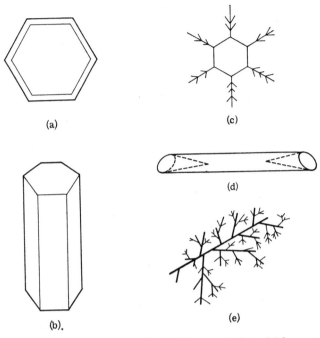

FIG. 3.12. Typical forms of ice crystals. (a) Hexagonal plate, (b) hexagonal column, (c) stellar crystals, (d) needle, (e) dendrite.

Experiment sheds valuable light on the phenomenon. If a thread is suspended in a chamber in which the air temperature increases upward from, say, −30°C to 0°C and water vapor is introduced into the chamber, ice crystals form in a very curious fashion on the thread. Near the top of the chamber hexagonal plates form, further down there are prisms and needles, still further down plates and dendrites, and finally prisms again near the bottom. The separation between the various crystal forms is quite sharp and corresponds closely to the following temperature ranges:

$$
\begin{aligned}
&0 \text{ to } -3°C && \text{hexagonal plates} \\
&-3 \text{ to } -8°C && \text{prisms and needles} \\
&-8 \text{ to } -12°C && \text{plates and dendrites} \\
&-12 \text{ to } -16°C && \text{plates} \\
&-16 \text{ to } -25°C && \text{prisms}
\end{aligned}
$$

The temperature ranges also depend slightly on the humidity. These same forms of crystals are observed in the atmosphere, so that it must be concluded that temperature plays a crucial role in the specification of crystal form. The mechanism responsible for this remarkable relation is under intensive investigation in several laboratories.

3.12   Precipitation

In the preceding sections various microphysical processes responsible for the growth of cloud particles have been discussed. In trying to understand the relation of these processes to precipitation, it is important to realize that the microphysical processes which govern the behavior of cloud particles are dependent on the large scale or macrophysical processes, and therefore an account of the factors which govern precipitation is likely to be a much more complex affair than is an account of the microphysical factors. However, it is possible to give a reasonable explanation of the formation of precipitation with the aid of calculations on simple models of the air motion.

The liquid or solid water contained in a cloud at a particular time is rarely sufficient to provide more than a trace of precipitation, as may be verified by solving problem 6. Obviously, in order to produce an economically significant amount of precipitation an updraft must provide a continuous and plentiful supply of moisture to the cloud. Clouds which are supplied with adequate water vapor must process the vapor into precipitation. Condensation on liquid droplets, although important in the early stages of droplet growth, cannot alone produce raindrops. On the other hand, it has been shown in Section 3.10 that ice particles surrounded by supercooled water drops may grow extremely rapidly and under favorable conditions may accumulate sufficient mass to become raindrops after melting. Wegener in 1911 had already recognized the rapid growth of ice particles, but Bergeron in 1935 was the first to hold this mechanism responsible for the formation of rain. For some time after Bergeron's theory was published it was believed that precipitation could only be initiated in clouds which extend considerably above the height of melting. However, studies of showers have shown that clouds from which precipitation falls often do not extend to the height of 0°C. In these cases precipitation can form only by accretion. In clouds which extend well above the height of melting, both processes are present; the importance of the Wegener-Bergeron mechanism is likely to predominate in clouds with relatively weak updrafts (stratiform clouds), whereas the accretion mechanism is likely to predominate in clouds with relatively large updrafts (cumuliform clouds). In the following part of this section methods for calculating rate of precipitation by the two mechanisms will be discussed.

Precipitation by the Wegener-Bergeron Mechanism

Consider an active stratiform cloud in which a continuous supply of water vapor is provided by an updraft of the order of 10 cm sec^{-1}, and imagine that the updraft causes the temperature in the cloud to decrease at a uniform rate. When the temperature is slightly below 0°C condensation takes place on supercooled droplets but at a rather slow rate. At a temperature between −10°C and −20°C, ice nuclei present in the cloud may initiate ice crystal formation. Because of their rapid growth at the expense of the supercooled water droplets, the ice particles quickly attain sufficient size to fall with respect to the surrounding air. The size attained during their fall through the cloud depends on the number of ice particles formed. If many particles compete, the supercooled droplets evaporate altogether, and the vapor pressure attains a value somewhere between saturation pressure over ice and saturation vapor pressure over water. This reduces the rate of growth of the ice crystals. If only a few ice crystals are formed, they grow at a maximum rate.

As the crystal falls, it usually enters air which is increasingly warm and moist. Therefore, a crystal at a temperature colder than −12°C usually experiences an increased rate of growth by sublimation as it falls toward warmer air, whereas for a crystal which is warmer than −12°C the reverse is likely to be true.

The rate of vertical displacement of the ice crystal may be expressed by

$$\frac{dz}{dt} = u - v \tag{3.16}$$

where v represents the terminal velocity of the ice crystal and u the updraft velocity of the air in the cloud. The terminal velocity of hexagonal plates is mainly a function of the thickness of the plate, but not directly of the mass. For thickness of 15 μ the terminal velocity is about 40 cm sec^{-1} and for thickness of 40 μ the terminal velocity is about 70 cm sec^{-1}. If both u and v are considered constant, then Eq. (3.16) can be integrated easily over the thickness of the cloud (z_0) giving for the time required for the crystal to fall from the top of the cloud, where it began to grow, to the bottom

$$t_0 = \frac{z_0}{u - v} \tag{3.17}$$

If the temperature and the vapor pressure are known through the entire cloud layer, then Eq. (3.16) may be integrated numerically to yield the mass of the crystal as a function of time. Before the crystal reaches the ground it may have melted into a water droplet of the same mass. During its fall some evaporation may take place because the air under the cloud may not be saturated. For this reason drops may decrease in size between the cloud base and the ground, and in many cases do not reach the ground at all. Neglecting evaporation between cloud and ground, the rate of precipitation can be determined using the pseudo-adiabatic chart if the updraft and the thickness of the cloud are known. Problem 7 provides an exercise.

The Accretion Mechanism

At radii greater than about 20 μ growth by accretion exceeds growth by condensation as shown in Fig. 3.6, and at radii greater than about sixty microns only accretion is important.

Now consider a large droplet which enters the base of a cloud and is carried upward by an updraft of velocity v. The initial large droplet may result from chance accretion of smaller droplets, from condensation on giant salt nuclei or from the breakup of large drops falling from the upper part of the cloud. In any case such droplets are relatively rare, but may nevertheless be very important in producing raindrops. If Eq. (3.16) is combined with Eq. (3.9), the radius of the droplet is expressed as a function of height by

$$\frac{dr_0}{dz} = \frac{\sum_{i=1}^{n} \epsilon_i v_i' \, \Delta\rho_{di}}{4\rho_w(u - v)} \tag{3.18}$$

The accretion efficiency is uncertain; but the uncertainty may be minimized by considering only the large droplets for which ϵ is near unity. Further simplification may be achieved by considering that all collisions occur with particles of negligible fall speed, so that $v \approx v'$. Under these conditions Eq. (3.18) becomes

$$4\rho_w \int_{r_0}^{r_0'} \frac{(u - v) \, dr_0}{v} = \int_{z_0}^{z} \rho_d \, dz \tag{3.19}$$

Equation (3.19) may be applied to the case of a droplet which enters the cloud at the bottom, is carried aloft by the vertical current u until v exceeds u, and finally falls out the bottom of the cloud. If further simplification is achieved by assuming ρ_d and u independent of height, integration of Eq. (3.19) over the complete path gives

$$r_0' - r_0 = u \int_{r_0}^{r_0'} \frac{dr_0}{v} \tag{3.20}$$

In order to integrate Eq. (3.20) the terminal velocity must be known as a function of radius. Experimental results are shown in Fig. 3.7. Integration of Eq. (3.20) now gives the raindrop radius as a function of updraft velocity. Examples shown in Fig. 3.13 indicate that drops of 2.5-mm radius can develop in updrafts of from 3 to 4 m sec^{-1}. Drops larger than this are likely to break up. For a particular updraft the 20 μ droplets achieve larger size than the 30 or 40 μ droplets; this anomalous result occurs because the smaller droplets are carried higher into the cloud than the larger droplets,

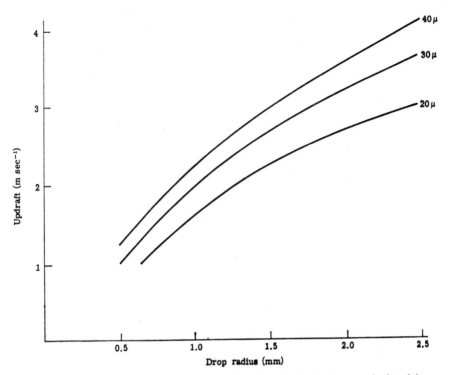

FIG. 3.13. Final raindrop radius as a function of updraft velocity as calculated from Eq. (3.20) for initial radii of 20, 30 and 40 μ [after F. H. Ludlam, *Quart. J. Roy. Meteorol. Soc.* **77**, 402 (1951)].

and therefore eventually grow to greater size. This of course requires a greater length of time.

The minimum vertical extent necessary for formation of showers by accretion has been calculated by Ludlam* in the following way. A droplet which grows to a radius greater than about 150 μ during its ascent through the cloud has a fall velocity of about one meter per second. Therefore, Eq. (3.19) can be written in the form

$$\int_{r_0}^{150\mu} \frac{u \, dr_0}{\epsilon v} - \int_{r_0}^{150\mu} \frac{dr_0}{\epsilon} = \frac{1}{4\rho_w} \int_0^h \rho_d \, dz \qquad (3.21)$$

where h represents the minimum cloud depth necessary for droplets to grow to 150 μ radius. The path followed by the drop during growth is illustrated in Fig. 3.14. If the liquid water concentration is assumed to be equal to the decrease in saturation mixing ratio which occurs for ascent along the wet adiabatic lines from the cloud base, graphical integration

* F. H. Ludlam, *Quart. J. Roy. Meteorol. Soc.* **77**, 402 (1951).

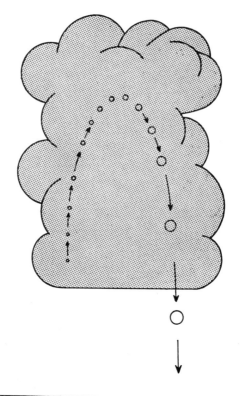

FIG. 3.14. Schematic diagram of the trajectory of a growing droplet in a convection cell.

using the pseudo-adiabatic diagram serves to determine the liquid water concentration. Results of calculations using an accretion efficiency of 0.85 are shown in Fig. 3.15. This figure shows that minimum thickness is roughly proportional to updraft velocity. For warm clouds the minimum depth for an updraft of 5 m sec^{-1} is 2 or 3 kilometers; the whole cloud therefore is well above freezing. On the other hand, for cold clouds the minimum depth is greater, so that the cloud tops are likely to be well below 0°C, and formation of ice crystals is likely.

Among the many simplifying assumptions which have been made in this discussion, the arbitrary choice of an accretion efficiency is probably most questionable. In spite of this, the results of the calculations discussed here are in rough agreement with the results of observation in clouds.

An analysis for the case of supercooled droplets and ice crystals has been carried out also by Ludlam.* The results show that growth by sublimation

* F. H. Ludlam, *Quart. J. Roy. Meteorol. Soc.* **78,** 543 (1952).

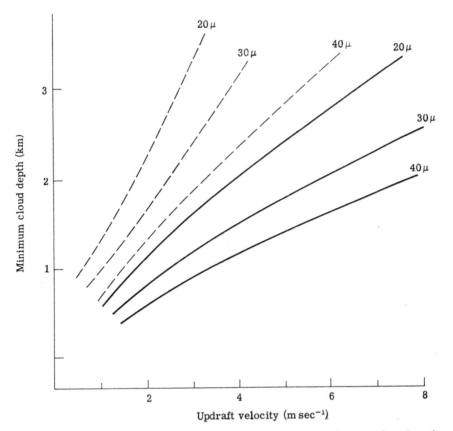

FIG. 3.15. Minimum cloud depth for drop formation by accretion as a function of updraft velocity. Results are based on Eq. (3.21) using accretion efficiency of 0.85 and cloud base temperatures of $-5°C$ (*dashed lines*) and 20°C (*solid lines*) [after F. H. Ludlam, *Quart. J. Roy. Meteorol. Soc.* **77**, 402 (1951)].

is, of course, most important in the early stage when the fall velocity of the crystals is small. Soon, however, the crystals reach a size such that they fall relative to the surrounding air, and subsequently the ice particles may grow primarily by accretion. If the temperature is low or the accretion slow, the supercooled droplets freeze upon coalescing with the crystals. But if the temperature is only a few degrees below zero and if accretion is rapid, the latent heat of melting may be released so rapidly that the temperature rises to 0°C and water remains unfrozen. The falling ice particles may shed liquid water which then may grow by the water accretion process. Under favorable conditions this may be an efficient method for developing the large drops necessary for shower formation.

3.13   Artificial Cloud Modification

The mechanism of natural precipitation may be incompletely efficient in two respects, and these inefficiencies give man the opportunity under special conditions to exert limited control over the mechanism. Both inefficiencies involve "trigger" mechanisms, processes within the total precipitation mechanism which exert a large influence over subsequent development. In Section 3.8 it was pointed out that naturally occurring ice nuclei vary widely in concentration and in effectiveness. Several possibilities for cloud modification follow from this observation. Addition of artificial nuclei active at temperatures below 0°C may initiate freezing of supercooled droplets and thus set off the rapid growth of ice crystals which may lead to precipitation. Or, if an abundance of artificial ice nuclei is added to a supercooled cloud, so many minute crystals may form that a supercooled cloud is transformed to an ice crystal cloud in which further growth of crystals is inhibited by competition for water vapor between the myriad crystals. This is referred to as "overseeding" a cloud. The second opportunity for exerting control over the precipitation mechanism lies in influencing the accretion process through the introduction of relatively large droplets or of hygroscopic particles which grow to large size in a short time.

Modification of supercooled clouds by adding ice nuclei has been successfully carried out many times. Although early experiments of potential importance were made by several people, the first carefully observed experiments were carried out by Schaefer[*] in 1946. He observed the dramatic effect of dropping a bit of dry ice (solid CO_2) into a supercooled laboratory cloud, thereby producing some 10^8 ice crystals in a volume of about 130 liters. Subsequently, a field observation demonstrated that dry ice dropped into a supercooled layer cloud can convert large volumes of the cloud into ice crystal cloud. In some cases precipitation fell, leaving clear air surrounded by cloud. The physical process in this case probably is the freezing of many droplets by contact with the dry ice surface (temperature -78.5°C). The frozen droplets may be shattered into tiny crystals by the internal pressure created by freezing; these then grow rapidly in the supercooled cloud. Ice crystal clouds also have been formed by dropping dry ice into air which is saturated with respect to ice but not with respect to water.

Soon after Schaefer's demonstration of the effect of dry ice on supercooled cloud Vonnegut[†] discovered that silver iodide crystals are effective in producing ice crystals in water clouds at temperatures below -5°C. Subsequently, silver iodide was shown to be effective in modifying natural supercooled clouds. The importance of Vonnegut's discovery is that, al-

---

[*] V. J. Schaefer, *Science* 104, 457 (1946).
[†] B. Vonnegut, *J. Appl. Phys.* 18, 593 (1947).

though silver iodide is somewhat less effective than dry ice, it presumably can be introduced into the atmosphere upwind of the target area and perhaps at the ground where the temperature may be above 0°C and carried into supercooled clouds by natural air currents; in this way Vonnegut's discovery enormously extended the feasibility of large-scale seeding operations.

Cloud modification by influencing the accretion mechanism was suggested in 1948 by Langmuir.* Subsequent development of the accretion theory of precipitation, as outlined in the preceding section, indicated that an effective procedure should be the injection of droplets of thirty-micron radius into the bottom of a cumulus cloud having a depth greater than one and a half kilometers. However, it is important to realize that the potential effectiveness of precipitation stimulation by accretion is much less than that by ice crystal formation due to the stubborn facts of the microphysical processes involved, so that it is easy to understand that observational evidence of seeding with water is somewhat equivocal at present.

Stimulation of accretion by adding giant hygroscopic nuclei to the air may be effective in cases where natural giant nuclei are deficient. A reliable test requires that the concentration of naturally occurring giant nuclei be counted in the cloud, but this has not been done. Without this information, it is not possible to say whether addition of hygroscopic material should have a stimulating or an inhibiting effect on precipitation.

There is a large step between proof that cloud modification is possible and proof that large scale control of precipitation is possible, and this step has not yet been taken. Extensive cloud seeding trials, using silver iodide generators (usually based on the ground) have been conducted by research groups and by private commercial companies, but the reports of these trials have been conflicting. The difficulty in determining the path of the silver iodide crystals is added to all the other unknown factors, so that even negative results are not entirely convincing.

In spite of the fact that large scale cloud seeding remains to be adequately tested, an understanding of the essential physical processes enables us to distinguish the soundly based possibilities from the fantastic. The hydrological cycle may be imagined as a powerful, complex, and smoothly operating engine continuously transforming enormous amounts of energy. The astute use of cloud seeding then may serve as a "quality" lubricant on one of the bearings. Viewed in worldwide terms cloud seeding must be of marginal effectiveness in increasing precipitation. It can probably redistribute precipitation to a distinctly limited extent, and it can probably increase or decrease total annual precipitation locally by something less than a factor of two in the most favorable areas. It is unlikely that the

* I. Langmuir, *J. Meteorol.* **5**, 175 (1948).

total earthwide precipitation can be changed appreciably by cloud seeding, and it is unlikely that precipitation in arid areas can be increased enough to make the desert bloom. The possibilities which remain are sufficiently exciting. There are many areas whose natural precipitation is marginal; the economies of these areas might be changed dramatically by adding 10–20% to the annual precipitation. And, the restriction of precipitation to prevent flooding and to reserve moisture for other regions downwind are reasonable possibilities which have not yet been thoroughly explored.

PART II: ELECTRICAL CHARGE GENERATION AND ITS EFFECTS

3.14 Elementary Principles of Electricity

Coulomb's Law

An electrical point charge (q) placed at a distance $(\mathbf{r})$ from a charge (Q) experiences a force given by Coulomb's Law in the form

$$\mathbf{F}_c = \frac{qQ}{4\pi\epsilon r^2}\frac{\mathbf{r}}{r}$$

(3.22)

where $\mathbf{r}$ is directed from Q to q and ϵ, the *permittivity*, is a characteristic of the medium. Permittivity is defined for vacuum by $\epsilon_0 = (4\pi)^{-1}$, and for air $\epsilon \approx \epsilon_0$. When $\mathbf{F}_c = 1$ dyn, $r = 1$ cm, $\epsilon = (4\pi)^{-1}$, and $q = Q$, then each charge is by definition equal to one *electrostatic unit of charge* or one *franklin** (Fr). The forces repel each other when the signs of the charges are the same and attract each other when the signs are opposite. Coulomb's Law is fundamental in the same sense as Newton's laws or the Law of Conservation of Energy.

The Electric Field

The force on a unit test charge due to another charge Q may be associated with an *electric field* which is proportional to Q. This bears close analogy to the gravitational field discussed in Chapter I. The electric field is defined as the limit of the force per unit test charge as the test charge diminishes to zero, and is written

$$\mathbf{E} \equiv \lim_{q\to 0}\frac{\mathbf{F}}{q}$$

(3.23)

---

* The franklin (Fr) has been adopted here as the electrostatic unit of charge following the suggestion of various authors, e.g., Guggenheim, Fleury, de Boer. Electrical units are listed in Appendix II.C.

Equation (3.22) shows that the *electrostatic field* may be expressed by

$$E_c \equiv \frac{Q}{4\pi\epsilon r^2}\frac{\mathbf{r}}{r} \qquad (3.24)$$

The electric field may arise from other sources than the electrostatic, and in Chapter VI it will be necessary to discuss this subject again and in more detail.

The potential energy of a unit charge in the electrostatic field is defined by the work necessary to bring the unit of charge from $r = \infty$ to $r = r_0$ against the electrostatic field. Thus, the *electrostatic potential* or simply the *potential* at r_0 is expressed by

$$V_0 = \int_\infty^{r_0} \frac{Q}{4\pi\epsilon r^2}\frac{\mathbf{r}}{r} \cdot d\mathbf{r} = \frac{Q}{4\pi\epsilon r_0} \qquad (3.25)$$

Equations (3.24) and (3.25) together show that the potential and the electrostatic field vector are related by

$$-\frac{\mathbf{r}}{r}\frac{\partial V}{\partial r} = E_c$$

Surfaces of equal potential may be imagined which surround the charge Q; the test charge may be moved on a potential surface without work just as masses may be moved on a geopotential surface without work. It follows that the surface of a charged conductor must be an equipotential surface if the charges are at rest, and the potential must represent the work done in bringing a charge Q from infinite distance to the conductor. Therefore, the potential of a conductor is proportional to the charge, and the *capacitance* (C) may be defined by

$$V \equiv \frac{Q}{C} \qquad \text{or} \qquad Q = CV \qquad (3.26)$$

Combining Eq. (3.25) with (3.26) yields for capacitance of a sphere in vacuum

$$C = r \qquad (3.27)$$

Ohm's Law

Experiment shows that electrical charge flows along a conductor at a rate which is proportional to drop in potential. Thus

$$\Delta V = R\frac{dQ}{dt} = RI \qquad (3.28)$$

where the *resistance* (R) is characteristic of the conductor. Equation (3.28) is the well-known Ohm's law; it is one of several equations which are analogous to the heat flux equation [Eq. (2.73)].

3.15   Charge Separation in Clouds

Clouds which extend well above the height of 0°C and in which there are present strong updrafts act as powerful electrostatic generators, the upward moving air carrying small particles of one sign (usually positive) and the falling precipitation elements charge of the other sign (usually negative). Experiment shows that separation of charge occurs during freezing at the interface between liquid water and ice or between two ice surfaces when there is a difference in temperature or in concentration of dissociated contaminant. A suggested explanation is based on the fact that (a) the concentration of dissociated ions increases rapidly with increasing temperature or increasing dissolved contamination and (b) the *mobility* (the average drift velocity under a potential gradient of one volt per centimeter) of the positive hydrogen ion is an order of magnitude greater than the mobility of the negative hydroxyl ion. The consequence of (a) is that a temperature gradient is accompanied by a concentration gradient of both positive and negative ions. The consequence of (b) is that diffusion of positive ions is more rapid than diffusion of negative ions, and a positive charge therefore develops in the colder region. Estimates made by Latham and Mason[*] of the charge separation and the potential difference developed by this process amount to $5 \times 10^{-5} \, dT/dx$ Fr cm^{-2} and $2\Delta T$ millivolts (mv), respectively.

To understand how these results apply to the process of charge separation in clouds, both the micro- and macrophysical properties of clouds must be understood, and here we encounter once again, while walking rapidly, a slowly opening door. At least two processes may be effective. Latham and Mason[†] have suggested that most of the charge separation results from the following mechanism. Small supercooled water droplets are carried upward by an updraft and collide with falling soft hail pellets. Collision results in freezing of part of the droplet and in raising the temperature of the whole droplet to 0°C by the release of the latent heat of fusion. Heat is then lost to the surrounding cold air by conduction, with the result that a radial temperature gradient is established as shown in Fig. 3.16. Freezing proceeds from the outside, but as the interior freezes and expands the outer shell is shattered. The positively charged fragments are carried upward by the air current while the falling pellet carries its negative charge downward. Charging occurs efficiently for radii between about 20 and 50 μ but decreases sharply beyond these limits. The lower limit reflects failure of very small droplets to shatter upon freezing, and the upper limit reflects ejection or splashing of water from the negatively charged interior of large droplets

---

[*] J. Latham and B. J. Mason, *Proc. Roy. Soc.* **A260**, 323 (1961).
[†] J. Latham and B. J. Mason, *Proc. Roy. Soc.* **A260**, 537 (1961).

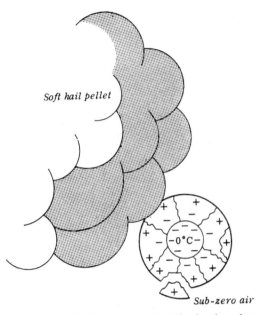

FIG. 3.16. Temperature gradient and charge distribution in a freezing droplet upon adhering to a soft hail pellet.

into the air stream. Laboratory measurements have shown that a droplet of 40 μ radius in an environment of $-15°C$ separates a charge of 4×10^{-6} Fr, and this value appears to be consistent with the theory of preferential ion diffusion described above.

On the other hand, Reynolds et al.* had earlier suggested that particles which grow rapidly by coalescence with supercooled droplets (soft hail pellets) are warmer than their environment by several degrees. Collision of these particles with colder small ice crystals brings about separation of charge, the small crystals being carried upward with newly developed positive charges and the large ice particles falling with negative charge.

The rate of charge separation may now be estimated for the droplet freezing process. Consider a cloud with supercooled droplet concentration, n, and pellet concentration N. If ϵ_c represents the collision efficiency of the pellets with radius r, v' the relative fall velocity of the pellets and droplets, and q the charge separated per collision, the rate of charging per unit volume is expressed by

$$\frac{dQ}{dt} = \epsilon_c \pi r^2 n v' N q \qquad (3.29)$$

Although the terms on the right have not been accurately determined, and

* S. E. Reynolds, M. Brook, and M. F. Gourley, J. Meteorol. 14, 426 (1957).

although they undoubtedly vary from one case to another, the following are conservative but realistic estimates for convective clouds:

$$r: \quad 0.2 \text{ cm}$$
$$\epsilon_c: \quad 1$$
$$v': \quad 1 \text{ m sec}^{-1}$$
$$N: \quad 10^{-3} \text{ cm}^{-3}$$
$$n: \quad 1 \text{ cm}^{-3}$$
$$q: \quad 4 \times 10^{-6} \text{ Fr}$$

These values yield a charging rate of one coulomb per cubic kilometer per minute. For a thunderstorm of 50 km³ volume 10³ coulombs could be generated in 20 minutes, and this rate of charging is consistent with the discharge rate observed in lightning.

An estimate of rate of charging by the ice crystal collision process may be made using Eq. (3.29) if n now represents concentration of ice crystals, and q represents charge separation per collision. Experimental measurements of charge separation by Reynolds et al. exceed corresponding measurements by Latham and Mason by five orders of magnitude! so that it is not possible unequivocally to assess the importance of the ice crystal collision process. The much higher value found by Reynolds et al. may result from the fact that supercooled droplets were collected by the ice crystal target during the experiment so that separation by freezing and shattering may have been occurring. In any case there is laboratory evidence for charge separation of magnitude sufficient to account for the observed charge separation in convective clouds. In order to understand the effects of these cloud generators other electrical properties of the atmosphere must be considered.

3.16   Origin and Distribution of Ions

For many purposes air acts as a good insulator; but when careful measurements are made, air turns out to have characteristic conductivity which varies with time, location, and especially with height. The conducting property of air has been known since the late eighteenth century when Coulomb called attention to the slow discharge of a charged body through air, but an adequate explanation was proposed only at the beginning of the twentieth century by Elster and Geitel in Germany and C. T. R. Wilson in England. They attributed the conductivity to the movement of positive and negative ions of molecular size through the air under an electrical field. The investigation of the source of these ions led to the discovery of cosmic rays by Hess in 1911, and to the development of cosmic-ray physics.

Primary cosmic rays are particles of very great energy, mostly protons, which enter the earth's atmosphere from all directions and produce other high energy particles by colliding with neutral air molecules. These products of collision are called secondary cosmic rays, and they are responsible for production of ions by collision with air molecules. The energy of cosmic radiation is comparable in magnitude to the light received from stars and is distributed over an enormous range. The primaries appear to have maximum total energy in the neighborhood of 10^{10} electron volts (ev) with occasional particles of 10^{12}, 10^{14}, even the staggering figure of 10^{19} electron volts. The latter figure is approximately one calorie! and it should be understood that each measurement of cosmic-ray energy is likely to be an underestimate. The lower energy particles are absorbed in the upper atmosphere, while the highest energy particles penetrate all the way to the earth, producing multiple high energy particles each of which leaves a trail of ionized air molecules. The average number of ion pairs produced at sea level by cosmic rays is about 1.5 cm^{-3} sec^{-1}; ion production increases with latitude due to the effect of the earth's magnetic field in deflecting the charged primary rays, thereby shielding the region of the geomagnetic equator more effectively than the higher latitudes. Ion production increases with height up to about three hundred per cubic centimeter per second at thirteen kilometers and decreases above that height as shown in Fig. 3.17. Primary cosmic rays of the greatest energies may produce cascades or showers of as many as 10^8 high energy particles; on these occasions the ion density may be suddenly increased by more than an order of magnitude throughout a conical volume of diameter at the ground as large as a kilometer or more.

Near the ground ions are also produced by radioactive decay of elements in the soil and in the air; ion production by this process is variable depending on air currents, static stability, and atmospheric pressure as well as on proximity to radioactive rocks and the condition of the surface. Over land radioactivity produces roughly eight ion pairs per cubic centimeter per second on an average, but radioactivity decreases to negligible intensity at a height of a few kilometers. Over the sea radioactivity is not important in producing ions.

Near the outer limit of the atmosphere cosmic rays are ineffective ionizers because only the primary particles are present in the cosmic radiation and these are few in number. Ions are produced, however, in this region by absorption of ultraviolet and X-radiation from the sun. At the low densities present above, say two hundred kilometers, the electrons ejected from the neutral oxygen and nitrogen atoms can remain for long periods as free electrons. Under the daytime sun's radiation electron densities of 10^6 cm^{-3} occur, and these decrease by only one order of magnitude at night. The

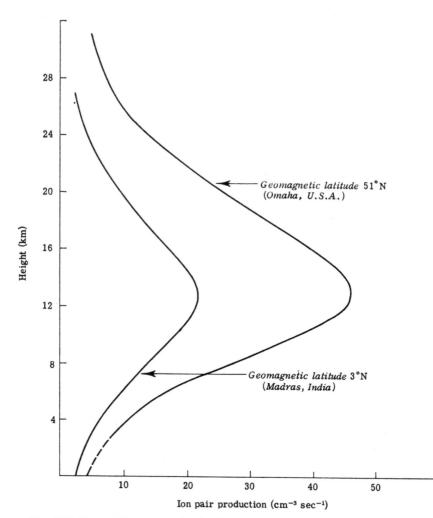

Fig. 3.17. Rate of ion pair production by cosmic rays as a function of height [after O. H. Gish, *Compendium Meteorol.*, p. 101 (1951)].

atmosphere, therefore, contains a spherical shell of high ion density, called the *ionosphere*, which extends from a height of about 80 kilometers to above 300 kilometers. Inside this shell, ion density produced by cosmic rays amounts to only hundreds per cubic centimeter. Properties of the ionosphere will be discussed in more detail in Chapters IV and VI.

Ionization, whether produced by cosmic ray or ultraviolet light, results in creation of a negative electron and a much heavier positive ion. Near sea level the electrons, with average lifetimes of 10^{-5} seconds, very quickly become attached to neutral molecules; the resulting negative ions as well as the positive ions have lifetimes of roughly 10^2 seconds. These ions may attract a group of neutral molecules but their size remains essentially molecular, and they are therefore called *small* ions. Other ions become attached to aerosol particles which are very large compared to molecules; they are called *large* or *Langevin* ions and their average life is roughly 10^3 seconds.

In an electrostatic field ions are accelerated, negatives toward the center of positive potential and positives toward the center of negative potential. However, the mean free path in air at sea level is only about 10^{-7} centimeter, so that an ion experiences a series of accelerations interrupted by abrupt collisions; it therefore moves through the air with an average drift velocity which depends on mean free path, potential gradient, charge of the ion, and mass of the ion. The mobility of small ions in air at sea level is roughly 1 to 2 cm $\sec^{-1}$ (volt $\text{cm}^{-1})^{-1}$, and of large ions is 10^{-4} to 10^{-2} cm $\sec^{-1}$ (volt $\text{cm}^{-1})^{-1}$.

3.17 Conductivity

Small ions are constantly produced at a rate (p) which depends upon the processes described above, and they are constantly destroyed by recombination with ions of the opposite sign and taken out of circulation by attachment to large particles. Recombination of each ion is proportional to ion density so that the rate of recombination is proportional to the square of the ion density. Similarly, rate of ion attachment is proportional to the density of small ions and to the density of large particles. Under conditions of equilibrium between ion production and destruction

$$p = \alpha n^2 + \beta nN \qquad (3.30)$$

where n and N represent, respectively, the number of small ions and large particles per unit volume, α is called the *recombination coefficient* for small ions and has the rough value 1.6×10^{-6} $\text{cm}^3\,\sec^{-1}$, and β is called the *combination coefficient* for small ions and large particles and has the rough value 3×10^{-6} $\text{cm}^3\,\sec^{-1}$. The individual terms of Eq. (3.30) are only formal descriptions of the creation and destruction of ions; no additional physics is represented by the equation. However, if α, β, N, and p are known from experiment and observation, the concentration of small ions can be calculated. Near cities where N may be 5×10^3 cm^{-3} or even larger, small ion concentration is determined mostly by attachment to large particles, but above a height of a kilometer or over the ocean recombination is the controlling process. The concentration of small ions increases with height from roughly 6×10^2 cm^{-3} at sea level to a maximum of 5×10^3 cm^{-3} at a height of 15 km. Equation (3.30) is applied in problem 8.

The typical lifetime of small ions may be estimated in the following way. Imagine ion production within a cubic centimeter at the rate p, and allow this production to continue for a time τ until the cubic centimeter contains n ions. Therefore, $p\tau = n$. Then allow destruction of ions to proceed at the rate p. Ion concentration subsequently will remain at n and under equilibrium conditions the average lifetime will be equal to τ and expressed by

$$\tau = \frac{n}{p} \tag{3.31}$$

Problem 9 provides an application of Eq. (3.31).

The *specific conductivity* of air may be defined by Ohm's law in the form

$$\lambda \equiv \frac{i}{dV/dz} \tag{3.32}$$

where i represents electrical current per unit area. Equation (3.32) is a special form of Eq. (3.28). Because the mobility (w) may be expressed by the identity

$$i \equiv new \frac{dV}{dz} \tag{3.33}$$

where n represents number of ions per unit volume and e charge on each ion, Eq. (3.32) may be written

$$\lambda = new \tag{3.34}$$

Equations (3.32)–(3.34) may be written separately for ions of different mobilities, charges and concentrations; but this refinement will be omitted here. Because of upward movement of negative ions under the atmospheric potential gradient, near the ground there are more positive than negative ions. However, the resulting positive space charge is small and is not essential to the purpose of this discussion. The mobility of small ions greatly exceeds that of large ions, so that conductivity depends mainly on the number of small ions per unit volume.

One may expect ion mobility to be inversely proportional to air density; and, although it also depends on other factors, the density dependence is the only one to be considered here. Upon introducing the density dependence into Eq. (3.34), the conductivity may be expressed in terms of conductivity at a base height where the density is ρ_0 by the equation

$$\lambda = \frac{ne\rho_0}{n_0 e_0 \rho} \lambda_0$$

Between sea level and a height of 15 km n increases by an order of magnitude while ρ decreases by an order of magnitude. Consequently, conductivity increases by two orders of magnitude over this height range. The increase in cosmic radiation with geomagnetic latitude results in an increase in n between geomagnetic equator and high latitudes by a factor of about three. Balloon measurements of air conductivity and related calculations are shown in Fig. 3.18.

The chief importance of electrical conductivity and of its vertical distribution in the atmosphere is the fact that a vertical current is permitted

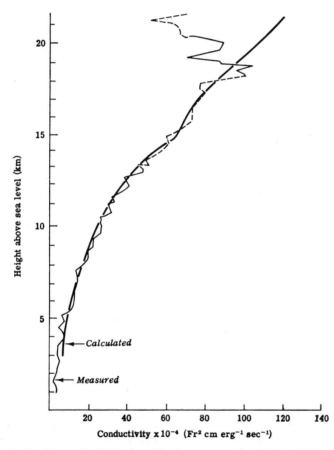

FIG. 3.18. Specific conductivity of positive ions measured on balloon flight of Explorer II, 11 November 1935 and calculated from cosmic-ray ionization [after O. H. Gish, *Compendium Meteorol.*, p. 101 (1951)].

to flow readily from the upper parts of charged clouds to the ionosphere whereas current between the lower part of the cloud and the ground is greatly inhibited by the low conductivity of the atmosphere near sea level.

3.18   The Lightning Discharge

In Section 3.15 the charge separation process which may occur in supercooled convective clouds has been described. In an active convection cell the large ice particles (soft hail pellets) carry negative charges toward the bottom of the cloud while small ice particles carry positive charges upward. In this way an active convection cell may develop the distribution of charge shown in Fig. 3.19. This figure shows that the electrical charge is

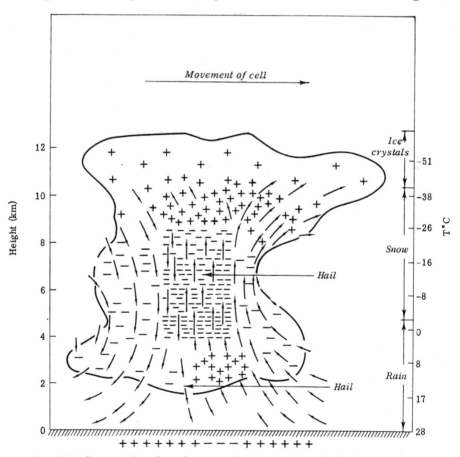

Fig. 3.19. Cross section through very active convection cell showing temperature, vertical air velocity and charge distribution (after L. B. Loeb, *Modern Physics for the Engineer*, p. 330. McGraw-Hill, New York, 1954).

associated with the region of strong updraft, the negative cell extending between heights of 4 and 9 kilometers and the positive cell being centered above 10 kilometers. The diameter of the cell is usually less than 4 kilometers. The small positive concentration near the bottom of the cell is characteristic of thunderstorm cells, but its origin and significance is uncertain. Within this volume of about 75 km^3 there may be stored electrical charge amounting to 10^3 coulombs or more. All but about 50 to 100 coulombs are present at any one time in an unorganized state. Discharge by lightning occurs between the positive and negative centers and between the cloud and the induced positive ground charge. Following discharge the upper positive and lower negative centers are recharged from the intermediate region in about 20 seconds.

In order to understand the phenomenon of lightning it is helpful to first understand the spark discharge in air between the plates of a charged capacitor. The spark may be initiated by a single ion pair in the space between the plates. The negative ion is accelerated toward the positive plate and the positive ion toward the negative plate, and each achieves a certain kinetic energy before it collides with a neutral particle. If this kinetic energy exceeds the ionization potential,the neutral particle yields a new ion pair, which in turn is accelerated in the electric field. In this way the number of ions grows by geometric progression until an "avalanche" is produced which provides a conducting path for discharge of the plates. The conditions required for spark discharge are that the accelerating potential gradient be so related to the mean free path between air molecules that the ions can attain kinetic energy equal to the ionization potential of the air molecule. If the potential across a capacitor at sea level is gradually increased, spark discharge occurs in dry air when the potential gradient reaches 30×10^3 volts cm^{-1}. If droplets are present, the "breakdown" potential is lower, and it has been estimated that the critical potential required within clouds in which droplets of one millimeter radius are present is about 10×10^3 volts cm^{-1}.

The analogy between the spark discharge and lightning is assuredly close, but the sequence of events which occurs in lightning has not been predicted by theory or explained in detail. Therefore only a descriptive account can be given of the common lightning flash which originates in the negative cloud base and transfers negative charge to the earth. Cloud to cloud flashes develop in a similar manner, but are not directly effective in transferring charge to the ground and therefore are of less interest here.

The potential difference between convective clouds and the ground and within the clouds, as indicated by the few observations available, is of the order 10^2 to 10^3 volts cm^{-1} over most of the volume. However, within the cloud the field fluctuates locally over a wide range as turbulent air motions bring the low-lying mass of positive ions closer or farther from the mass of negative ions shown in Fig. 3.19. In this way the breakdown potential gradient of 10^4 volts cm^{-1} may suddenly develop over a region hundreds of meters in length. There follows a lightning stroke within the cloud which transfers negative charge to the base of the cloud, neutralizing the small positive charge and charging the base of the cloud negatively. Breakdown potential is not achieved in the column between cloud and ground except immediately adjacent to the negative base. Where breakdown potential is achieved negative charge advances downward and forms an ionized path of about 10 centimeters radius, called the *pilot leader*. This path grows toward the earth at a rate of 10^7 to 10^8 cm sec^{-1} for a distance of roughly 10^2 meters length. This may be thought of as the length limited by the quantity of charge in the head of the pilot leader which can maintain breakdown potential without a resupply of electrons from the cloud base. Advance of the pilot leader produces a conducting path of 10^{13} to 10^{15} ion pairs per centimeter of path. Down this path surges a negative charge which revives the potential at the head of the pilot leader, and the series of events and another extension of the pilot leader is repeated. Finally there is created a conducting path extending from the base of the cloud to within a short distance above the earth. The potential gradient in the neighborhood of sharp points connected to the earth is now high enough that the breakdown potential is reached, and a positive streamer advances from such a point to meet the pilot leader at a height of 5 to 50 meters above the ground. When the pilot leader and streamer meet, earth and cloud are joined by a conducting path roughly 10 centimeters in radius, and up this path rushes the wave of ionizing potential called the *return stroke*. It advances at about 10^{10} cm sec^{-1} and practically fully ionizes the channel. Now the negative current in the cloud base rushes earthward through the brilliantly luminous channel and discharges roughly the lowest kilometer of the cloud. Following the discharge the conductivity of the channel decreases, and concurrently the potential of the cloud base gradually recovers. After about 1/20 second the potential is sufficient for development of the *dart leader*, a new character in the drama which serves to reactivate the conducting path. The number of strokes varies widely about the average value of three to four, and each of these strokes discharges successively higher regions of the negatively charged portion of the cloud. The total charge transferred from cloud to ground also varies widely. The average value appears to be about 20 coulombs which, as shown in Section 3.15, can be generated in a thunderstorm of 75 cubic kilometer volume in about fifteen seconds.

Observations made in aircraft flying above thunderstorms show that there is an average (positive) current of nearly one ampere flowing upward from cloud to ionosphere; other observations suggest that, at least occasionally, lightning may occur between cloud and ionosphere.

3.19 The Mean Electric Field

The thunderstorms which are always active somewhere in the atmosphere may be imagined as serving as huge Van de Graaff generators providing a potential difference between ionosphere and earth amounting to about 4×10^5 volts. Because both earth and ionosphere are excellent conductors, charge is conducted readily in the horizontal direction, and the two spherical surfaces are each at uniform potential. Therefore, positive current flows downward through the atmosphere over the entire earth except for the regions of thunderstorms; this current amounts to only about 4×10^{-16} ampere cm^{-2} or 2×10^3 amperes over the whole earth.

The vertical distribution of potential is represented by Eq. (3.32) if the current and the conductivity is known as a function of height. Evidently, since conductivity increases sharply with height, the potential gradient decreases equally sharply with height. If Eq. (3.32) is integrated between the ground and the ionosphere, the potential difference can be calculated

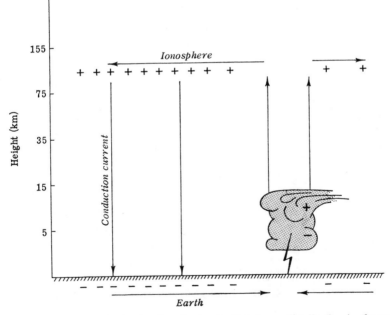

FIG. 3.20. Charge generation by thunderstorms, the charge distribution in the earth and ionosphere and the positive conduction current.

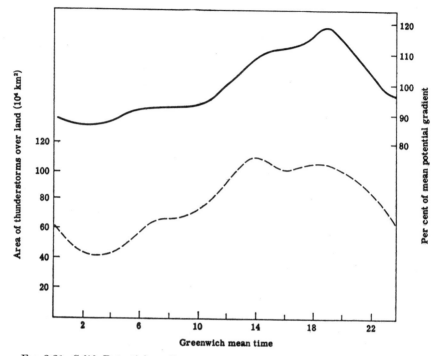

FIG. 3.21. *Solid:* Potential gradient over the sea. *Dashed:* Area of observed thunderstorms as a function of time of day (after J. A. Chalmers, *Atmospheric Electricity*, p. 204. Pergamon, New York, 1957).

from observations of the vertical current and the distribution of conductivity. In practice, the potential gradient near the earth is more easily measured than is the vertical current, so that Eq. (3.32) is first used to calculate the current and then is used in integrated form to calculate the potential difference. An exercise is provided in problem 10. The generation of potential difference and the current flow between earth and ionosphere are illustrated in Fig. 3.20.

Observations of vertical potential gradient over the sea show that there occurs an average diurnal variation of about 30%. Maxima and minima occur, respectively, at the same time all over the globe. This daily variation is in close agreement with the diurnal variation of the number of observed thunderstorms as is shown in Fig. 3.21. In fact it is this evidence which first demonstrated to Whipple* the role played by thunderstorms in maintaining the mean vertical potential gradient of the atmosphere.

---

* F. J. W. Whipple, *Quart. J. Roy. Meteorol. Soc.* **55**, 1 (1929).

List of Symbols

List of Symbols (Continued)

		First used in Section
η	Viscosity	3.6
λ	Penetration range, specific electrical conductivity	3.6, 3.17
ρ	Air density	3.6
ρ_w	Water density	3.2
ρ_d	Mass of liquid water per unit volume of air	3.6
σ	Surface tension	3.1
τ	Average lifetime of small ions	3.17

Subscripts

c	Curved surface, Coulomb
h	Solute
i	Ice
s	Saturation
v	Vapor
w	Water

Problems

1. Calculate the equilibrium vapor pressures over spherical droplets for the following radii: 10^{-1}, 10^{-2}, 10^{-3}, 10^{-4} cm if the vapor pressure is 6.1 mb, temperature is 0°C, and surface tension is 75 erg cm^{-2}. Plot e_c as a function of radius.

2. Lord Kelvin derived Eq. (3.1) by considering the following thought experiment. A closed box contains a tray filled with water. A capillary tube is placed vertically in the tray. The air is pumped out of the box and water evaporates until equilibrium is reached at the capillary water surface and the flat water surface. Show how Eq. (3.1) can be deduced from this experiment.

3. From Eq. (3.4) find the radius which requires the greatest supersaturation for equilibrium for specified values of mass of solute, temperature, pressure, etc.

4. (a) Show that the penetration range can be expressed by

$$\lambda = \frac{2}{9} \frac{\rho_w r^2 v'}{\eta}$$

where v' represents the initial velocity of the large droplet with respect to the small droplet.

(b) Show that Langmuir's inertia parameter can be expressed by

$$K \equiv \frac{\lambda}{r_0} = \frac{4}{81} \frac{g\rho_w^2}{\eta^2} r^2 r_0$$

where r and r_0 are, respectively, the radii of the small and large droplets.

5. If an ice crystal of one microgram mass falls from a cloud into an isothermal region where the temperature is -10°C and the vapor pressure is 2.0 mb, how long will it fall before the crystal is reduced to 10^{-7} gram? Assume that the diffusion capacitance may be expressed by $(m/4)^{1/3}$ cm.

6. Calculate the greatest possible depth of precipitation which could fall from a cloud which extends from 700 mb where the temperature is -5°C to 500 mb where the temperature is $-23°C$ if the average liquid water content is 0.3 g m^{-3} and if there are no updrafts within the cloud.

7. For the conditions of problem 6 calculate the precipitation in six hours under steady state conditions if the updraft is 1.0 m sec^{-1}.

8. Calculate the small ion concentration under conditions of equilibrium between ion production and recombination for the following cases: ($\alpha = 1.6 \times 10^{-6}$ cm^3 sec^{-1}, $\beta = 3 \times 10^{-6}$ cm^3 sec^{-1}).

(a) $N = 5 \times 10^3$ cm^{-3}; $p = 10$ cm^{-3} sec^{-1}
(b) $N = 5 \times 10^2$ cm^{-3}; $p = 1.5$ cm^{-3} sec^{-1}
(c) $N = 10^2$ cm^{-3}; $p = 10$ cm^{-3} sec^{-1}

9. Calculate the average lifetimes of small ions for the cases given in problem 8.

10. (a) If the vertical potential gradient near the ground is measured as 150 volts meter^{-1} and the conductivity is 1.5×10^{-16} (ohm cm)$^{-1}$, find the vertical conduction current.

(b) Find the potential difference between the earth and the atmosphere at a height of 20 km for the following representative observations of conductivity.

Height (km)	λ (10^{16} ohms^{-1} cm^{-1})	Height (km)	λ (10^{16} ohms^{-1} cm^{-1})
0	2.5	12	38
2	3	14	60
4	6	16	70
6	10	18	85
8	20	20	110
10	26		

(c) If the conductivity is assumed uniform between 20 and 80 kilometers, find the potential difference between the ground and the lower ionosphere (80 km).

General References

Hirschfelder, Curtiss, and Bird, *Molecular Theory of Gases and Liquids*, provides a detailed and authoritative account of intermolecular forces.

Mason, *The Physics of Clouds*, provided the first unified and comprehensive account of cloud physics. It is an invaluable reference for deeper penetration of the subject.

Fletcher, *The Physics of Rainclouds*, provides the best account of nucleation. His book is an excellent text for the student, but does not include the electrical aspects of cloud physics.

McDonald, *The physics of cloud modification* (*Advances in Geophys.* **5**), considers the complex, interacting processes which determine cloud modification. It is particularly good in interpretation and in illuminating the distinctions between problems with quantitative answers and those which remain matters of individual judgment or bias.

Junge, *Atmospheric chemistry* (*Advances in Geophys.* 4), presents an organized but detailed account of nuclei, their constitution, distribution, and physical processes.

Halliday and Resnick, *Physics for Students of Science and Engineering*, and Sears, *Principles of Physics*, Vol. 2, give good accounts of elementary electrical principles.

CHAPTER IV

Solar and Terrestrial Radiation

"Common sense is a docile thing.
It sooner or later learns the ways of science." HENRY MARGENAU

PART I: PRINCIPLES OF RADIATIVE TRANSFER

HAVING DESCRIBED in some quantitative detail the physical properties of the atmosphere, we now turn to an examination of why these properties and their interrelations are so characteristically ever-changing. The answer lies deeply embedded in an understanding of atmospheric energy sources and the processes of energy transfer within the atmosphere. These will be explored in this and the following chapter.

The most important of the processes responsible for energy transfer in the atmosphere is electromagnetic radiation which travels in wave form through vacuum at 2.997930×10^{10} cm sec^{-1} and through air at very nearly this speed. All wavelengths are possible and the totality of all wavelengths is called the electromagnetic spectrum. Various parts of this spectrum have been given names which are listed in Fig. 4.1 together with their approximate boundaries.

In Chapter VII the properties and behavior of electromagnetic waves are discussed in detail. Here, only those properties which are necessary to an understanding of energy transfer are considered.

4.1 Definitions and Concepts

Radiant energy is energy in transit. The amount of radiant energy passing an area per unit time is called *the radiant flux*; and the radiant flux per unit area, the *radiant flux density*, is represented by

$$F = \frac{d^2E}{dA\,dt}$$

where E represents radiant energy. The flux density is usually expressed in w cm^{-2}, or cal cm^{-2} min^{-1}.

Radiant energy may be propagated uniformly in all directions, or it may be directionally dependent. The radiant energy per unit time coming from a specific direction and passing through unit area perpendicular to

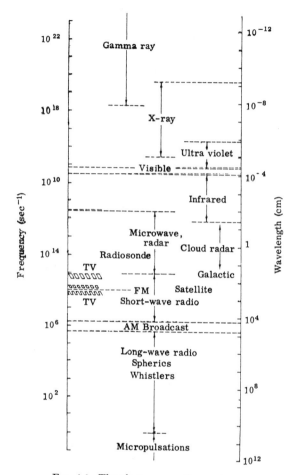

FIG. 4.1. The electromagnetic spectrum.

that direction is called the *intensity*. The intensity and flux density are related by

$$I = \frac{dF}{d\omega \cos \theta} \tag{4.1}$$

where $d\omega$ represents the differential of solid angle and θ the angle between the beam of radiation and the direction normal to the surface (usually horizontal) on which the intensity is measured. The definition of solid angle is given in Appendix I.H.

Equation (4.1) can also be written in the integral form

$$F = \int_0^{2\pi} I \cos \theta \, d\omega \tag{4.2}$$

Figure 4.2 illustrates the effect on flux density of orientation of the radiant beam with respect to the surface. Radiation whose intensity is independent of direction is called *isotropic radiation*. In this case because $d\omega = 2\pi \sin\theta\, d\theta$, as shown in Appendix I.H, Eq. (4.2) may be integrated to yield

$$F = \pi I \qquad (4.3)$$

Since the radiant energy is distributed over various wavelengths and also because the absorptive or emissive properties of materials are a function

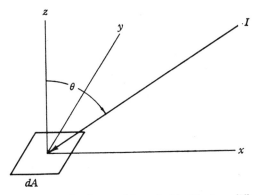

FIG. 4.2. Radiation of intensity I at zenith angle θ incident on differential area dA.

of wavelength, it is useful to define *monochromatic flux density* (F_λ) and *monochromatic* or *specific intensity* (I_λ) by

$$F_\lambda \equiv \frac{d(F_\lambda)}{d\lambda} \quad \text{and} \quad I_\lambda \equiv \frac{d(I_\lambda)}{d\lambda} \qquad (4.4a)$$

Intensity and flux density are also often expressed per unit *frequency* interval. In this case

$$F_\nu \equiv \frac{d(F_\nu)}{d\nu} \quad \text{and} \quad I_\nu \equiv \frac{d(I_\nu)}{d\nu} \qquad (4.4b)$$

The transformation from wavelength to frequency follows from

$$\nu = c/\lambda \qquad (4.5)$$

where ν represents frequency, the number of waves that pass a given point in one second, and c represents the speed of light. From Eqs. (4.4) and (4.5) it follows that

$$F_\lambda = -\frac{c}{\lambda^2} F_\nu \quad \text{and} \quad I_\lambda = -\frac{c}{\lambda^2} I_\nu$$

Certain properties of radiation may be understood as depending on the characteristics of a continuous series of waves; other properties require that radiation be treated as a series of discrete particles or *quanta*. Some of the properties associated with waves are discussed in Chapter VII, while some of the properties associated with quanta are discussed in Sections 4.3 and 4.5.

4.2   Absorption and Emission of Radiation

Radiation is absorbed in passing through matter, and the fraction absorbed is a specific characteristic of the material. The ratio of the absorbed to the incident radiation at a certain wavelength is called the *monochromatic absorptivity* (a_λ) and is usually a function of the wavelength. A body with absorptivity equal to unity for all wavelengths is called a *black body*. Perfect black bodies do not exist in nature, but they may be very closely approximated especially in the infrared or long wave range. Of the incident radiation that is not absorbed part is reflected and part is transmitted. The ratio of the reflected to the incident radiation is called the *monochromatic reflectivity* (r_λ), and the ratio of the transmitted to the incident radiation is called the *monochromatic transmissivity* (τ_λ). The three ratios are related by

$$a_\lambda + r_\lambda + \tau_\lambda = 1$$

For a black body $r_\lambda = \tau_\lambda = 0$, and $a_\lambda = 1$ for all wavelengths.

Kirchhoff's Law

A molecule which absorbs radiation of a particular wavelength also is able to emit radiation of the same wavelength. The rate at which emission takes place is a function of the temperature and the wavelength. This is a fundamental property of matter of the greatest importance which leads to a statement of Kirchhoff's law.

In order to develop Kirchhoff's law consider a perfectly insulated enclosure with black walls, and consider that this system has reached an equilibrium state characterized by uniform temperature and isotropic radiation. Because the walls are black, all the radiation striking a surface is absorbed, and because there is equilibrium the same amount of radiation that is absorbed by a surface area is also emitted by that area. Furthermore, because the black body absorbs the maximum possible radiation, it also emits the maximum. If it would emit more, equilibrium would be impossible, and this violates the Second Law of Thermodynamics. The radiation inside the system is called *black-body* radiation. The intensity of black-body radiation clearly is a function of temperature only.

Consider now the same system with the exception that the absorptivity of one of the walls is less than unity. Under equilibrium conditions the radiation in this system is also black-body radiation, because each wall must absorb just as much as it emits. The wall with absorptivity a_λ absorbs only a_λ times the black-body radiation and therefore also emits a_λ times the black-body radiation. If the monochromatic emissivity (ϵ_λ) is now defined as the ratio of the emitted radiation to the maximum possible emission at the same temperature at wavelength λ, the result is Kirchhoff's law in the form

$$a_\lambda = \epsilon_\lambda \qquad (4.6)$$

Beer's Law

The experiment depicted in Fig. 4.3 shows that absorption by a homogeneous medium is directly proportional to thickness of the absorbing layer

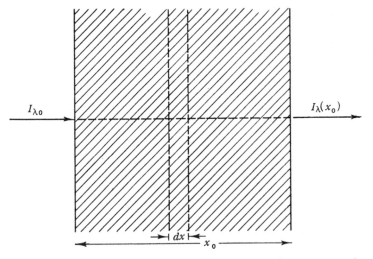

FIG. 4.3. Depletion of radiation intensity in traversing a homogeneous absorbing layer.

and to the intensity of the incident radiation. This result is expressed by the equation

$$dI_\lambda = -k_\lambda \rho I_\lambda \, dx \qquad (4.7)$$

where k_λ represents the *absorption coefficient* of the layer and ρ represents the density of the absorbing substance. Integration of Eq. (4.7) yields

$$I_\lambda = I_{\lambda 0} e^{-k_\lambda \int_0^x \rho \, dx} \qquad (4.8)$$

which is known as Beer's law. It is convenient to replace the coordinate by the *optical thickness* defined by

$$u \equiv \int_0^x \rho \, dx \qquad (4.9)$$

which is the mass of absorbing gas of the unit column through which the radiation passes. If the total optical thickness of the layer is represented by u_0, then the monochromatic transmissivity may be expressed by

$$\tau_\lambda \equiv \frac{I_\lambda(x_0)}{I_{\lambda 0}} = e^{-k_\lambda u_0}$$

Similarly

$$a_\lambda = 1 - e^{-k_\lambda u_0}$$

Schwarzschild's Equation

Consider that a beam of intensity I_λ passes through a layer where absorption and also emission takes place. According to Eqs. (4.7) and (4.9), the radiation absorbed over an optical thickness du can be expressed by $k_\lambda I_\lambda \, du$. Kirchhoff's law requires that the emission in that layer is $k_\lambda I_\lambda^* \, du$, where I_λ^* represents the black-body monochromatic intensity. Therefore, the net change in intensity is given by

$$\frac{dI_\lambda}{du} = -k_\lambda(I_\lambda - I_\lambda^*) \tag{4.10}$$

Equation (4.10), known as Schwarzschild's equation, can be integrated if the temperature and the absorption coefficient along the path of the beam is known.

Because the transmitted and reflected components are simply additive, when reflection occurs the reflected radiation must be subtracted before Beer's law is applied. An example is provided by problem 1.

4.3*   Theory of Black-Body Radiation

The theory of the energy distribution of black-body radiation was developed by Planck and was first published in 1901. He postulated that energy can be emitted or absorbed only in discrete units or photons defined by

$$u = h\nu \tag{4.11}$$

where the constant of proportionality h has the value 6.625×10^{-27} erg sec and is known as *Planck's constant*. Although the Planck postulate was introduced arbitrarily, some intuitive support may be added if one considers each atom to be analogous to a mechanical oscillator (tuning fork) which may oscillate at only discrete characteristic frequencies. Einstein subsequently showed that the energy of the photon is expressible by mc^2, where m represents the equivalent mass of the photon. Therefore the momentum of the photon is given by

---

* This section contains advanced material which is not essential for understanding the following sections and chapters.

$$p = \frac{h\nu}{c} \tag{4.12}$$

Consider now a system filled with photons which move in all directions with the speed c. To find the distribution of the photons in phase space, or more specifically, in the frequency interval $\Delta\nu$, it is necessary to consider the most probable distribution of photons. This is analogous to the calculation of most probable distribution of molecular velocities which was discussed in Section 2.15. Equations (2.57), (2.58), and (2.60) are still valid in this case. However, photons are created and destroyed at each interaction, so that there is no constraint on the number N, and Eq. (2.59) is no longer valid.

Lagrange's method of undetermined multipliers applied to Eqs. (2.58) and (2.60) now yields for the average number of photons per cell in the interval $\Delta\nu_i$

$$N_i = (e^{\beta u_i} - 1)^{-1} \tag{4.13}$$

Combining Eq. (4.13) with Eqs. (2.64) and (4.11) yields the Bose-Einstein distribution in the form

$$N_i = (e^{h\nu_i/kT} - 1)^{-1}$$

The fact that β has the same value in this case as is indicated by Eq. (2.64) can be checked by substituting Eq. (4.13) back into Eq. (2.58), with the result

$$\delta \ln W = \sum \beta u_i \, \delta N_i = \beta \, \delta u$$

Upon combining this result with Eq. (2.66)

$$\delta s = \frac{k}{Nm} \delta \ln W = k\beta \, \delta u$$

But for a system with constant volume

$$ds = T \, du$$

which leads again to the conclusion that

$$\beta = 1/kT \tag{2.64}$$

To find the number of photons per frequency interval it may be recognized from Eq. (4.12) that a shell of phase volume corresponding to $\Delta\nu$ is represented by $4\pi h^3 V \nu^2 \, \Delta\nu/c^3$, where V represents the total volume in ordinary space. Upon dividing this volume in phase space by H, the volume of a single cell, the number of cells contained in the increment $\Delta\nu$ is represented by

$$\Delta n = \frac{4\pi V h^3 \nu^2 \, \Delta\nu}{c^3 H}$$

This is the number of cells for photons of a single transverse degree of freedom; since photons exhibit two transverse degrees of freedom, the total number of cells is twice that given here. The number of photons in $\Delta\nu$ is then

$$\Delta N = 2\bar{N}_i \Delta n = \frac{8\pi V h^3 \nu^2 \Delta\nu}{c^3(e^{h\nu/kT} - 1)H}$$

and the energy density per unit frequency interval is

$$\frac{h\nu \, \Delta N}{V \, \Delta\nu} \equiv u_\nu = \frac{8\pi h\nu^3 h^3}{c^3(e^{h\nu/kT} - 1)H} \tag{4.14}$$

Although Planck had originally intended to allow the cell size to become zero, it is clear from Eq. (4.14) that this is not possible. The smallest possible cell size is determined by Heisenberg's uncertainty principle, which states that if the position of a photon is known to within Δx, the momentum cannot be known more accurately than within $h/\Delta x$. Therefore, if the precise position of the photon is known, its momentum is completely unknown; and if the momentum is known precisely, its position is completely uncertain. So, clearly, h^3 is the smallest possible cell size in which a photon can be specified.

Because for black-body radiation the photons travel in all directions with speed c

$$I_\nu^* = \frac{u_\nu c}{4\pi} \tag{4.15}$$

Upon combining Eqs. (4.14) and (4.15) Planck's law of black-body radiation intensity per unit frequency interval may be written in the form

$$I_\nu^* = \frac{2h\nu^3}{c^2(e^{h\nu/kT} - 1)} \tag{4.16}$$

This law has been experimentally verified to a high degree of accuracy.

4.4   Characteristics of Black-Body Radiation

If the intensity is expressed in terms of wavelength, Eq. (4.16) may be written

$$I_\lambda^* = 2hc^2\lambda^{-5}(e^{hc/k\lambda T} - 1)^{-1} \tag{4.17}$$

Figure 4.4 shows that the black-body radiation intensity increases markedly with temperature and that the wavelength of maximum intensity decreases with increasing temperature.

The total radiation intensity of a black body is found by integrating Eq. (4.17) over all wavelengths. Upon substituting $x \equiv hc/\lambda kT$, the radiation intensity is given by

$$I^* = \int_0^\infty I_\lambda^* \, d\lambda = \frac{2k^4 T^4}{c^2 h^3} \int_0^\infty \frac{x^3 \, dx}{e^x - 1}$$

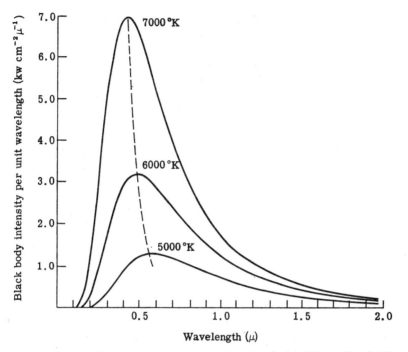

FIG. 4.4. Black-body intensity per unit wavelength calculated from Eq. (4.17) for temperatures 5000°, 6000°, and 7000°K.

The integral $\int_0^\infty x^3 (e^x - 1)^{-1} dx$ can be shown to have the value $\pi^4/15$, so that

$$I^* = \frac{2\pi^4 k^4}{15c^2h^3} T^4 \equiv bT^4 \qquad (4.18)$$

This is the *Stefan-Boltzmann law*, according to which the intensity emitted by a black body varies as the fourth power of the absolute temperature.

Since black-body intensity is independent of the direction of emission, Eq. (4.3) shows that the flux density emitted by a black body is

$$F^* = \pi b T^4 \equiv \sigma T^4 \qquad (4.19)$$

where σ represents the Stefan-Boltzmann constant. It has the value 5.67×10^{-5} erg cm^{-2} sec^{-1} °K^{-4} (0.817×10^{-10} cal cm^{-2} min^{-1} °K^{-4}).

The wavelength of maximum intensity for black-body radiation may be found by differentiating Planck's law with respect to the wavelength, equating to zero, and solving for the wavelength as required in problem 3. The result is *Wien's displacement law*, which may be written

$$\lambda_m = \alpha/T \qquad (4.20)$$

where $\alpha = 0.290$ cm deg. This relation makes it possible to compute the temperature of a black body by measuring the wavelength of maximum monochromatic intensity. If the body is only approximately black, computation using Eq. (4.20) yields an approximate temperature called the *color temperature* of the body. In problem 4 it is required to obtain the relation between $I^*_{\lambda\text{max}}$ and the temperature.

4.5   The Line Spectrum

When inspected in detail with high resolution, the absorption and emission spectrum of a particular material consists of a large number of individual and characteristic spectral lines. The relationship of spectral lines to atomic structure can be understood on an elementary basis by use of the model of the hydrogen atom proposed by Niels Bohr in 1913. He postulated that the electron revolves in a circular orbit, its motion being governed by the balance between the electrostatic and the apparent centrifugal force, with the restriction that its angular momentum is a multiple of $h/2\pi$. Consequently, transition from one orbit to another orbit cannot take place continuously, but occurs in jumps each of which is accompanied by a change in energy. The energy expressed by Eq. (4.11) is emitted by the atom as a photon of electromagnetic radiation; and the atom may absorb a photon of the same energy. The energy of the quantum jump is analogous to the change in kinetic plus potential energy required to shift an earth satellite from one orbit to another, as expressed by Eq. (1.13).

Each quantum jump between fixed energy levels results in emission or absorption of characteristic frequency or wavelength. These quanta appear in the spectrum as emission or absorption lines. For the simple hydrogen atom the line spectrum is correspondingly simple, whereas the spectra of water vapor and carbon dioxide are considerably more complex.

Spectral lines have a characteristic shape suggestive of the Maxwellian distribution function. The finite width of the line is, in part, a result of molecular collisions; the interaction between atoms when they are very close together causes a slight change in the energy levels of the electrons and in the energy of an absorbed photon. This effect is called the pressure broadening of the line. A less important source of line broadening is the Doppler shift associated with the velocity distribution of the molecules. Those molecules moving toward the approaching photons experience a shift toward higher frequency, and vice versa. It is therefore clear that line width should increase with temperature, as well as with pressure. The Doppler shift is discussed in Section 7.14.

The shape of a single spectral line in the lowest 80 kilometers of the atmosphere may be expressed by

$$k_\nu = \frac{k_l}{\pi} \frac{\alpha}{(\nu - \nu_0)^2 + \alpha^2} \qquad (4.21)$$

where α represents the half-width of the line, and the *normalized line intensity* is expressed by

$$k_l = \int_{-\infty}^{\infty} k_\nu \, d\nu$$

Equation (4.21) follows from the discussion of refractive index in Section 7.6. The monochromatic transmissivity or absorptivity may be calculated by substituting Eq. (4.21) into the appropriate forms from Sections 4.1 and 4.2. The absorptivity is plotted in Fig. 4.5 for three values of the

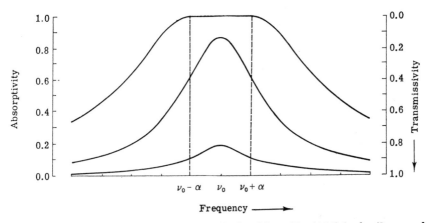

FIG. 4.5. Absorptivity for a spectral line calculated from Eq. (4.21) for $k_l u / 2\pi\alpha$ equal to 0.1, 1.0, and 5 [after W. M. Elsasser, *Meteorol. Monographs* **4** (23), 5 (1960)].

optical thickness expressed in the dimensionless form $k_l u / 2\pi\alpha \equiv X$. For small values of X the absorption is proportional to k_l. For large values of X the absorptivity in the center of the line is unity, and further increase in optical thickness has no effect. However, in the line wings the absorptivity continues to increase with further increase in optical thickness.

The actual absorption spectrum of the atmosphere contains innumerable lines some of which are organized into absorption bands of limited spectral interval. In this case the wings of the various lines overlap, and the combined absorption has to be taken into account. For small absorptivity the individual line absorptivities can be added directly to give the combined absorptivity. For strong absorptivity the combined effect is more complicated, and is always less than the sum of the individual line absorptivities.

PART II: RADIATION OUTSIDE THE ATMOSPHERE

The energy source for nearly all the physical processes discussed in earlier chapters—expansion of air, precipitation, generation of ionospheric potential—is, of course, the sun. This statement is equally true for many phenomena outside the normal scope of atmospheric physics: biological processes, formation of soils, and the oxidation of automobile tires, for example. In each of these examples energy originally incident on the top of the atmosphere in the form of radiation has undergone a series of transformations, finally culminating in the phenomenon of special interest. The link between solar energy and atmospheric phenomena is particularly close, so that it is of the greatest importance to understand how the energy transformations occur.

4.6   The Sun

In spite of its importance to virtually all processes on earth, the sun is by no means unique. Among the billions of stars in our galaxy it is about average in mass but below average in size, and our galaxy is, of course, only one among millions of galaxies. The sun's importance results from its closeness; it is 150×10^6 kilometers from the earth, whereas the next closest star is 3×10^5 times as far away.

The sun is a gaseous sphere with a diameter of 1.42×10^6 kilometers and a surface temperature of about 6×10^3 °K. The temperature increases toward deeper layers until temperatures high enough to sustain nuclear reactions are reached. The source of solar energy is believed to be fusion of four hydrogen atoms to form one helium atom, and the slight decrease in mass which occurs in this reaction accounts for the energy released in the solar interior. This energy is transferred by radiation and convection to the surface, and is then emitted as both electromagnetic and particulate radiation.

Each square centimeter of the solar surface emits on the average with a power of about 6.2 kilowatts or 9.0×10^4 cal min^{-1}. This immense intensity is maintained by the generation of only 3.8×10^{-6} cal cm^{-3} min^{-1}, whereas a corresponding value for burning coal is typically 10^9 times as large. The solar spectrum is shown in Fig. 4.6.

The sun's energy is radiated uniformly in all directions, and nearly all of the energy disappears into the vastness of the universe. As is illustrated in

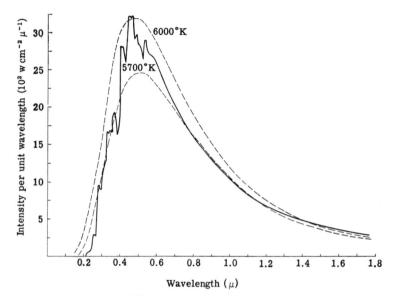

FIG. 4.6. Observed solar spectrum (*solid line*) and black-body intensities for temperatures of 6000° and 5700°K (*dashed lines*) [after F. S. Johnson, *J. Meteorol.* **11,** 431 (1954)].

problem 5, only a minute fraction of the output of the sun is intercepted by the earth. Although there is no *a priori* reason to assume that the sun emits radiation at a constant rate, observations have indicated so far no measurable variations.

The Solar Spectrum

The distribution of electromagnetic radiation emitted by the sun approximates black-body radiation for a temperature of about 6000°K as shown in Fig. 4.6. The similarity between the solar and black-body spectra provides the basis for estimates of the temperature of the visible surface layer of the sun. Because the sun does not radiate as a perfect black body, different comparisons are possible with different results. The Stefan-Boltzmann law may be used in combination with the solar constant, as in problem 6, to calculate a temperature of 5780°K. Or, the wavelength of maximum intensity, 0.4750 μ, combined with Wien's displacement law [Eq. (4.20)] may be used to calculate a temperature of 6080°. Although it is not possible from the two radiation laws to determine a unique temperature for the solar surface, the uncertainty amounts to less than 10^3 out of 6×10^3 degrees. This range of temperature may result from the fact that different parts of the spectrum are emitted by different layers of the sun.

Other Features of the Sun

Although to the eye armed with only a smoked glass, the sun seems to have a uniform brightness, inspection of the solar surface (*photosphere*) with a telescope reveals a granular structure. The granulae are bright areas with a diameter averaging between three hundred and fifteen hundred kilometers. They have a lifetime of the order of a few minutes, suggesting a violent convection of tremendous scale. Figure 4.7 is a picture of the sun's

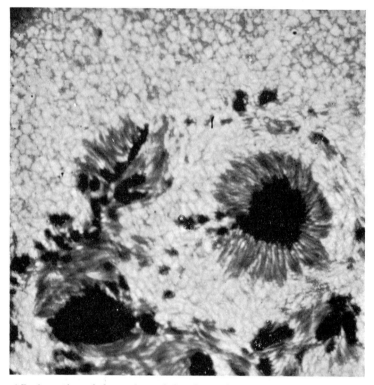

Fig. 4.7. A portion of the surface of the photosphere at 10.01 CST 17 August 1959 taken by Project Stratoscope. The radii of the large sunspots are roughly equivalent to the diameter of the earth. (Published by courtesy of M. Schwarzschild.)

surface showing clearly the granular structure. The resemblance to some types of stratocumulus clouds or to *Benard-cells* is particularly challenging. Figure 4.7 also shows the structure of a sunspot. Clearly visible are the dark core (the umbra) and the radiant structure (the penumbra) surrounding the core. The origin and dynamics of sunspots are still uncertain, although an important role must be played by magnetic fields near the surface. Because the solar gas consists of charged particles, a magnetic field exerts a force on the gas particles moving in the field. This force may prevent the development of convection and the transport of hot interior matter upward to the photosphere. This may account for the fact that sunspots appear as comparatively dark areas showing that they are cooler than the surrounding photosphere. The sunspots are much more stable than the granulae and have a lifetime varying from a few days to more than a month. Sometimes, usually in the vicinity of sunspots, very bright areas, called *faculae* or *flares*, become visible.

The photosphere is surrounded by a spherical shell, about 1.5×10^4 kilometers in thickness, called the *chromosphere*. It is composed of thin gases and may be considered the solar atmosphere. The spectrum of the chromosphere is characterized by many rather sharp absorption lines, called the *Fraunhofer lines* after their discoverer. Although these lines have been intensively studied as clues to properties of the chromosphere, it is likely that the region is so far from thermodynamic equilibrium that Kirchhoff's law does not apply, and therefore the Fraunhofer lines do not provide a reliable indication of temperature. The temperature is consequently uncertain, but has been estimated to be in the range from 5×10^4 to $10^5 °K$. Violent disturbances in the chromosphere result in large masses of ionized hydrogen being emitted from the sun. Beyond the chromosphere the corona extends many millions of miles into space. The weak continuous spectrum of the corona indicates that it consists of an extremely thin gas of a very high temperature ($10^6 °K$).

The period of rotation of the sun, as determined from sunspot observations, is about 24 days at the equator, 26.5 days at 25° latitude, and is not well known near the poles.

The frequency of sunspots exhibits cycles of roughly 22 years. A cycle begins with a few sunspots appearing at about 35° latitude. Gradually the number of sunspots increases and at the same time the latitude of initial detection decreases. Five to six years after the beginning of a cycle a maximum in sunspot activity is reached. After that the number of sunspots decreases and the latitude of initial detection also decreases until at about eleven years, a minimum in activity occurs. Before the last sunspots disappear at about 5° latitude, already new sunspots have appeared at high latitudes. Although this suggests a cycle of roughly eleven years, closer observation shows that the new sunspots have a reversed magnetic field with respect to the old ones. Therefore the total cycle of the sunspots consists of two periods, each of about eleven years.

4.7 Determination of the Solar Constant

Even though solar radiation is attenuated by scattering and absorption in passing through the atmosphere, the intensity of solar radiation at the top of the atmosphere, the *solar constant*, may be calculated from measurements made at the earth's surface. If the atmosphere is assumed to consist of a series of homogeneous plane parallel layers, the optical thickness of the atmosphere at zenith angle θ is $u_\infty \sec \theta$ where $u_\infty \equiv \int_0^\infty \rho \, dz$. This assumption introduces an error which becomes important only for zenith angles greater than 70°. The monochromatic intensity may be expressed by Beer's law [Eq. (4.8)] in the form

$$I_\lambda = I_{\lambda 0} \exp\left(-\sigma_\lambda u_\infty \sec \theta\right)$$

or

$$\ln I_\lambda = \ln I_{\lambda 0} - \sigma_\lambda u_\infty \sec \theta \tag{4.22}$$

where $I_{\lambda 0}$ represents the monochromatic intensity at the top of the atmosphere and σ_λ represents the combined effects of absorption and scattering and is called the *extinction* coefficient. Observations of I_λ are made for several zenith angles during a single day. If for each observation σ_λ has been the same, then a plot of I_λ against $\sec \theta$, as shown in Fig. 4.8, may be

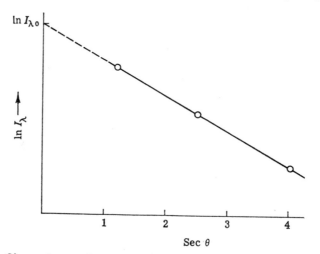

FIG. 4.8. Observed monochromatic solar intensities (I_λ) plotted as a function of zenith angle (θ).

extrapolated to $\sec \theta = 0$, and $I_{\lambda 0}$ may be read from the ordinate. Because $\sec \theta$ cannot be less than unity, the point determined in this way may be understood as representing the top of the atmosphere $(u = 0)$.

A complication is introduced because the *spectrobolometer* used to measure I_λ is not capable of making absolute determinations of intensity. Instead, monochromatic intensities (I_λ') are measured relative to a reference intensity. The corresponding relative intensities at the top of the atmosphere are then obtained by extrapolation as illustrated in Fig. 4.8. The relative intensities may now be plotted against wavelength, and the area under the curve expressed by the integrals

$$\int_0^\infty I_\lambda' \, d\lambda \quad \text{and} \quad \int_0^\infty I_{\lambda 0}' \, d\lambda$$

may be measured graphically. However, because the atmosphere is opaque for wavelengths shorter than 0.34 and for wavelengths longer than 2.5 μ, these integrals are evaluated only between 0.34 and 2.5 μ. Corrections, amounting to about 8%, are added for the omitted ranges.

In order to convert the above integrals to energy units, the total solar flux density is measured by an instrument called the *pyrheliometer* which is illustrated in Fig. 4.9. The black surface at the far end of the instrument is

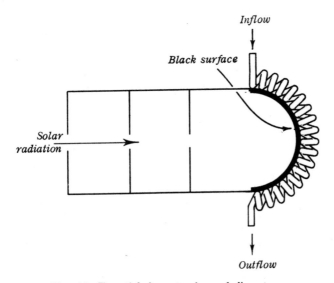

FIG. 4.9. Essential elements of a pyrheliometer.

exposed to solar radiation by directing the instrument toward the sun and is kept within a small temperature range by contact with circulating water. Water flow is maintained at a constant rate, so that the heat flow is proportional to the temperature difference between inflow and outflow. The radiation emitted through the opening of the instrument and other minor neglected effects are accounted for by appropriate corrections.

Now, assuming that the ratio I_λ'/I_λ is independent of wavelength, the solar flux density outside the atmosphere may be expressed by

$$F_0 = F \frac{\int_0^\infty I'_{\lambda 0}\, d\lambda}{\int_0^\infty I_\lambda'\, d\lambda} \tag{4.23}$$

Finally, in order to determine the solar constant from F_0, it is necessary to recognize that F_0 changes as the distance of the earth from the sun changes. The relation between F_0 and the solar constant is obtained by assuming that all the solar energy which passes through a sphere of radius R_0 also passes through a sphere of radius $\overline{R}$. Therefore, where $\overline{F}_s$ represents the solar constant

$$4\pi R_0^2 F_0 = 4\pi \overline{R}^2 \overline{F}_s \tag{4.24}$$

The accuracy of the determination of the solar constant is limited by errors introduced in interpolation between measured wavelengths, and, particularly at short wavelengths, by correction for the ultraviolet portion of the spectrum. This portion of the spectrum is completely absorbed by the atmosphere, so that appreciable uncertainty is unavoidable by this method. Recent rocket measurements have extended spectral observations into the ultraviolet wavelengths, and have reduced the uncertainty in the solar constant. The International Radiation Committee adopted in 1957 the value 1.98 cal cm^{-2} min^{-1} (138.2 mw cm^{-2}) for the solar constant. Because the uncertainty of this value exceeds 1%, the solar constant properly may be used as 2.00 cal cm^{-2} min^{-1}.

4.8   Distribution of Solar Energy Intercepted by the Earth

The flux of solar radiation per unit horizontal area at the top of the atmosphere depends strongly on zenith angle of the sun and much less strongly on the variable distance of the earth from the sun. If the zenith angle is assumed to be constant over the solid angle subtended by the sun, the flux density on a horizontal surface may be expressed by

$$F = F_0 \cos \theta$$

The mean value of the distance of earth from the sun $(\overline{R})$ is 149.5×10^6 kilometers. R_0 varies from 147.0×10^6 kilometers about January 3 (perihelion) to 152.0×10^6 kilometers about July 5 (aphelion).

The zenith angle depends on latitude, on time of day and on the tilt of the earth's axis to the rays of the sun, that is, on the celestial longitude or the date. The total energy received per unit area per day (q_0) may be calculated by integrating over the daylight hours. Thus, if R_0 is assumed constant during a single day

$$q_0 = F_s\left(\frac{\overline{R}}{R_0}\right)^2 \int_{\text{sunrise}}^{\text{sunset}} \cos\theta\, dt \qquad (4.25)$$

Equation (4.25) has been evaluated for a variety of latitudes and dates, with the results summarized in Fig. 4.10.

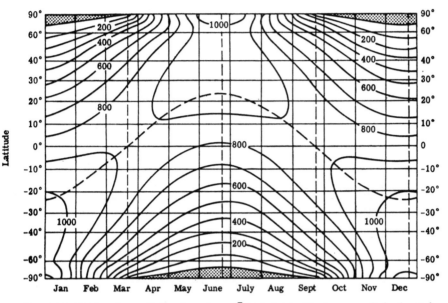

FIG. 4.10. Daily solar radiation in cal cm^{-2}(24 hr)$^{-1}$ incident on a unit horizontal surface at the top of the atmosphere as a function of latitude and date [after R. J. List (ed.), "Meteorological Tables," 6th ed., p. 417. Smithsonian Inst., Washington, D.C., 1951].

Figure 4.10 shows that maximum insolation occurs at summer solstice at either pole. This is a result of the long solar day (24 hours). The maximum at the southern hemisphere is higher than at the northern hemisphere because the earth is closer to the sun during the northern winter than during the northern summer. However, the reader is invited to prove in problem 7 that the total annual insolation is the same for corresponding latitudes in northern and southern hemispheres.

4.9 Short- and Long-Wave Radiation

The solar radiation received by the earth is partly reflected and partly absorbed. Over an extended period energy equivalent to that absorbed must be emitted to space by the earth and the atmosphere; otherwise, significant changes in temperatures of the earth would be observed.

Although the flux absorbed by the earth and atmosphere is closely equal to that emitted, and although both distributions are very roughly black in character, the spectral curves of absorbed and emitted radiation overlap almost not at all. In Fig. 4.11 black-body radiation is plotted for a body

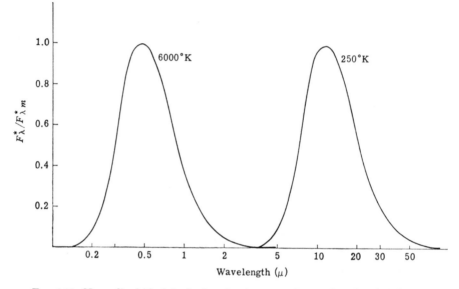

FIG. 4.11. Normalized black-body flux density per unit wavelength calculated from Eqs. (4.3) and (4.17) for temperatures of 6000° and 250°K.

of 6000°K corresponding to the radiation emitted by the sun and for a body of 250°K corresponding roughly to the radiation emitted by the earth and the atmosphere. The two curves are clearly separated into two spectral ranges above and below about four microns; for this reason it is customary to call the radiation received from the sun *short-wave radiation*, and the radiation emitted by the earth and atmosphere *long-wave radiation*. This distinction makes it possible to treat the two types of radiation separately and thereby to reduce the general complexity of transfer of radiant energy.

4.10   Radiation Measurements from Satellites

It is inherent in the discussion of Section 4.7 that determination of the radiation flux density received from the sun requires several observations under rather special conditions and that a single determination is subject to considerable uncertainty. Consequently, these observations cannot be used to provide accurate values of the solar constant at frequent intervals in time. Instruments mounted in satellites which circle the earth above the absorbing atmosphere can provide these data as well as the equally important data concerning radiation emitted by the earth and atmosphere. The objective of these observations is to determine three separate radiation fluxes: direct solar radiation, solar radiation reflected by the earth and atmosphere, and long-wave radiation emitted by the earth and atmosphere. V. E. Suomi* has developed a simple device which solves this problem. Two spherical sensors, one black and one white are exposed to the three fluxes, and in a short time each achieves radiative equilibrium. Equilibrium for the black sensor is expressed by

$$4\pi r^2 a_b \sigma T_b{}^4 = \pi r^2 a_b (F_0 + F'_{sr} + F'_l)$$

where the subscript b refers to the black sensor. The absorptivity a_b is assumed to be the same for short- and long-wave radiation. The flux densities F'_{sr} (short-wave reflected radiation) and F'_l (long-wave radiation) are defined for spherical sensors by

$$F'_{sr} = \int_0^\omega I_{sr}\, d\omega$$

and   (4.26)

$$F'_l = \int_0^\omega I_l\, d\omega$$

where ω is the solid angle by which the satellite "sees" the earth. I_{sr} and I_l are, respectively, the intensities of reflected and emitted radiation received from the earth. Equations (4.26) express the fact that a spherical sensor is equally sensitive to radiation coming from all directions. To reconcile these equations with Eq. (4.2) is left as problem 8.

Radiative equilibrium for the white sensor is represented by

$$4 a_{wl} \sigma T_w{}^4 = a_{ws}(F_0 + F'_{sr}) + a_{wl} F'_l$$

where the subscript w refers to the white sensor. The absorptivity of white paint at short wavelengths is small (about 0.1) and at long wavelengths is large (about 0.9) and is assumed to be the same as for the black sensor.

The equilibrium equations for the black and white sensors may be used to solve for the sum of the short-wave flux densities $(F_0 + F'_{sr})$ and the long-wave flux density (F'_l). The result is, upon introducing $a_l = a_b = a_{wl}$ and $a_s = a_{ws}$

$$F_0 + F'_{sr} = \frac{4\sigma a_l}{a_l - a_s}\,(T_b{}^4 - T_w{}^4)$$ (4.27)

and

---
* V. E. Suomi, *NASA Tech. Note* D, 608 (1961).

$$F'_l = \frac{4\sigma}{a_l - a_s} (a_l T_w{}^4 - a_s T_b{}^4) \tag{4.28}$$

F_0 may be measured directly by a pyrheliometer in the satellite or it may be estimated from Eq. (4.24).

In order to convert the measured flux densities into intensities emitted and reflected by the earth, it may be recalled that F'_{sr} and F'_l depend on the solid angle ω under which the earth is seen by the satellite sensor. Figure 4.12 indicates that, where ψ represents half the plane angle subtended by the earth

$$\frac{R}{R + h} = \sin \psi$$

and Appendix I.H shows that the solid angle may be expressed in terms of ψ by

$$\omega = 2\pi \int_0^\psi \sin \psi' \, d\psi' = 2\pi(1 - \cos \psi)$$

Upon combining these two equations

$$\omega = 2\pi \left(1 - \frac{(2Rh + h^2)^{1/2}}{R + h} \right) \tag{4.29}$$

Thus, the solid angle depends only on the height of the satellite above the earth; and as the height decreases to zero, the solid angle approaches 2π, as we should expect.

The intensity of the reflected solar radiation depends on the reflectivity of earth and atmosphere (*albedo*) and on the zenith angle (θ) of the incoming solar radiation. If the reflected radiation is diffuse, then the flux density of reflected radiation is equal to πI_{sr}, and this must be equal to the radiation

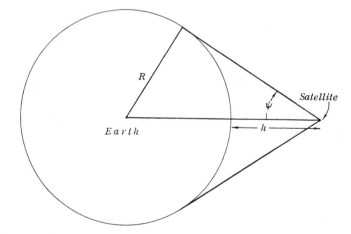

Fig. 4.12. Cross section through the earth in the plane of an earth satellite.

received per unit area ($F_0 \cos \theta$) multiplied by the albedo. Thus the albedo (A) may be calculated from

$$\pi I_{sr} = A F_0 \cos \theta$$

Substituting this equation into (4.26) yields

$$F'_{sr} = \frac{F_0}{\pi} \int_0^\omega A \cos \theta \, d\omega$$

For a satellite at fairly low altitude, $\cos \theta$ does not vary much over the observed area, and so may be taken outside the integral. And if the average albedo is defined by

$$\bar{A} = \frac{1}{\omega} \int_0^\omega A \, d\omega$$

the reflected solar radiation may be expressed by

$$F'_{sr} = \frac{F_0}{\pi} \bar{A} \omega \cos \theta \qquad (4.30)$$

Upon introducing Eq. (4.29), the average albedo becomes

$$\bar{A} = \frac{F'_{sr}}{2 F_0 \left\{ 1 - \dfrac{(2Rh + h^2)^{1/2}}{R + h} \right\} \cos \theta} \qquad (4.31)$$

Thus, determination of the average albedo requires measurements of the solar flux density, the reflected solar flux, zenith angle, and height of the satellite above the earth.

The long-wave radiation emitted by the earth and atmosphere may be represented by black-body radiation at an effective temperature T_e. This means that a black body with temperature T_e radiates with an intensity I_l, and the effective temperature is defined by

$$I_l \equiv \frac{\sigma T_e^{4}}{\pi} \qquad (4.32)$$

Because the satellite responds to an average over the solid angle intercepted by the earth, it is convenient to introduce an average effective temperature defined by

$$\overline{T_e}^{4} \equiv \frac{1}{\omega} \int_0^\omega T_e^{4} \, d\omega$$

It follows from Eqs. (4.26), (4.29), and (4.32) that

$$\overline{T_e}^{4} = \frac{\pi F'_l}{\sigma \omega} = \frac{F'_l}{2\sigma \left\{ 1 - \dfrac{(2Rh + h^2)^{1/2}}{R + h} \right\}} \qquad (4.33)$$

The distribution of outgoing long-wave radiation has been observed by Explorer VII *; and maps have been prepared for selected days. These maps exhibit large scale organization with systematic variations amounting to as much as 0.1 cal cm^{-2} min^{-1} (7 mw cm^{-2}).

World-wide averages of outgoing long-wave radiation ($\sigma \overline{T}_e{}^4$) and $\overline{A}$ representative of a fairly short period can be obtained from a well-designed satellite observational program. If these averages are sufficiently accurate, it is possible to determine the net rate of gain or loss of energy by the earth and atmosphere. An exercise is provided in problem 9. This information has never been available from any source, and we have very little idea of the magnitudes or the changes which may occur in the rate of gain or loss. This challenging information will soon be available.

PART III: EFFECTS OF ABSORPTION AND EMISSION

4.11 Absorption

Attenuation of radiation results from absorption, scattering and reflection. In this section the major absorbing properties of atmospheric gases will be described. Oxygen and nitrogen, which together constitute 99% of the atmosphere by volume, absorb strongly that fraction of solar radiation having wavelengths shorter than about 0.3 micron. Consequently, ultraviolet radiation from the sun is absorbed in the very high atmosphere. The very short wavelengths of great energy ionize the gas molecules. Other wavelengths in the ultraviolet are absorbed by ozone in the layer between 20 and 50 kilometers above the earth. The photochemical processes associated with these radiations are discussed in Part IV of this chapter, and other aspects of the ionosphere are discussed in Chapters VI and VII.

The absorption spectra for atmospheric gases are shown in Fig. 4.13. In the infrared a large fraction of the radiation is strongly absorbed by water vapor and carbon dioxide. Ozone, nitrous oxide, and methane absorb lesser but appreciable energy. The very rapid change in absorptivity (and therefore emissivity) with small change in wavelength which is evident in the expanded portion of Fig. 4.13 makes calculation of the total energy absorbed or transmitted by the atmosphere very difficult, and atmospheric absorption spectra are still not known with sufficient detail for highly accurate calculations. The atmosphere is evidently nearly transparent between 0.3 and 0.8 micron, but the intensity of terrestrial and atmospheric radiation is very weak in this spectral range. Within the 8 to 12 micron range, except for strong absorption by ozone at 9.6 microns, terrestrial radiation escapes directly to space. This is the wavelength interval of

* V. E. Suomi and W. C. Shen, *J. Atmospheric Sci.* **20**, 62 (1963).

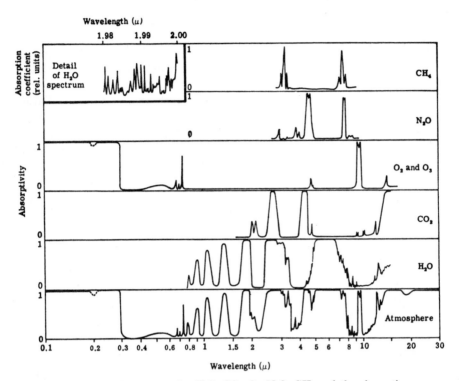

FIG. 4.13. Absorption spectra for H_2O, CO_2, O_2, N_2O, CH_4 and the absorption spectrum of the atmosphere [after J. N. Howard, *Proc. I.R.E.* **47**, 1451 (1959), and R. M. Goody and G. D. Robinson, *Quart. J. Roy. Meteorol. Soc.* **77**, 153 (1951)].

strongest terrestrial radiation, and the radiation is said to escape through the open atmospheric "window."

4.12 The "Atmosphere" Effect

Figures 4.11 and 4.13 taken together emphasize an important characteristic of the atmosphere. The atmosphere is much more transparent for short-wave radiation than for long-wave radiation. The result is that solar energy passes through the atmosphere and is absorbed at the earth's surface. On the other hand, most of the long-wave radiation emitted by the earth's surface is absorbed by the atmosphere. Consequently, solar energy is trapped by the earth and atmosphere.

This effect is sometimes referred to as the *greenhouse effect* because in a similar manner glass covering a greenhouse transmits short-wave solar radiation but absorbs the long-wave radiation emitted by the plants and earth inside the greenhouse. However, it is not commonly recognized that, whereas the absorbing effect of the atmosphere results in temperatures well above what they would be without an atmosphere, the high temperatures in a greenhouse are not to be attributed to absorption of long-wave radiation by the glass. In 1909 R. W. Wood carried out an experiment with two small model greenhouses of which one was covered with glass and one with rock salt. Rock salt is transparent for both short- and long-wave radiation and therefore does not "trap" the radiation in the sense mentioned above. Both model greenhouses reached about the same high temperatures, proving that the effectiveness of greenhouses in growing plants is not the result of absorption of long-wave radiation by glass. Greenhouses reach much higher temperatures than the surrounding air because the glass cover of the greenhouse prevents the warm air from rising and removing heat from the greenhouse. This effect is four to five times as important as the absorption of long-wave radiation by the glass.

Trapping of radiation by the atmosphere is typical of the atmosphere and therefore may be called the *"atmosphere effect."* In order to make an estimate of this effect, the average temperature of the earth may be compared with the radiative equilibrium temperature. For an albedo of 0.35 and solar constant of 2.0 cal cm^{-2} min^{-1} the equilibrium temperature is 248°K, whereas the average observed temperature near the earth's surface is 283°K, 35°K above the radiative equilibrium temperature. If all long-wave radiation is assumed to be absorbed in a single layer of the atmosphere, the result of problem 10 shows that the average temperature of the earth's surface is 296°K. It may be concluded that the atmosphere forms a protective blanket for life on earth although, as shown in problem 11, the "effective absorptivity" is less than unity. A more detailed discussion of radiative transfer of energy is given in the following sections.

4.13 Transfer of Radiation between Two Parallel Surfaces

Radiative exchange between two parallel surfaces with negligible absorption in the intervening space provides one of the simplest models of radiation transfer. This model is approximated by the earth's surface and a cloud layer or by two cloud layers if the water vapor content in the intervening air is very low or the distance between the surfaces is small.

Assume that the radiating surfaces are *gray*, that is, that they have absorptivities a_1 and a_2, respectively, which are independent of wavelength. The question is, how much energy is transferred from surface (1) to surface (2)?

If F_1 represents the flux density in the direction of surface (2) and F_2 the flux density in direction of surface (1), then the net transfer of energy per unit area and unit time from surface (1) to surface (2) is $F_1 - F_2 = F_n$, the *net flux density* or *net radiation*. F_1 consists of radiation emitted by surface (1) plus the radiation reflected from the same surface, so

$$F_1 = a_1 \sigma T_1^4 + (1 - a_1)F_2$$

and similarly

$$F_2 = a_2 \sigma T_2^4 + (1 - a_2)F_1$$

Solution of these equations yields

$$F_1 = \frac{1}{1 - (1 - a_1)(1 - a_2)} \{a_1 \sigma T_1^4 + (1 - a_1)a_2 \sigma T_2^4\}$$

$$F_2 = \frac{1}{1 - (1 - a_1)(1 - a_2)} \{a_2 \sigma T_2^4 + (1 - a_2)a_1 \sigma T_1^4\}$$

and

$$F_1 - F_2 = F_n = \frac{a_1 a_2}{1 - (1 - a_1)(1 - a_2)} \sigma(T_1^4 - T_2^4)$$

When $a_1 = a_2 = 1$, this equation reduces to

$$F_n = \sigma(T_1^4 - T_2^4)$$

which represents the transfer between black surfaces. In problem 12 a possible case of transfer between the surface of the earth and a cloud layer is given.

The problem becomes quickly very complicated if the energy transfer between arbitrarily shaped bodies is considered. These cases fortunately are not of great interest in atmospheric problems.

4.14 Transfer of Long-Wave Radiation in a Plane Stratified Atmosphere

The problem is to express the upward and downward radiation fluxes through unit horizontal area as a function of the vertical distribution of temperature and of absorbing substance. Temperature is assumed independent of the x and y directions. The procedure to be followed is (1) to express the contribution made by an arbitrary volume element to the monochromatic flux density at the reference level and (2) to integrate over all wavelengths the separate contributions of all volume elements above and below the reference level.

To obtain the contribution from a volume element choose an element of vertical optical thickness du at an optical height u above the reference level, as illustrated in Fig. 4.14. The monochromatic intensity (dI_λ) emitted by the element in the direction θ is given by Kirchhoff's law in the form $k_\lambda I_\lambda^* \, du \sec \theta$. This radiation is attenuated according to Beer's

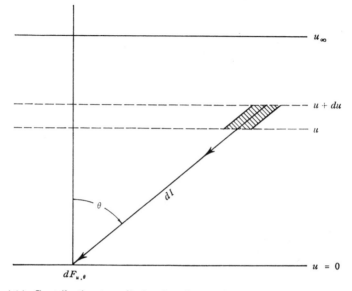

FIG. 4.14. Contribution to radiation flux density $(dF_{u,\theta})$ at $u = 0$ made by a differential element at optical thickness u and zenith angle θ.

law, so that the fraction of the intensity increment dI_λ that arrives at the reference level is

$$dI_\lambda \exp\left(-k_\lambda u \sec \theta\right) = k_\lambda I_\lambda{}^* \exp\left(-k_\lambda u \sec \theta\right) du \sec \theta$$

All volume elements seen at zenith angle θ contribute an equal increment to the flux density at dA. Remembering that $d\omega = 2\pi \sin \theta \, d\theta$ and that the relation between intensity and flux density is given by Eq. (4.1), the contribution to flux density at the surface made by a circular ring of the differential layer is recognized to be

$$dF_{\lambda,u,\theta} = 2\pi k_\lambda I_\lambda{}^* \exp\left(-k_\lambda u \sec \theta\right) \sin \theta \, d\theta \, du$$

Integration with respect to θ gives the contribution to the monochromatic flux density from the entire layer in the form

$$dF_{\lambda,u} = 2\pi k_\lambda I_\lambda{}^* \, du \int_0^{\pi/2} \sin \theta \exp\left(-k_\lambda u \sec \theta\right) d\theta \qquad (4.34)$$

The integral is a function of $k_\lambda u$ only. If the transformation $y \equiv \sec \theta$ is made, the integral takes the form

$$\int_1^\infty \frac{e^{-k_\lambda u y}}{y^2} \, dy \equiv H_2(k_\lambda u) \qquad (4.35)$$

where $H_2(k_\lambda u)$ is called a Gold function of the second order. This function has been tabulated for various $k_\lambda u$ values. Since $\pi I_\lambda^* = F_\lambda^*$, Eq. (4.34) may be written

$$dF_{\lambda u} = 2k_\lambda F_\lambda^* H_2(k_\lambda u) \, du$$

The contribution to the flux density made by all wavelengths emitted by the slab du may be expressed by

$$dF = 2 \left\{ \int_0^\infty k_\lambda F_\lambda^* H_2(k_\lambda u) \, d\lambda \right\} du \tag{4.36}$$

The integral on the right-hand side is a function of temperature and optical thickness and can be evaluated for each pair of values of T and u if the absorption coefficient is known. Despite the complexity of Eq. (4.36), the reader may be assured that he has not undertaken the task of Sisyphus; he will have to toil to the top of the mountain only one more time.

One further integration, this time over the entire optical thickness of the atmosphere, yields the total downward directed flux density. A similar integration describes the upward directed flux density. However, because evaluation of k_λ depends on difficult laboratory measurements which are limited in resolution and accuracy, the integrals are less than exact. For this reason, it is appropriate to simplify the procedure by expressing the contribution to flux density made by the integral in Eq. (4.36) as proportional to black-body emission. In order for this procedure to give the radiation flux density accurately, the proportionality coefficient must be independent of temperature over the range encountered in the vertical column; that is, the proportion of the area under the black-body envelope which represents energy absorbed in an atmospheric layer must not change as the black-body envelope changes with change in temperature. This condition is at least approximately fulfilled for water vapor and carbon dioxide for the atmospheric temperature range, as may be inferred from Fig. 4.13. The integral in Eq. (4.36) may now be expressed by

$$2 \int_0^\infty k_\lambda F_\lambda^* H_2(k_\lambda u) \, d\lambda = \gamma \sigma T^4$$

where γ represents the constant of proportionality which is only a function of optical thickness. Equation (4.36) now may be written

$$dF = \sigma T^4 \gamma \, du \tag{4.37}$$

From this equation the product $\gamma \, du$ may be recognized as the contribution of the layer to the total or *flux emissivity* of the atmosphere. If ϵ_f represents flux emissivity

$$d\epsilon_f = \gamma \, du \tag{4.38}$$

Equation (4.38) shows that γ is the rate of change of flux emissivity with optical thickness.

The total downward flux density is obtained by integrating Eq. (4.37) from the reference level ($u = 0$) to the top of the atmosphere ($u = u_\infty$). This yields for a cloudless atmosphere

$$F \downarrow = \int_0^{u_\infty} \gamma \sigma T^4 \, du \tag{4.39}$$

The upward flux density consists of a contribution from the atmosphere below the reference level plus a contribution from the black-body radiation emitted by the earth's surface. The latter is given by multiplying the radiation emitted by the surface by the flux transmissivity of the atmosphere below the reference level. The upward flux density therefore can be written as

$$F \uparrow = \int_0^{u_0} \gamma \sigma T^4 \, du + (1 - \epsilon_{f0}) \sigma T_0^4 \tag{4.40}$$

where the index 0 refers to the earth's surface and the flux transmissivity is replaced by $(1 - \epsilon_{f0})$, following Kirchhoff's law. The flux emissivity (ϵ_{f0}) is found by integrating Eq. (4.38) from the reference level to the earth's surface; therefore

$$\epsilon_{f0} = \int_0^{u_0} \gamma \, du$$

Black-body radiation from cloud layers may be in principle accounted for in a similar way.

Equations (4.39) and (4.40) can be evaluated when γ is known as a function of optical thickness. In the following section a discussion of the experimental determination of the flux emissivity and γ is given. Because the temperature is an arbitrary function of the optical thickness, Eqs. (4.39) and (4.40) cannot be integrated analytically. Instead, graphical integration may be performed by dividing the atmosphere into finite layers and approximating the integrals by summation over the layers. It is convenient to transform the optical thickness to pressure as the vertical coordinate. If only the contribution of water vapor to optical thickness is considered

$$u = \int_{z_0}^{z} \rho_v \, dz = \frac{1}{g} \int_p^{p_0} q \, dp \tag{4.41}$$

where q is the specific humidity. After substitution of Eq. (4.41), Eqs. (4.39) and (4.40) may be written as summations in the form

$$F \downarrow = \frac{\sigma}{g} \sum_{i=1}^{n_\infty} T_i^4 q_i \gamma_i \, \Delta p_i$$

and $\hspace{6cm}$ (4.42)

$$F \uparrow = (1 - \epsilon_{f0}) \sigma T_0^4 + \frac{\sigma}{g} \sum_{i=1}^{n_0} T_i^4 q_i \gamma_i \, \Delta p_i$$

where n_∞ represents the number of layers that are taken from the reference level to the top of the atmosphere and n_0 the number of layers between reference level and the earth's surface.

To evaluate the flux density it is only necessary to determine mean values of T, q, and γ for each layer, multiply by the appropriate Δp and add. This procedure is illustrated in problem 13.

The contribution to total atmospheric radiation made by carbon dioxide may be treated in a similar manner if care is taken to avoid counting twice the contributions from overlapping CO_2 and H_2O bands. A simpler approximate method is to draw up separate emissivity curves for various ratios of CO_2 to H_2O and to use γ in Eqs. (4.42) for the appropriate ratio. Other gases emit very much less strongly than H_2O and CO_2 and are neglected in these calculations.

4.15   Experimental Determination of Flux Emissivity

In order to evaluate Eqs. (4.42) it is necessary to know γ (or ϵ_f) as a function of optical thickness. This information can be derived from the results of laboratory measurements of transmissivity. A tube is filled with a known amount of the absorbing gas, a radiation source of constant intensity is placed at one end of the tube, and the transmitted intensity is measured at the other end. The emissivity is then found by dividing the difference in intensities at the ends of the tube by the intensity of the source. By changing the amount of absorbing gas in the tube or by using mirrors inside the tube, the optical thickness may be varied, and the emissivity may be determined as a function of optical thickness.

The parallel beam emissivity determined as described here may be converted to flux emissivity by integrating over the hemisphere shown in Fig. 4.14. By analogy with Eq. (4.2) the flux emissivity is found to be

$$\epsilon_f(u) = 2 \int_0^{\pi/2} \epsilon(u \sec \theta) \sin \theta \cos \theta \, d\theta \qquad (4.43)$$

where u represents the optical thickness of the slab for which the flux emissivity is to be determined. From Eq. (4.43) it may be recognized that in order to obtain ϵ_f for a layer of thickness u, ϵ must be known for substantially greater optical thicknesses.

In general the flux emissivity exceeds the parallel beam emissivity for the same optical thickness. The limits are easily recognized. When optical thickness is very small, absorption by the intervening atmosphere is negligible, so that flux emissivity is directly proportional to optical thickness and $\epsilon(u \sec \theta) = \epsilon(u) \sec \theta$, and Eq. (4.43) reduces to

$$\epsilon_f(u) = 2\epsilon(u) = \epsilon(2u)$$

On the other hand, if the optical thickness is very large, the radiation emitted by the slab becomes black-body radiation for all directions and $\epsilon_f(u) = \epsilon(u) \approx 1$. For a wide range of optical thickness in between the two extremes Eq. (4.43) can be approximated by

$$\epsilon_f(u) = \epsilon(1.60 \, u)$$

Figure 4.15 shows that this equality holds within 20% over the range of $k_\lambda u$ from 10^{-2} to beyond 10. This range represents virtually all the atmospheric absorption.

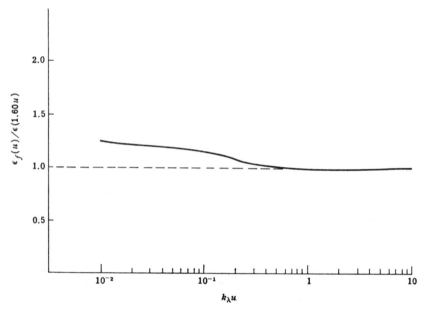

FIG. 4.15. Ratio $\epsilon_f(u)/\epsilon \, (1.60u)$ as a function of $k_\lambda u$ computed from Eq. (4.43)

By using this equation or Eq. (4.43) where higher accuracy is necessary, experimental observations of the parallel beam emissivity may be converted to flux emissivity. Experimental results obtained in this way for water vapor and for carbon dioxide are shown as functions of optical thickness in Fig. 4.16. Also shown in the figure is the flux emissivity for a typical mixture of water vapor and carbon dioxide. At very small optical thickness (less than 10^{-5} g cm^{-2} of water vapor as indicated by Elsasser*) emissivity increases linearly with optical thickness. Beyond 10^{-5} g cm^{-2} the emissivity increases more and more slowly with increasing thickness as a result of absorption along the path. Emissivity of unity (complete blackness) is

---

* W. M. Elsasser, *Meteorol. Monographs* **4** (23), 43 pp. (1960).

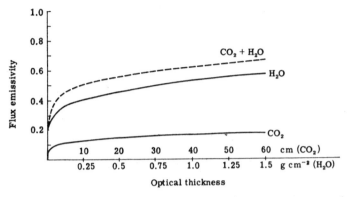

FIG. 4.16. Experimental observations of emissivity of pure water vapor, of carbon dioxide and of an atmospheric mixture of CO_2 (0.032%) and H_2O (mixing ratio 5 g/kg) as a function of optical thickness for temperature of 10°C and pressure 1013 mb [after W. M. Elsasser, *Meteorol. Monographs* **4** (23), 43 pp. (1960)].

reached at optical thickness of infinity. The optical thickness in the clear atmosphere ranges between about one and ten grams per square centimeter, and the corresponding emissivity of the atmosphere ranges from about 0.65 to 0.85.

The rate of change of emissivity with path length (γ) decreases monotonically with path length from roughly 10^3 cm^2 g^{-1} at 10^{-5} g cm^{-2} to about 10^{-2} cm^2 g^{-1} at 10 g cm^{-2} optical thickness. The decrease is very rapid between 10^{-5} and 10^{-3} g cm^{-2}. A detailed graph of this function for water vapor is given in Fig. 4.17.

4.16 Divergence of Net Radiation

The cooling or warming experienced by a layer of air due to radiation may be calculated from the principle of conservation of energy. If the net radiation flux density, defined by $F_n \equiv F\uparrow - F\downarrow$, is smaller at the top of the layer than at the bottom, the difference must be used to heat the layer. Thus

$$c_p\rho \left(\frac{\partial T}{\partial t}\right)_R \Delta z = -F_n(z + \Delta z) + F_n(z) = -\frac{\partial F_n}{\partial z} \Delta z$$

or, after introducing Eq. (4.41),

$$\left(\frac{\partial T}{\partial t}\right)_R = -\frac{1}{c_p\rho} \frac{\partial F_n}{\partial z} = -\frac{q}{c_p} \frac{\partial F_n}{\partial u} \tag{4.44}$$

The term $\partial F_n/\partial u$ is referred to as the *divergence of net radiation* or simply *flux divergence*.

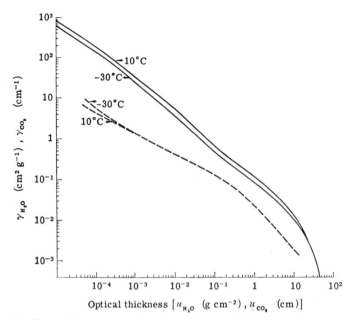

FIG. 4.17. Rate of change of emissivity with optical thickness (γ) as a function of optical thickness for H_2O (*solid lines*) and for CO_2 (*dashed lines*) (after W. M. Elsasser *Meteorol. Monographs* **4** (23), 43 pp. (1960))].

After substituting Eqs. (4.39) and (4.40) into (4.44) and utilizing a balanced mixture of mathematical technique and physical insight, the rate of radiational warming may be expressed by

$$\left(\frac{\partial T}{\partial t}\right)_R = \frac{\sigma q}{c_p}\left\{4\int_{T_r}^{T_\infty}\gamma T^3\,dT - T_\infty^4\gamma(u_\infty) + 4\int_{T_r}^{T_0}\gamma T^3\,dT\right\}\quad(4.45)$$

where T_r represents the temperature at the reference level and T_∞ represents the temperature at the height above which the water vapor becomes negligible. Radiational temperature change in the cloudless atmosphere may be evaluated from Eq. (4.45) by dividing the atmosphere into a sufficiently large number of layers and summing over the layers as required in problem 14.

Calculations of radiative temperature change using Eq. (4.45) are straightforward although somewhat tedious, particularly if detail is required. Results depend, of course, on the vertical distributions of temperature and humidity, but throughout most of the troposphere, where temperature decreases with height at 4 to 8°C km^{-1}, the air cools by radiation at 1 to 2°C day^{-1}. The excess of temperature over radiational equilibrium temperature indicates that internal energy is supplied to the troposphere by other processes. These are discussed in Chapter V.

Because γ decreases sharply beyond optical path lengths corresponding to vertical distances of a few centimeters, radiational temperature change depends strongly on the temperature distribution in the immediate neighborhood of the reference point. At the base of temperature inversions, above which temperature increases with height, radiational warming at a rate of several degrees per hour may be calculated from Eq. (4.45).

Errors in radiation calculations are to be attributed to inaccuracies in determination of flux emissivity and, especially, to errors in humidity observations. The latter are very serious at low temperatures and low humidities where the humidity element carried aloft by the radiosonde is insensitive. Radiative flux or flux divergence in the cloudless regions above or below cloud layers may be calculated if the cloud layers are treated as black bodies at known temperatures. This is not very satisfactory, however, because "blackness" of clouds depends on thickness and on the existing drop size spectrum, and because infrared radiation is scattered by cloud droplets in a complex manner. For these reasons, it is important that radiation calculations be supplemented by direct observations.

4.17 Direct Measurement of Flux Divergence

Flux divergence may be measured by the radiometersonde, a protected flat plate radiometer which is carried aloft by a balloon. As shown in Fig. 4.18, one plate faces upward and the other downward, and each receives radiation from the appropriate hemisphere. The polyethylene cover is transparent for most of the short- and long-wave ranges of wavelengths.

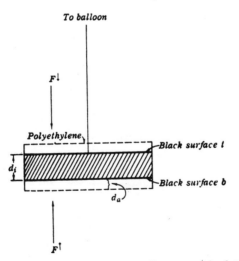

FIG. 4.18. Essential elements of a flat plate radiometer with plates t and b shielded from ventilation by a polyethylene cover.

In order to develop a simple relation between rate of radiational temperature change and temperatures observed in the radiometersonde, the small absorptivity of the polyethylene shield is neglected. Under conditions of radiative equilibrium the upward and downward radiation fluxes in the instrument can be derived with the method developed in Section 4.13 and utilized in problem 15. The result considering only long-wave radiation is for the top plate

$$F\downarrow = \sigma T_t^4 + \frac{1 - r(1 - a)}{a(1 - r)} L_t \qquad (4.46)$$

where r represents the reflectivity of polyethylene, a the absorptivity of the black surface, T_t temperature of the top plate, and L_t loss from top plate due to conduction to the surrounding air and to the other plate. Similarly, the equilibrium condition for the bottom plate is given by

$$F\uparrow = \sigma T_b^4 + \frac{1 - r(1 - a)}{a(1 - r)} L_b \qquad (4.47)$$

where the subscript b refers to the bottom plate. The losses are assumed to be proportional to temperature gradient. Under steady state conditions this may be expressed by

$$L_t = (k_a/d_a)(T_t - T_a) + (k_i/d_i)(T_t - T_b)$$

where k_a represents the thermal conductivity of the air trapped between the plate and the polyethylene, d_a vertical separation of plate and polyethylene, T_a air temperature, k_i thermal conductivity of insulation between the plate, and d_i separation of the plates. In a similar manner

$$L_b = (k_a/d_a)(T_b - T_a) + (k_i/d_i)(T_b - T_t)$$

T_t^4 may be expanded in a binomial series about T_a with the result

$$T_t^4 = T_a^4 + 4T_a^3 (T_t - T_a) + 6T_a^2 (T_t - T_a)^2 + \cdots$$

For $T_t - T_a \ll T_a$, all terms beyond the first two on the right may be neglected. Similarly, T_b^4 may be expanded with the result

$$T_b^4 = T_a^4 + 4T_a^3(T_b - T_a) + 6T_a^2(T_b - T_a)^2 \cdots$$

Finally, when these equations are substituted into Eqs. (4.46) and (4.47) and then into (4.44), the linear terms yield

$$\left(\frac{\partial T}{\partial t}\right)_R = \frac{1}{c_p\rho}\left[12\sigma T_a^2(T_t - T_b)\frac{\partial T_a}{\partial z} + \left\{4\sigma T_a^3 + \left(\frac{k_a}{d_a} + \frac{2k_i}{d_i}\right)\frac{1 - r(1 - a)}{a(1 - r)}\right\}\frac{d}{dz}(T_t - T_b)\right] \qquad (4.48)$$

The coefficient on the right is slowly varying with air temperature and pressure and may be determined with reasonable accuracy by calibration.

The rate of radiative temperature change, therefore, is easily evaluated from observations of the rate of change with height of the temperature difference between upper and lower plates of the instrument as illustrated in problem 16. In order to obtain the greatest accuracy k_i and r should be as small as possible. The results of observations using a radiometersonde designed by Suomi and Kuhn* are shown in Fig. 4.19.

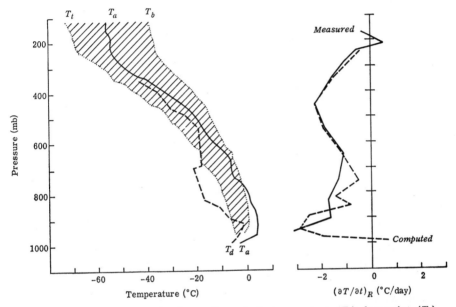

FIG. 4.19. Radiometersonde observations of air temperature (T_a) dew point (T_d), radiometer temperatures $(T_t$ and $T_b)$, rate of radiational temperature change measured using Eq. (4.48) and computed from Eq. (4.45) [after V. E. Suomi, D. O. Staley, and P. M. Kuhn, *Quart. J. Roy. Meteorol. Soc.* **84**, 134 (1958)].

PART IV: PHOTOCHEMICAL PROCESSES

4.18 Dissociation of Oxygen and Its Consequences

Absorption of radiation by a gas takes place when a photon strikes a molecule and the energy of the photon is transferred to the molecule. The molecule may gain so much energy in this way that it becomes unstable and splits into two new particles. Because the change in structure is produced by a photon, it is called a photochemical reaction.

An important photochemical reaction that occurs in the upper atmosphere is the dissociation of molecular oxygen by absorption of a photon. The reaction is expressed by

---

* V. E. Suomi and P. M. Kuhn, *Tellus* **10**, 161 (1958).

$$C_2 + h\nu \rightarrow O + O \tag{4.49}$$

The minimum energy necessary for this reaction corresponds to a wavelength of 0.2424 μ. The rate of dissociation of O_2 is proportional to the concentration of O_2 molecules and the amount of absorbed radiation integrated over all wavelengths less than 0.2424 μ. Therefore for the reaction described by (4.49)

$$\left(\frac{dn_2}{dt}\right)_{\text{diss}} = -n_2 \int_0^{0.2424} \beta_{\lambda 2} k_{\lambda 2} F_{\lambda 2} \, d\lambda$$

where n_2 is the number of O_2 molecules per cubic centimeter and β_λ is the fraction of the number of O_2 molecules dissociated per unit energy of radiation absorbed.

The last equation can be evaluated only approximately because $\beta_{\lambda 2}$ and $k_{\lambda 2}$ are known only approximately. The radiation flux density may be expressed using Beer's law in the form

$$F_{\lambda 2} = F_{\lambda 20} \exp\left(-\frac{m_2}{N_0} \int_z^\infty k_{\lambda 2} n_2 \sec \theta \, dz\right) \tag{4.50}$$

where m_2 is the molecular mass of O_2 and N_0 is Avogadro's number. Equation (4.50) shows that $F_{\lambda 2}$ depends on the distribution of n_2 with height. But this distribution of n_2 is just what we should like to determine. Also, because of pressure broadening of the absorption bands, k_λ depends on height. For these reasons determination of the vertical distribution of the various constituents participating in photochemical processes is not a simple task.

Following formation of atomic oxygen as described by reaction (4.49), a series of further reactions may occur, of which the most important are

$$O + O + M \rightarrow O_2 + M \tag{4.51}$$

$$O + O_2 + M \rightarrow O_3 + M \tag{4.52}$$

and

$$O_3 + O \rightarrow 2O_2{}^* \tag{4.53}$$

where the asterisk indicates an excited state.

Reactions (4.51) and (4.52) require three-body collisions, and the third body, represented by M, may be any particle capable of absorbing the extra energy released by the reaction. When two particles, say O and O collide, the O_2 formed would be unstable and dissociate again unless it were able to release its excess energy to a third particle within a very short time.

Reaction (4.52) represents the process which is chiefly responsible for the creation of ozone in the atmosphere. Ozone is decomposed again by the reaction (4.53) and by the additional important reaction

$$O_3 + h\nu \rightarrow O_2 + O \tag{4.54}$$

In this reaction the wavelength of the photon must be less than 1.1340 μ. The following equations may now be written for the rates of change of the concentrations n_1 of O, n_2 of O_2, and n_3 of O_3:

$$-2K_1n_1{}^2n_M - K_2n_1n_2n_M - K_3n_1n_3 + 2n_2 \int_0^{0.2424} \beta_{\lambda2}k_{\lambda2}F_{\lambda2}\,d\lambda$$

$$+ n_3 \int_0^{1.1340} \beta_{\lambda3}k_{\lambda3}F_{\lambda3}\,d\lambda = \frac{dn_1}{dt}$$

$$K_1n_1{}^2n_M - K_2n_1n_2n_M + 2K_3n_1n_3 - n_2 \int_0^{0.2424} \beta_{\lambda2}k_{\lambda2}F_{\lambda2}\,d\lambda$$

$$+ n_3 \int_0^{1.1340} \beta_{\lambda3}k_{\lambda3}F_{\lambda3}\,d\lambda = \frac{dn_2}{dt}$$

$$-K_3n_1n_3 + K_2n_1n_2n_M - n_3 \int^{1.1340} \beta_{\lambda3}k_{\lambda3}F_{\lambda3}\,d\lambda = \frac{dn_3}{dt}$$

where n_M is the number of M particles per unit volume, and K_1, K_2, and K_3 are the reaction coefficients of reactions (4.51), (4.52), and (4.53), respectively. The distribution of the three oxygen constituents with height is now determined by the above equations together with two equations of the form (4.50), the hydrostatic equation and the equation of state. The fact that the absorbed radiation increases the temperature, which in its turn affects the chemical reaction coefficients, complicates the calculations. It turns out that at the height where the ozone-forming reaction [(4.52)] occurs little of the $F_{\lambda2}$ radiation penetrates because most of this radiation is absorbed at greater heights. Therefore, very little dissociation of O_2 takes place, and even in the *ozonosphere* (the layer where ozone is formed and decomposed) the concentrations of O and O_3 are small compared to the concentration of O_2. Thus n_2 (the concentration of O_2) may be considered independent of n_1 and n_3; and if uniform temperature is assumed, n_2 may be determined by Eq. (2.21) in the form

$$n_2 = n_{20} \exp\left(-\Phi/R_M T\right) \tag{2.21a}$$

where n_{20} is the number of O_2 molecules per unit volume at geopotential $\Phi = 0$. Furthermore, it usually is assumed that the atmosphere is in *photochemical equilibrium*, that is

$$\frac{dn_1}{dt} = \frac{dn_2}{dt} = \frac{dn_3}{dt} = 0$$

Under these simplifying assumptions the set of equations has been solved yielding vertical distributions of ozone in fair agreement with the observations shown in Fig. 4.20.

Photochemical equilibrium is approached when the major formation and decomposition processes are so rapid that the time required to form the equilibrium reservoir is short compared with diurnal variations. The photochemical processes (4.49) and (4.54) are most active above the height where the respective radiation flux densities ($F_{\lambda 2}$ and $F_{\lambda 3}$) are depleted by absorption. In the lower part of the ozonosphere the active

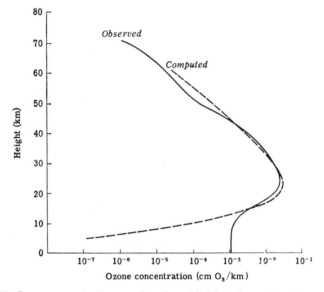

FIG. 4.20. Ozone concentration as a function of height, observed and computed for photochemical equilibrium by Paetzold [after H. K. Paetzold and E. Regener, *in* "Handbuch der Physik" (S. Flügge, ed.), Vol. 47, p. 394. Springer-Verlag, Berlin, 1957].

radiation is almost completely absorbed, and consequently the time to restore equilibrium is very long. Therefore, as shown in Fig. 4.20, the observed concentration in the lower ozonosphere may differ markedly from the corresponding equilibrium concentration.

Above the ozonosphere reactions (4.49) and (4.51) are dominant, and photochemical equilibrium is expressed by

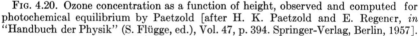

$$K_1 n_1^2 n_M - n_2 \int_0^{0.2424} \beta_{\lambda 2} k_{\lambda 2} F_{\lambda 2}\, d\lambda = 0 \qquad (4.55)$$

The distribution of n_1 for the case $n_1 \ll n_2$ is determined by this equation in combination with Eqs. (2.21a) and (4.50). If dissociation is so large that n_1 is comparable to n_2, the hydrostatic equation cannot be integrated to the simple form of Eq. (2.21a); instead it must be left in the more general differential form

$$\frac{d(n_1 + n_2 + n_N)}{dz} = -\frac{g(n_1 m_1 + n_2 m_2 + n_N m_N)}{kT} \qquad (4.56)$$

where n_N is the number of nitrogen molecules per unit volume; m_N is the mass of a nitrogen molecule and k is Boltzmann's constant. Equations (4.50), (4.55), and (4.56) determine the vertical distributions of n_1 and n_2 under conditions of photochemical equilibrium and uniform temperature. Although this set of equations is simpler than the set appropriate to the ozonosphere, it is still a considerable task to solve them. The vertical distribution of O_2 and O above 80 kilometers under conditions of photochemical equilibrium is given in Fig. 4.21.

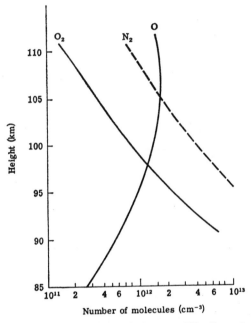

FIG. 4.21. Number of molecules of atomic oxygen (O), diatomic oxygen (O_2) and diatomic nitrogen (N_2) per cubic centimeter as a function of height [after M. Nicolet, *in* "Physics of the Upper Atmosphere" (J. A. Ratcliffe, ed.), p. 34. Academic Press, New York, 1960].

Recombination of monatomic oxygen, as expressed by reaction (4.51), releases 5.1 ev per atom pair. Kellogg* has suggested that this exothermic reaction accounts for the fact that the upper atmosphere is markedly warmer in winter than in summer. He has shown that subsidence of air in polar regions at less than 1 cm sec^{-1} may carry monatomic oxygen downward to pressures where recombination occurs efficiently; the heat released is sufficient to heat the air by 10°C day^{-1}.

* W. W. Kellogg, *J. Meteorol.* 18, 373 (1961).

Evidence of auroral displays indicates that some atomic nitrogen is present above heights of about 100 kilometers, but solar radiation is not effective in its dissociation. Mitra* has shown that dissociation may be caused by recombination of N_2^+ ions and electrons according to the reaction

$$N_2^+ + e^- \rightarrow 2N$$

Ionization results from absorption of ultraviolet radiation, as will be discussed in the next section, so that this reaction is a secondary effect of radiation.

4.19   Photoionization

Ionization of molecules and atoms by ultraviolet radiation can be treated relatively simply when it is assumed that absorption takes place in a narrow band ($\Delta\lambda$). Equation (4.50) may then be written in the form

$$F_\lambda = F_{\lambda 0} \exp\left(-k_\lambda \sec \theta \int_z^\infty \rho \, dz \right) \qquad (4.57)$$

where ρ is the density of the absorbing gas. If again uniform temperature is assumed, this equation may be combined with Eq. (2.20), yielding

$$F_\lambda = F_{\lambda 0} \exp\left[-k_\lambda \rho_0 \frac{R_m T}{g} \sec \theta \exp\left(-\frac{\Phi}{R_m T} \right) \right] \qquad (4.58)$$

The production rate of ions is proportional to the amount of radiation absorbed per unit volume, that is

$$p = \beta_\lambda k_\lambda \rho F_\lambda \Delta\lambda$$

Substitution of Eqs. (2.20) and (4.58) into this equation yields

$$p = \beta_\lambda k_\lambda F_{\lambda 0} \Delta\lambda \rho_0 \exp\left[-\frac{\Phi}{R_m T} - k_\lambda \rho_0 \frac{R_m}{g} \sec \theta \exp\left(-\frac{\Phi}{R_m T} \right) \right] \qquad (4.59)$$

The height of maximum ion production rate is found by differentiation to be

$$z_m = \frac{R_m T}{g} \ln\left[\frac{R_m T}{g} k_\lambda \rho_0 \sec \theta \right] \qquad (4.60)$$

and the maximum production rate is

$$p_m = \frac{g \beta_\lambda F_{\lambda 0} \Delta\lambda \cos \theta}{R_m T} \exp(-1) \qquad (4.61)$$

Figure 4.22 illustrates schematically the vertical distribution of ionizing radiation, density and ion production described by Eqs. (4.57), (2.20), and (4.59). Equations (4.60) and (4.61) show that for a particular reaction

---

* S. K. Mitra, *Nature* 1 67, 897 (1951).

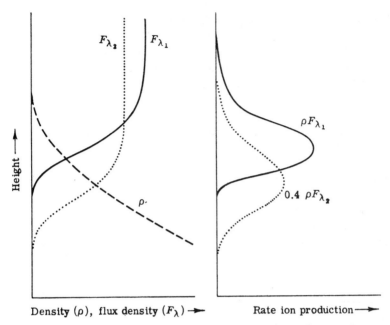

FIG. 4.22. *Left:* Schematic representation of vertical distribution of air density (ρ) and of radiation flux density for two wavelengths, $F_{\lambda 1}$ and $F_{\lambda 2}$. *Right:* Products $\rho F_{\lambda 1}$ and $\rho F_{\lambda 2}$ which are proportional to rate of ion production.

the height z_m and the production rate p_m depend only on the angle of incidence of the radiation.

The ion density resulting from this production rate can now be computed for photochemical equilibrium from Eq. (3.30). In the upper atmosphere electron attachment to large ions may be neglected, and decay occurs only by recombination. Under these conditions

$$p = \alpha n^2 \tag{4.62}$$

where α is the recombination coefficient. Combination of Eq. (4.62) with (4.59) yields the vertical distribution of n in the form

$$n = \left(\frac{\beta_\lambda k_\lambda F_{\lambda 0} \Delta \lambda \rho_0}{\alpha} \right)^{1/2} \exp \left[-\frac{\Phi}{2R_m T} - k_\lambda \rho_0 \frac{R_m T}{2g} \sec \theta \exp \left(-\frac{\Phi}{R_m T} \right) \right] \tag{4.63}$$

The maximum ion concentration is found by combining Eq. (4.62) with Eq. (4.61) to be

$$n_m = \left(\frac{g \beta_\lambda F_{\lambda 0} \Delta \lambda \cos \theta}{\alpha R_m T} \right)^{1/2} \exp \left(-\tfrac{1}{2} \right)$$

A region where the ion distribution is in agreement with Eq. (4.63) is called a simple Chapman region* in honor of the geophysicist who developed the preceding theory. The D layer at a height of 60 kilometers most nearly fulfills the assumptions required by the Chapman theory. Extension of the theory to include continuous absorption and height variation of temperature and composition renders the equations rather complex and requires numerical or graphical techniques for solution.

Although the Chapman theory has been extremely valuable in the intuitive understanding of the formation of ionospheric layers, the specific reactions responsible for the various regions of the ionosphere are still uncertain. Table 4.1 gives a summary of the photochemical effects of solar radiation from which it is evident that each of the various layers may be caused by a number of processes. Recent rocket observations have indicated that the intensities of X-rays and the Lyman series of the hydrogen spectrum are sufficient to play a role in the formation of the ionosphere.

Further discussion of the ionosphere and its effects is given in Sections 6.7, 6.8, and 7.15.

4.20   Airglow

The energy absorbed in reactions (4.49) and (4.54) may result in raising the temperature of the gas (as in the ozonosphere) or it may be re-radiated in the form of photons. The latter process is illustrated by the reactions

$$2O \rightarrow O_2 + h\nu$$

and

$$O + NO \rightarrow NO_2 + h\nu$$

These and many other photon emitting reactions occurring in the upper atmosphere result in a very faint luminescence of the sky, called the *airglow*. This relatively constant radiation should be distinguished from the highly variable radiation of the aurora which is emitted at somewhat greater heights and is associated with the influx of solar particles in the upper atmosphere.

Until recently airglow had been observed only during night and twilight conditions, but rockets equipped with suitable instruments make airglow observations possible during the day.

Airglow at night, also called *night glow*, contributes on a moonless night between 40 and 50% of the total illumination by the night sky, slightly more than the radiation received from the stars. Spectroscopic observations reveal many emission lines and bands and a continuum which extends from below 0.4 μ well into the infrared.

---

* S. Chapman, *Proc. Phys. Soc.* (*London*) **43**, 26, 483 (1931).

TABLE 4.1[a]

SUMMARY OF EFFECTS OF SOLAR RADIATION ON UPPER ATMOSPHERIC GASES

Spectral region (μ)	Reaction	Height (km)	Remarks
0.300–0.210 (Hartley absorption bands)	$O_3 + h\nu \rightarrow O_2 + O^*$ (excited)	50–60	Strong absorption by ozone. Although the reaction takes place for absorbed radiation with wavelengths $<1.1340\ \mu$, ozone absorbs strongly only from 0.300 to 0.210 μ.
0.1925–0.1760 (Runge Schumann absorption bands)	$O_2 + h\nu \rightarrow O_2^*$ (excited) $\rightarrow 2O$ $O_2^* + O_2 \rightarrow O_3 + O$ $O + O_2 + M \rightarrow O_3 + M$	50–80	Comparatively weak absorption. Sequence of ozone formation.
0.1751–0.1200 (Runge Schumann continuum)	$O_2 + h\nu \rightarrow O + O^*$ (excited)	80–110	Strong absorption. Dissociation of O_2.
0.12157 (Lyman α)	$NO + h\nu \rightarrow NO^+ + e$	60–90	Formation of D region?
0.10247 (Lyman β)	$O_2 + h\nu \rightarrow O_2^+ + e$	90	Contribution to base of E region.
0.1012–0.0910	$O_2 + h\nu \rightarrow O_2^+ + e$	50–80	Weak absorption. Contribution to D region.
0.0910–0.0795	$O + h\nu \rightarrow O^+ + e$	> 200	Very strong absorption. Ionization of O contributes to F_1 and F_2 regions?
0.0795–0.0755	$N_2 + h\nu \rightarrow N_2^+ + e$	140–160	Comparatively weak absorption. Contribution to E_2 region?
0.0744–0.0661	$O_2 + h\nu \rightarrow O_2^+$ (excited) $+ e$	90–120	Strong absorption. Contribution to E_1 region?
0.0661–0.0585	$N_2 + h\nu \rightarrow N_2^+$ (excited) $+ e$	200	Very strong absorption. Contribution to F_1 region.
2×10^{-2}–1.5×10^{-3} (X-rays)	General ionization	90–450	Contribution to E and F regions.
$\sim 2.5 \times 10^{-4}$	General ionization	60–90	Contribution to D region?

[a] Based on S. K. Mitra, *Compendium Meteorol.*, p. 245 (1951), and H. Friedman, in *Physics of the Upper Atmosphere* (J. A. Ratcliffe, ed.), p. 133. Academic Press, New York, 1960.

The constituents responsible for the lines and bands of the airglow spectrum include molecular and atomic oxygen, nitrogen, hydroxyl, and possibly hydrogen, but many of the lines and bands have not been conclusively identified. The excitation of these particles is a result of various processes of which the most important probably is the recombination of ions. Other possibilities of minor importance are collisions with incoming particles and the electrical discharge caused by potential differences generated by ionospheric winds. The continuum may result from the photochemical processes described in the beginning of this section.

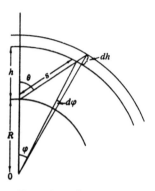

FIG. 4.23. Cross section through the earth and the atmosphere showing the contribution to the radiation received at the surface from a horizontal layer in the upper atmosphere.

The processes responsible for the airglow probably occur at various heights. If a single process which occurs in a sharply defined layer can be isolated, then observations may be used to determine the height of the layer. To do this the intensity of the relevant line or band is measured as a function of zenith angle as is illustrated in Fig. 4.23. This diagram shows that the intensity emitted in the direction θ by the layer dh is given by

$$I_\lambda(\theta) = I_\lambda(0) \sec (\theta - \phi) \qquad (4.64)$$

where $I_\lambda(0)$ is the intensity emitted in the vertical direction. If no absorption takes place between the layer and the earth's surface, this intensity is also measured at the surface.

For a given zenith angle the angle ϕ is a function of height of the layer. When h is increased by dh, the angle ϕ is increased by $d\phi$. From the small triangle in the layer dh of Fig. 4.23 it can be deduced that

$$dh^2 \sec^2 (\theta - \phi) = dh^2 + (R + h)^2 d\phi^2$$

or

$$\frac{d\phi}{dh} = \frac{\left(\sec^2 (\theta - \phi) - 1 \right)^{1/2}}{R + h} = \frac{\sin (\theta - \phi)}{(R + h) \cos (\theta - \phi)}$$

where R is the radius of the earth. This equation can be integrated to

$$\frac{R + h}{R} = \frac{\sin \theta}{\sin (\theta - \phi)}$$

or

$$\sec (\theta - \phi) = \left\{ 1 - \left(\frac{R \sin \theta}{R + h} \right)^2 \right\}^{-1/2}$$

Substitution of this equation into (4.64) yields

$$I_\lambda = I_\lambda(0) \left\{ 1 - \left(\frac{R \sin \theta}{R + h} \right)^2 \right\}^{-1/2} \tag{4.65}$$

For $h \ll R$ the right-hand side of Eq. (4.65) approaches $I_\lambda(0) \sec \theta$. In Fig. 4.24 $I_\lambda/I_\lambda(0)$ is plotted for a few values of h/R as a function of θ. This

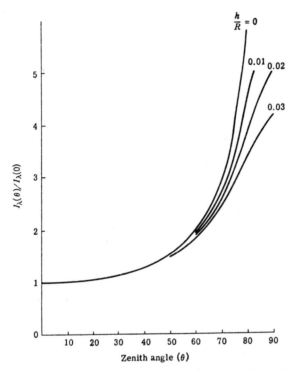

FIG. 4.24. Ratio of intensity emitted by shallow layer at zenith angle θ to intensity at zero zenith angle as calculated from Eq. (4.65).

method for determining the height of the emitting layer is called the van Rhijn method.[*] Because this height is a small fraction of the earth's radius, very accurate measurements of the intensity are required. Unfortunately, the radiation is weak and variable in space and time, so that accurate observations are very difficult to achieve.

In case the emitted radiation is partly absorbed on the way down, the intensity is reduced according to Beer's law over the path s, and Eq. (4.65) is modified to

$$I_\lambda = I_\lambda(0) \left\{ 1 - \left(\frac{R \sin \theta}{R + h} \right)^2 \right\}^{-1/2} \exp \left[-k_\lambda \int_0^s \rho \, ds \right]$$

[*] P. J. van Rhijn, *Publ. astr. Lab. Groningen* No. 31, 1 (1921).

where k_λ is the absorption coefficient and s is a function of h for constant θ given by

$$s = R \cos \theta \left\{ \left(1 + \frac{2Rh + h^2}{R^2 \cos^2 \theta} \right)^{1/2} - 1 \right\}$$

The heights found by various investigators using this method have varied between 70 and 300 kilometers, indicating that most of the radiation is emitted by the upper mesosphere and lower thermosphere. Direct measurements of airglow emission using rocket-borne photometers indicate that most of the airglow originates between heights of 80 and 120 kilometers.

List of Symbols

		First used in Section
a	Absorptivity	4.2
A	Area, albedo	4.1, 4.10
b	Stefan-Boltzmann constant for intensity	4.4
c	Speed of light	4.1
c_p	Specific heat at constant pressure	4.16
d	Thickness	4.17
E	Radiant energy	4.1
F	Radiant flux density	4.1
$\overline{F}_s$	Solar constant	4.7
F'	Spherical radiant flux density	4.10
g	Force of gravity per unit mass	4.18
h	Planck's constant, height	4.3, 4.10
H	Volume of cell in phase space	4.3
$H_n(x)$	Gold function of nth order	4.14
I	Intensity	4.1
k	Absorption coefficient, Boltzmann's constant, thermal conductivity	4.2, 4.3, 4.17
K	Reaction coefficient for chemical reactions	4.18
m	Equivalent mass of photon, mass of molecule	4.3, 4.18
n	Number of cells per frequency interval per degree of freedom, number of molecules per unit volume	4.3, 4.18
N	Number of photons per frequency interval	4.3
N_i	Average occupation number per cell in phase space	4.3
N_0	Avogadro's number	4.18
p	Momentum of photon, pressure, production rate of ions	4.3, 4.14, 4.19
q	Specific humidity	4.14
q_0	Energy received per unit horizontal area per day at top of the atmosphere	4.8
r	Reflectivity	4.2
R	Radius	4.8
R_m	Specific gas constant for gas with molecular mass M	4.19
s	Specific entropy, path length	4.3, 4.20
t	Time	4.1
T	Absolute temperature	4.3

		First used in Section
u	Optical thickness, specific internal energy	4.2, 4.3
V	Volume	4.3
W	Thermodynamic probability	4.3
x, y, z	Cartesian coordinates	4.2, 4.16
X	Dimensionless optical thickness $(k_l u/2\pi\alpha)$	4.5
α	Wien's displacement constant, half-width of spectral line, recombination coefficient	4.4, 4.5, 4.19
β	Lagrange multiplier, fraction of molecules participating in photochemical reaction per unit absorbed energy	4.13, 4.18
γ	Rate of change of flux emissivity with optical thickness	4.14
ϵ	Emissivity	4.2
ϵ_f	Flux emissivity	4.14
θ	Zenith angle	4.1
λ	Wavelength	4.1
ν	Frequency	4.1
ρ	Density	4.2
σ	Stefan-Boltzmann constant for flux	4.4
σ_λ	Extinction coefficient	4.7
τ	Transmissivity	4.2
ϕ	Zenith angle as seen from center of earth	4.20
Φ	Geopotential	4.18
ψ	Half the plane angle subtended by the earth from a satellite	4.10
ω	Solid angle	4.1

Subscripts (other than those derived from defined symbols)

a	Air
b	Black, bottom
e	Effective
i	Index
l	Pertaining to spectral line, long-wave
m	Maximum
n	Net
0	Pertaining to earth's surface or other base
r	Reflected, reference
s	Short-wave
t	Top
v	Water vapor
w	White

Problems

1. If the reflectivity (r_λ) is equal to 0.2 at the surfaces $x = 0$ and $x = x_0$ as shown in Fig. 4.3, find an expression for $I_\lambda(x_0)$ in terms of $I_{\lambda 0}$, x_0, k_λ, and r_λ.

2. Express Planck's law in terms of $I_\nu{}^*(\nu, T)$ instead of $I_\lambda^*(\lambda, T)$.

3. Derive Wien's displacement law as expressed in Eq. (4.20).

4. Show that $I_{\lambda\,\max}^* = \text{const } T^5$.

5. If the average output of the sun is 6.2 kw cm^{-2}, the radius of the sun is 0.71×10^6 km, the distance of the sun from the earth is 150×10^6 km, and the radius of the earth is 6.37×10^3 km, what is the total amount of energy intercepted by the earth in one second?

6. The solar constant is 0.1382 w cm^{-2} (1.98 cal cm^{-2} min^{-1}). Using the information given in problem 5 together with the Stefan-Boltzmann law, find the temperature of the solar surface.

7. Show that the total annual insolation at the top of the atmosphere is the same for corresponding latitudes in northern and southern hemispheres.

8. The quantities F'_{sr} and F_l' introduced in Eq. (4.26) are not flux densities in the strict sense of the definition because the spherical sensors are equally sensitive for all directions. Find an expression for the actual flux densities F_{sr} and F_l passing through a horizontal plane at the position of the satellite in terms of F_{sr}' and ω, or F_l' and ω.

9. Using the solar constant as 138.2 mw cm^{-2}, compute the average effective temperature of the earth for average albedos of 0.3 and 0.4.

10. If one assumes that the atmosphere and earth reflect 35% of the incident solar radiation and that the atmosphere acts as a single isothermal layer which absorbs all long-wave radiation falling on it, find the radiative equilibrium temperatures of the atmosphere and the earth's surface.

11. If the average surface temperature of the earth is $288°K$ and the average albedo of the earth and atmosphere for solar radiation is 35%, find the "effective absorptivity" of the atmosphere for long-wave radiation.

12. Suppose that a cloud layer whose temperature is $7°C$ moves over a snow surface whose temperature is $0°C$. What is the maximum rate of melting of the snow by radiation from the cloud if the absorptivity of the cloud is 0.9 and of the snow is 0.95?

13. Find the downward radiative flux from water vapor in the atmosphere for the following vertical distribution of temperature and specific humidity.

Pressure (mb)	Temperature (°C)	Specific humidity (‰)
1010 (surface)	12.5	8.5
1009	12.8	8.7
1000	12.0	8.2
950	10.3	4.9
900	8.1	4.3
800	0.1	4.1
700	−7.5	2.0
500	−16.0	0.4

14. For the data given in problem 13, find the rate of radiational temperature change at a reference height corresponding to 950 mb.

15. Derive Eqs. (4.46) and (4.47).

16. An ideally designed radiometersonde is one in which absorptivity of the black surface is unity and reflectivity of the shield and conductivity of the insulation are negligible. If such an instrument is raised by a balloon at three hundred meters per minute and the temperature difference between upper and lower plate changes from $+2.0$ to $+1.0$ degree Celsius in one minute, find the mean radiational temperature change for the following conditions:

Pressure: 700 mb
Air temperature: $-7.5°C$
Conductivity of air: 0.18×10^4 erg cm^{-1} °K^{-1} sec^{-1}
Separation of plates and shield: 1.0 cm

General References

Charney, "Radiation," in *Handbook of Meteorology*, gives a concise introduction to the principles of radiative transfer.

Haltiner and Martin, *Dynamical and Physical Meteorology*, complements Parts I to III of this chapter. The organization is similar and the treatment is at about the same level as presented here.

Chandrasekhar, *Radiative Transfer*, is a classical text providing a clear and comprehensive account of the problems associated with radiative transfer on a more advanced level.

Kourganoff, *Basic Methods in Transfer Problems*, gives an introductory discussion similar to that of Chandrasekhar but with special emphasis on methods.

Elsasser, *Atmospheric Radiation Tables*, provides essential information concerning the absorbing constituents of the atmosphere.

Möller, "Strahlung in der Unteren Atmosphäre," in *Handbuch der Physik*, gives a detailed account of radiative transfer in the lower atmosphere including scattering.

Goody, *The Physics of the Stratosphere*, gives an authoritative discussion of radiative transfer from 10 to 90 km.

Mitra, *The Upper Atmosphere*, gives a comprehensive account of the various physical aspects of the atmosphere, in particular the photochemical processes. The book is rapidly becoming out of date.

Chamberlain, *The Physics of the Aurora and Airglow*, contains a thorough and up to date account of the various processes responsible for airglow.

Bates, "The airglow," in *Physics of the Upper Atmosphere* (J. A. Ratcliffe, ed.), complements Chamberlain's treatment. Differences of opinion among experts in this field make the use of several references desirable.

Proceedings of the I.R.E., *Special Issue on Nature of the Ionosphere*, provides ionization and airglow data from the IGY (1957-58) and extensive interpretation.

CHAPTER V

Transfer Processes and Applications

*"The scientific method, so far as it is a
method, is nothing more than doing one's
damndest with one's mind, no holds barred."* P. W. BRIDGEMAN

OF THE RADIANT ENERGY absorbed by the earth and its atmosphere most
is absorbed by the earth's surface; here it is transformed into internal
energy with the result that large vertical and horizontal temperature
gradients are developed. As a consequence, the immediate source of the
energy which drives the atmosphere may be identified with the earth's
surface and, specifically, with its internal energy distribution. A variety of
processes may bring about subsequent energy transformations: evaporation, conduction into the earth, long-wave radiation, and upward conduction and convection of heat into the atmosphere, each depending on
physical properties of the earth's surface and the atmosphere. These
processes and some of their interactions are discussed in this chapter.

5.1 Energy Transfer near the Earth's Surface

Consider a thin layer of the earth in contact with the atmosphere. The
principle of conservation of energy requires that the rate of net inflow of
energy through the two surfaces equals the rate of increase of total energy
within the layer. If it is assumed that each form of energy transfer is horizontally uniform, this principle may be expressed by the equation

$$-F_n(0) + F_g(-1) - F_E(0) - F_h(0) = \int_{-z_1}^{0} \rho_g c_g \frac{\partial T}{\partial t} \, dz \qquad (5.1)$$

where F_n represents the net radiation flux density at $z = 0$, $F_g(-1)$ the
energy flux density by conduction upward from the earth through the lower
boundary $(-z_1)$, $F_E(0)$ and $F_h(0)$ the energy flux densities upward from
the earth to the atmosphere by evaporation and conduction of heat,
respectively, ρ_g the density of the ground, and c_g the specific heat of the
ground. Equation (5.1) may also be applied to the surface layer of water if
the net radiation flux density at the lower boundary is taken into account.

If the layer is allowed to shrink to very small thickness, the integral becomes negligible and Eq. (5.1) reduces to

$$-F_n(0) + F_g(0) = F_E(0) + F_h(0) \qquad (5.2)$$

Because evaporation and conduction of heat are particularly difficult to measure, Eq. (5.2) is often used to measure their sum when the remaining terms are known.

In a similar manner the principle of conservation of energy may be applied to the layer of air in contact with the earth's surface in the form

$$F_h(0) = F_h(1) + \int_0^{z_1} \frac{\partial F_n}{\partial z}\, dz + \int_0^{z_1} \rho c_p \frac{\partial T}{\partial t}\, dz - \int_0^{z_1} L \frac{\partial \rho_d}{\partial t}\, dz \qquad (5.3)$$

where $F_h(1)$ represents rate of upward heat transfer across surface z_1, L latent heat of condensation, ρ_d mass of water in the form of droplets per unit volume, $F_n(0)$ and $F_n(1)$ the net radiative flux densities at $z = 0$ and $z = z_1$, respectively. The three integrals may be determined by measurement or by calculation, but for layers only ten meters or so in thickness these terms are usually small compared to $F_h(0)$ and $F_h(1)$. In fact, it is useful to define a surface boundary layer by the requirement that within this layer the rate of transfer of heat is essentially independent of height. Within this layer

$$F_h(0) = F_h(1) = F_h(2), \text{ etc.} \qquad (5.4)$$

because F_h is approximately constant with height. Whereas $F_h(0)$ represents heat flux by conduction, $F_h(1)$ and $F_h(2)$ represent transport (or *convection*) of heat by moving elements of fluid.

Although the terms in Eq. (5.3) which represent condensation and absorption or emission of radiation have been neglected because these terms are small compared to the individual terms representing heat flux density, it does not follow that these processes are unimportant. These processes will be discussed later in this chapter in connection with fog formation.

The contributions to energy transfer near the ground surface which are made by photosynthesis and by oxidation have not been considered in the above discussion. In general these contributions may be considered negligible, but there are conditions in which they become important or even dominant, e.g., the photosynthesis of a cornfield in the late afternoon in early summer or the release of chemical energy by a forest fire.

Equations (5.2) and (5.4) are important statements of energy conservation, but they tell very little about how energy transfer occurs. In order to understand the dependence of the transfer processes on properties of the earth and of the atmosphere the terms will be discussed separately.

5.2   Heat Conduction into the Earth

The rate of heat conduction into or out of the solid earth is an application of Eq. (2.73) and may be expressed by

$$F_g(0) = -\left(k_g \frac{\partial T}{\partial z}\right)_{-0} \tag{5.5}$$

where k_g represents the thermal conductivity of the earth at the surface where $F_g(0)$ is measured. Thermal conductivity ranges from about 5×10^{-3} cal cm^{-1} sec^{-1} °C^{-1} for wet soil or ice to about 1×10^{-3} for dry sand and about 0.25×10^{-3} for new snow. Consequently, the heat conduction may vary between rather wide limits.

The principle of conservation of energy may be applied to the layer of earth into which heat penetrates in a certain interval of time, say a day or a year. The heat conduction during this time must equal the total internal energy change within the layer. Therefore

$$\int_0^t \left(k_g \frac{\partial T}{\partial z}\right)_{-1} dt = -\int_{-z_1}^{-\infty} \rho_g c_g \, \Delta T \, dz \tag{5.6}$$

where ΔT represents the change of temperature during the time interval. If ρ_g, c_g, and k_g are known, temperature profiles suffice to calculate the heat conduction. An example is given in problem 1. Figure 5.1 clearly shows the downward progress of the diurnal temperature wave. This wave continues to advance downward and at the same time decreases in amplitude until it can no longer be identified below a depth of about 80 centimeters. A similar annual temperature wave is identifiable to much greater depths. The details of course depend strongly on the density and specific heat as well as on the rate of heat conduction at the upper surface.

The ideas developed above are applicable at least in a formal sense to water and air as well as to the solid earth. However, water and air motions result in much more efficient transfer of heat downward into the water and upward into the air, so that temperature changes are normally very small (tenths of degrees in water and degrees in air as compared to tens of degrees in the solid earth), and the layer of heat storage is likely to extend to great depths in water and to great heights in air.

5.3   Turbulent Transfer

The atmosphere and the ocean are free to transfer heat or other properties by motion as well as by molecular conduction. These motions are so effective in transfer that conduction is normally neglected except very near boundaries where motions are constrained. Consider the vertical

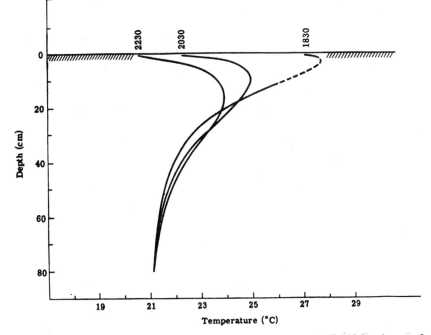

FIG. 5.1. Temperature profiles measured in the ground at O'Neill, Nebraska on 8 August 1953 (after H. Lettau and B. Davidson, "Exploring the Atmosphere's First Mile," p. 398. Pergamon, New York, 1957).

transport of a property of air such as dust concentration or specific humidity. If s represents the concentration of the property per unit mass, ρ the density of air, and w vertical velocity, the rate of vertical transport is expressed by $\rho w s$. Now, if this transport is measured a few meters above a horizontal rigid surface, upward and downward currents are found to be distributed in space and in time in the chaotic and random fashion which is referred to as turbulence. The net vertical transport of air, however, must be zero if the total mass below the surface is constant. The average value of ρw over the area is zero; but it is not true that the average value of $(\rho w)s$ is zero, for the upward currents may systematically carry higher values of s than the downward currents. This may be recognized explicitly by expressing s by $\bar{s} + s'$ where the bar and the prime represent, respectively, the average value and the departure from the average (called the *eddy* value). Upon treating ρw in the same way, the rate of average vertical transport is expressed by

$$F_s = \overline{\rho w s} = \overline{(\overline{\rho w} + (\rho w)')(\bar{s} + s')} = \overline{(\rho w)'s'} \qquad (5.7)$$

Equation (5.7) shows that accurate continuous observations of ρw and s

serve to determine the turbulent or eddy flux of s in the vertical. These observations can be made for various important properties, specific humidity and specific momentum, for example; but rather sophisticated instrumentation is necessary.

Turbulent heat flux (F_h) requires special consideration. In order to express the transport of energy by turbulence past a reference plane at height z_1 consider the volume enclosed by the earth's surface, the horizontal reference plane and vertical walls as shown in Fig. 5.2. It is assumed that

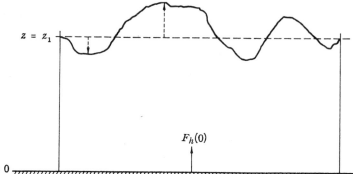

FIG. 5.2. Heat flux at the earth's surface ($F_h(0)$) into a layer of air of semi-infinite horizontal extent bounded at the top at time $t = 0$ by the dashed line and at the later time by the solid line.

the surface is the only heat source for this system and that the average properties of the system are constant in time. It follows from Eq. (5.3) that heat flux is independent of height. Because local mass transports take place through the upper boundary of the system, it is called an *open system*. Consider now a closed system that at time $t = 0$ coincides with the open system. The First Law of Thermodynamics requires for the closed system that the heat added equals the work done by the system plus the change in internal energy. The work is done at the upper surface and is expressed per unit area and unit time by pw. Because the internal energy of the open system does not change, the change in internal energy of the closed system must be equal to the net amount of internal energy passing through the reference level. Upon applying the First Law to the closed system, the heat entering the system at the ground may be expressed by

$$\left(\int F_h(0)\, dA \right) \Delta t = \left(\int F_h(z_1)\, dA \right) \Delta t$$

$$= \left\{ \int \rho w u\, dA + \int pw\, dA \right\}_{z_1} \Delta t$$

where A represents horizontal area and u specific internal energy. Upon taking the space average and omitting the height designation, the turbulent heat flux density at z_1 is expressed by

$$F_h = \overline{\rho w(u + p\alpha)} \tag{5.8}$$

and from Eqs. (2.10), (2.38), and (2.41)

$$F_h = \overline{c_p(\rho w)'T'} \tag{5.9}$$

Because $c_p T$ for an ideal gas is equivalent to the specific *enthalpy*, the turbulent flux of heat is properly referred to as the turbulent flux of enthalpy.*

We may gain insight into the mechanism of turbulent transfer by treating simplified models of turbulent motions. Models are introduced in cases in which the detailed description of a phenomenon is unknown or in which the details of a mechanism are extremely complex. Examples of the use of models are, (a) the introduction in Chapter II of the elementary molecule made possible development of a fundamental relation between temperature and kinetic energy, and (b) the energy of spectral lines may be derived from a simple model of the atom. Of course if these models are to be valuable, theoretical relations must agree with observations; but one must not assume that agreement of theory and observation necessarily implies that the model provides a complete account of the phenomenon.

Some of the important properties of atmospheric turbulence may be developed from a model in which one imagines a parcel of air, variously called an *eddy*, a *turbulon*, or *blob*, which may have any size and shape, which is homogeneous in the property s, and which moves as a coherent unit. Imagine also that this eddy passes through a horizontal test surface after originating a distance l (the *mixing length*) above or below the surface. At the point of origin of the eddy the property s is assumed to have been equal to the average value of s at that height. It is recognized then, that if s is conserved during vertical displacement

$$s' = -l\frac{\partial \bar{s}}{\partial z} \tag{5.10}$$

and if for a group of eddies of various values of mixing length it is assumed that $\overline{\rho w} = 0$, it follows that

$$F_s = -\overline{\rho w l}\frac{\partial \bar{s}}{\partial z} = -\overline{\rho w l}\frac{\partial \bar{s}}{\partial z} - \overline{w\rho l}\frac{\partial \bar{s}}{\partial z} \tag{5.11}$$

The second term on the right of Eq. (5.11) may be neglected compared to the first because density varies locally by only about a per cent whereas w is highly correlated with l. A *turbulent transfer coefficient* now may be defined by

---

* The derivation of Eq. (5.9) is due to R. B. Montgomery, *J. Meteorol.* **5**, 265 (1948).

$$K \equiv \overline{wl} \tag{5.12}$$

so that Eq. (5.11) becomes

$$F_s = -\rho K \frac{\partial \bar{s}}{\partial z} \tag{5.13}$$

By replacing s successively by specific enthalpy, latent heat, and momentum, equations may be written representing the turbulent transfer of these quantities. Thus

$$F_h = -\rho c_p K_h \left(\frac{\partial \overline{T}}{\partial z} + \Gamma \right) \tag{5.14}$$

$$F_E = -\rho L K_E \frac{\partial \bar{q}}{\partial z} \tag{5.15}$$

$$F_m = -\rho K_m \frac{\partial \bar{v}}{\partial z} \tag{5.16}$$

By analogy with the molecular transfer coefficients discussed in Section 2.16, the turbulent transfer coefficients K_h, K_E, and K_m are called, respectively, *eddy thermal diffusion coefficient*, *eddy diffusion coefficient*, and *eddy viscosity*. Verification of Eq. (5.14) is left for problem 2. The second term on the right of Eq. (5.14) represents the rate of change of temperature with height following an individual parcel as it moves vertically (the adiabatic lapse rate). Subscripts have been added to the individual transfer coefficients in Eqs. (5.14), (5.15), and (5.16) to indicate that they are not necessarily identical. This results from the fact that the behavior of the eddies may be biased in certain ways. For example, due to buoyancy warm eddies may tend to ascend, whereas cold eddies may tend to descend. Observational evidence suggests that the transfer coefficients are at least approximately equal for neutral static stability; for large vertical gradients of temperature of either sign the relation is still equivocal.

5.4 Turbulent Transfer in the Adiabatic Atmosphere

The simplest form of turbulent transfer takes place in the absence of buoyant forces, that is, in a layer characterized by the adiabatic lapse rate of temperature (neutral static stability). Although no transfer of heat takes place under these conditions, transfer of momentum and other properties may be quite important.

Consider the relation between vertical flux of momentum and the turbulent motions which accompany it in the normal case of increasing wind speed with height. At a reference height eddies with a deficit in momentum arrive from below and eddies with an excess in momentum arrive from above. Under steady conditions the mean velocity at the reference surface does not change, so the deficits and excesses of momentum of the various eddies add up to zero in the average. Formally, the average specific momentum from below and from above which arrives per eddy at the reference surface is given by

$$\bar{v} + l\,\frac{\partial \bar{v}}{\partial z}$$

A similar consideration for the kinetic energy of these eddies shows that they carry an excess of energy to the reference surface which is on the average

$$\overline{l^2}\left(\frac{\partial \bar{v}}{\partial z}\right)^2$$

This is the average specific kinetic energy per eddy which is converted from the mean flow to turbulent flow. Part of this turbulent energy is associated with the vertical velocity and part with the horizontal velocity. It is convenient, and it proves to be fruitful, to assume that the specific energy per eddy of the vertical velocity component is proportional to the input. Therefore

$$\overline{w^2} \propto \overline{l^2}\left(\frac{\partial \bar{v}}{\partial z}\right)^2 \tag{5.17}$$

It must be realized that under conditions of steady state the input of energy from the mean velocity field must be equal to dissipation of turbulent energy into heat. Therefore, the rate of dissipation must be proportional to the total turbulent energy.

Proportionality (5.17) may be combined with Eq. (5.12) and transformed to

$$K_m{}^2 \propto \overline{l^4}\left(\frac{\partial \bar{v}}{\partial z}\right)^2 \qquad \text{or} \qquad K_m \propto \overline{l^2}\,\frac{\partial \bar{v}}{\partial z} \tag{5.18}$$

The vertical momentum flux which is expressed by Eq. (5.16) therefore may be written

$$\frac{F_m}{\rho} \propto -\overline{l^2}\left(\frac{\partial \bar{v}}{\partial z}\right)^2 \tag{5.19}$$

and, because from Newton's Second Law time rate of change of specific momentum (F_m) per unit area must represent force per unit area (τ), this becomes

$$\frac{\tau}{\rho} \propto \overline{l^2}\left(\frac{\partial \bar{v}}{\partial z}\right)^2 \tag{5.20}$$

Now, assume that near the ground in conditions of neutral static stability mixing length is proportional to height above the ground. Thus, Eq. (5.20) may be written

$$\left[\frac{\tau}{\rho}\right]^{1/2} \equiv u_* = kz\,\frac{\partial \bar{v}}{\partial z} \tag{5.21}$$

where u_* represents the "friction velocity" and the constant of proportionality (k) is known as *von Karman's constant*. Measurements of stress and of vertical wind shear provide the data from which k may be calculated. Stress may be measured at the solid surface if it is assumed that close to the surface stress is independent of height. Wind tunnel observations have established the value of von Karman's constant as close to 0.40, and observations over natural surfaces under conditions of neutral stability are consistent with this value. Direct measurement of stress on natural surfaces is very difficult, so that values of k determined over these surfaces are subject to larger error than are wind tunnel observations.

Equation (5.21) may be integrated to give the wind profile equation for conditions of neutral stability. The resulting "adiabatic wind profile" is usually written in the form

$$\bar{v} = \frac{u_*}{k} \ln \frac{z}{z_0} \tag{5.22}$$

where z_0 is the constant of integration which is specified as the height where $\bar{v}$ vanishes. This does not imply that the real wind must vanish at this height, for the assumptions made in the derivation of Eq. (5.21) may well fail very close to the boundary. Values of z_0 (called the *roughness length*) vary from 10^{-3} centimeters for smooth mud flats to about 10 centimeters for tall grass or other surfaces of similar character. Observations over many types of surfaces have shown that the logarithmic profile appears with remarkable fidelity when the conditions of horizontal uniformity, steady state, and neutral static stability are fulfilled. Although this result strengthens faith in the validity of the assumptions on which (5.17) and 5.21) rest, it does not constitute proof, for other assumptions might also lead to Eq. (5.22). An example of the adiabatic wind profile is shown in Fig. 5.3.

The transfer coefficient for momentum (the eddy viscosity) may be expressed by combining Eqs. (5.16) and (5.19) in the form

$$K_m = k^2 z^2 \frac{\partial \bar{v}}{\partial z} \tag{5.23}$$

which may be transformed to

$$K_m = \frac{k^2 z(\bar{v} - \bar{v}_1)}{\ln z/z_1} \tag{5.24}$$

where $\bar{v}_1$ represents the average wind speed at height z_1 and $\bar{v}$ at height z. Equation (5.24) permits calculation of the transfer coefficient for momentum from the adiabatic wind profile. An example of stress calculation is given in problem 3.

The simple and illuminating relations just developed depend on the assumptions that turbulent stress is the only significant force acting in the

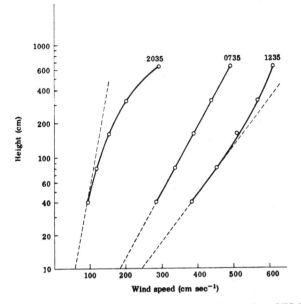

FIG. 5.3. Vertical distribution of average wind speed observed at O'Neill, Nebraska on 19 August 1953 under conditions of stability (2035 CST), neutral stability (0735 CST), and instability (1235 CST). The roughness length where the three straight lines converge is at a height of about 0.9 cm. (After H. Lettau and B. Davidson "Exploring the Atmosphere's First Mile," p. 444. Pergamon, New York, 1957).

horizontal and that stress is independent of height. What is the implication of the latter assumption, and what if it is not justified? Imagine a cubical element of air subjected to lateral stresses at the top and bottom as shown in Fig. 5.4. The stress τ_2 represents momentum carried downward through the upper face of the middle element. At the lower face momentum also is carried downward and exerts a stress on the air below the element. Newton's Third Law of Motion requires that an equal and opposite stress be exerted on the middle element by the air below the lower face, and this stress is represented by τ_1. If stress is independent of height, $\tau_1 = -\tau_2$; and if other forces are negligible, there is no increase of momentum within the element; and therefore the mean horizontal wind speed is constant with time. On the other hand, if more momentum enters through the upper face than leaves through the lower, $|\tau_2| > |\tau_1|$; and momentum increases with time within the element. Consequently, the assumption that stress is independent of height implies that the mean wind speed is constant in time. The stresses τ_1 and τ_2 also exert a torque on the element of air which may be visualized as producing rotational turbulent motions.

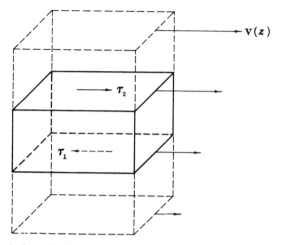

FIG. 5.4. Vertical profile of horizontal wind ($v(z)$) and the horizontal stresses, τ_1 and τ_2, acting *on* an element of air.

5.5 Turbulent Transfer in the Diabatic Atmosphere

Although the discussion of turbulent transfer in the adiabatic atmosphere has provided useful insight, the air near the earth's surface is seldom in a state of neutral equilibrium. More often, the air is either statically stable (temperature inversion) or statically unstable (super-adiabatic lapse rate); and these states are referred to as *diabatic*. In the diabatic case the turbulent energy is derived both from the kinetic energy of the mean wind and from buoyant energy associated with local temperature fluctuations. The discussion in Section 5.4 leading to proportionality (5.17) may now be extended for the diabatic case. The buoyant force per unit mass acting on an eddy whose temperature differs from its surroundings by T' is given by Newton's Law of Universal Gravitation in the form

$$\frac{g}{T} T'$$

When the eddy has traveled over a distance l and has arrived at the reference height z, the buoyant force has performed specific work proportional to

$$\frac{g}{T} T'l$$

Energy equivalent to this work enters directly into the vertical component of the turbulent velocity; on the other hand, energy produced by shear of the mean wind enters directly into one horizontal component but indirectly into the vertical component of turbulent velocity. It may now be recognized that proportionality (5.17) is a special form of a more general proportionality expressed by

$$\overline{w^2} \propto \overline{l^2}\left(\frac{\partial \overline{v}}{\partial z}\right)^2 + \alpha' \frac{g}{\overline{T}}\overline{lT'} \tag{5.25}$$

where α' is an empirical coefficient relating the work done by the buoyant force to the work done by vertical shear. Upon again multiplying by l^2, averaging, and using Eqs. (5.10) and (5.12), proportionality (5.25) is transformed to

$$K_m{}^2 \propto \overline{l^4}\left(\frac{\partial \overline{v}}{\partial z}\right)^2 - \alpha \frac{g}{\overline{T}}\overline{l^4}\left(\frac{\partial \overline{T}}{\partial z} + \Gamma\right)$$

which may be compared with proportionality (5.18) for the adiabatic case. Because of the statistical operation the coefficient α is probably not the same as α'. Upon eliminating K_m by use of Eq. (5.16), replacing the vertical momentum flux by the surface stress, and again assuming l proportional to z, this proportionality can be written as

$$u_*{}^4 = \left(\frac{\tau}{\rho}\right)^2 = \left(kz \frac{\partial \overline{v}}{\partial z}\right)^4 - \alpha \frac{g}{T}\left(\frac{\partial \overline{T}}{\partial z} + \Gamma\right)(kz)^4 \left(\frac{\partial \overline{v}}{\partial z}\right)^2$$

where k again represents von Karman's constant. The stress equation (5.21) now may be generalized to

$$u_*{}^2 = k^2 \left(\frac{\partial \overline{v}}{\partial(\ln z)}\right)^2 \left(1 - \frac{\alpha(g/\overline{T})(\partial \overline{T}/\partial z + \Gamma)}{(\partial \overline{v}/\partial z)^2}\right)^{1/2}$$

$$= k^2 \left(\frac{\partial \overline{v}}{\partial(\ln z)}\right)^2 (1 - \alpha \, \mathrm{Ri})^{1/2} \tag{5.26}$$

which shows clearly the effect of temperature lapse rate on the stress. The dimensionless combination

$$\mathrm{Ri} \equiv \frac{g(\partial \overline{T}/\partial z + \Gamma)}{\overline{T}(\partial \overline{v}/\partial z)^2}$$

is called the Richardson number. This number is a measure of the stability of the air. Equation (5.26) may be transformed to a simpler form by introducing the dimensionless variables

$$S \equiv \frac{kz}{u_*}\frac{\partial \overline{v}}{\partial z} \tag{5.27}$$

and

$$\zeta \equiv \frac{g}{\overline{T}}\frac{\partial \overline{T}/\partial z + \Gamma}{\partial \overline{v}/\partial z}\frac{kz}{u_*} \tag{5.28}$$

Equation (5.26) now may be written

$$S^4 - \alpha \zeta S^3 = 1 \tag{5.29}$$

This equation defines the diabatic wind profile in terms of ζ, a dimensionless height, and S, a dimensionless wind gradient. Under neutral conditions $\zeta = 0$, therefore $S = 1$, and we see that Eq. (5.21) is a special case of Eq. (5.29). Very near the surface ζ is very small, so that the diabatic profile approaches the adiabatic profile. Figure 5.5 shows that Eq. (5.29) agrees

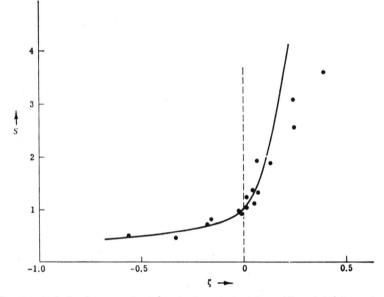

FIG. 5.5. Relation between the dimensionless wind gradient (S) and the dimensionless height (ζ) in the diabatic atmosphere. The line is plotted from Eq. (5.29) using $\alpha = 18$, and the points have been calculated from data given by H. Lettau and B. Davidson, "Exploring the Atmosphere's First Mile." Pergamon, New York, 1957.

well with observations under lapse or unstable conditions for an arbitrary choice of $\alpha = 18$. On the other hand, the great scatter of points on the positive or inversion side of the figure indicates that the theory may be seriously wrong in this region.

To plot the actual wind profile for a particular case, Eq. (5.29) may be integrated graphically or numerically as illustrated in problem 4.

Equation (5.29) has been developed in different ways by Ellison,[*] by Yamamoto,[†] and by Panofsky, Blackadar, and McVehil.[‡] However, in spite of twists in the derivation, the theory is incomplete because the arbitrary constant α has been used to absorb our ignorance concerning the

---

[*] T. H. Ellison, *J. Fluid Mech.* **2**, 456 (1957).

[†] G. Yamamoto, *J. Meteorol. Soc. Japan* [2] **37**, 60 (1959).

[‡] H. A. Panofsky, A. K. Blackadar, and G. E. McVehil, *Quart. J. Roy. Meteorol. Soc.* **86**, 390 (1960).

relative proportions of energy contributed by momentum and by buoyancy to the energy of the vertically moving eddies and also because mixing length may depend on stability.

Useful insights may be gained by relating the transfer coefficients to the parameters which characterize the atmosphere. From Eqs. (5.16) and (5.27) the eddy viscosity may be expressed by

$$K_m = u_* \frac{kz}{S} \tag{5.30}$$

which reduces for the adiabatic case to Eq. (5.23). Because Newton's Second Law, which is fundamental to Eq. (5.30), is not available for expression of the eddy coefficients for heat and water vapor, analogous relations for K_h and K_E cannot be written. For this reason, it is desirable to be able to relate K_h and K_E to K_m.

If Eqs. (5.14) and (5.15) are divided by (5.16)

$$\frac{K_h}{K_m} = \frac{F_h}{F_m} \frac{\partial \bar{v}/\partial z}{c_p \left(\partial \bar{T}/\partial z + \Gamma \right)} \qquad \frac{K_E}{K_m} = \frac{F_E}{F_m} \frac{\partial \bar{v}/\partial z}{L\, \partial \bar{q}/\partial z} \tag{5.31}$$

Because the flux-ratios may be assumed constant with height, variation in the ratios of the transfer coefficients is indicated by a dissimilarity between the wind, temperature, and humidity profiles. The numerical values of the ratios as determined by several investigators are of the order one. Problem 5 provides an exercise.

In the extreme case of free convection the mean wind speed is negligible, and the buoyant force is the only source of turbulent energy. For this case proportionality (5.25) reduces to

$$\overline{w^2} \propto -\alpha \frac{g}{T} l^2 \left(\frac{\partial \bar{T}}{\partial z} + \Gamma \right)$$

Substitution of this proportionality into Eq. (5.14) yields for the eddy thermal diffusion coefficient

$$K_h^2 = -\alpha'' \frac{g}{T} (kz)^4 \left(\frac{\partial \bar{T}}{\partial z} + \Gamma \right) \tag{5.32}$$

where α'' has been introduced as a new constant of proportionality. There is, of course, no a priori reason for supposing that $K_h = K_m$ or that $\alpha = \alpha''$. Combination of Eqs. (5.14) and (5.32) leads to

$$\left(\frac{\partial \bar{T}}{\partial z} + \Gamma \right) = \left(\frac{T}{\alpha'' g} \right)^{1/3} \left(\frac{F_h}{c_p \rho} \right)^{2/3} (kz)^{-4/3} \tag{5.33}$$

which defines the temperature profile in the free convection regime. Observations suggest that $\alpha'' \approx 32$. Now, if Eq. (5.26) is divided by $(\partial \bar{v}/\partial z)^2$ and is extrapolated to the free convection regime, the eddy viscosity may be expressed by

$$K_m{}^2 = -\alpha \frac{g}{T} (kz)^4 \left(\frac{\partial \overline{T}}{\partial z} + \Gamma\right)$$

Dividing Eq. (5.32) by this equation yields for the ratio of the eddy transfer coefficients

$$\frac{K_h}{K_m} = \left(\frac{\alpha''}{\alpha}\right)^{1/2} \approx 1.33$$

This result, together with the ratio quoted earlier, may indicate that K_h/K_m increases from unity for the neutral case to 1.33 for the free convection case. This increase may be gradual, but there is little observational evidence on this point. In any case, the search for a more general theory which will remove the necessity for arbitrary "constants" is a challenging one.

When the atmosphere is stably stratified, the term in Eq. (5.26) due to the buoyant force is negative, and the intensity of turbulence decreases rapidly with increasing stability. Turbulent transfer is slow, and the assumption that turbulent fluxes are independent of height is less valid than in the neutral or unstable case. A large scatter of the observed points should be expected, and this is revealed by Fig. 5.5. Large scatter of observations makes it difficult to conclude anything about the merits of the theory or the relative behavior of the transfer coefficients. Fortunately, under stable conditions the turbulent flux is usually small in magnitude.

5.6   Vertical Fluxes of Heat and Water Vapor

The mental excursions of the two previous sections should have provided the necessary conditioning for discussion of the methods for determining the magnitude of the flux densities of heat and water vapor. The fundamental method is, of course, to obtain simultaneous and continuous observations of the vertical wind velocity, temperature and humidity and then to determine the vertical transport directly by use of Eq. (5.7). This has been tried successfully by a number of investigators, but the method remains to be fully exploited.

The *aerodynamic method* makes the assumption that the eddy transfer coefficients for momentum, heat and water vapor are the same, that is, that $K_m = K_h = K_E$. Then Eqs. (5.14) and (5.15) may be written in expressions analogous to Eq. (5.26), yielding the flux density of heat and water vapor, respectively, in the form

$$F_h = -c_p \rho k^2 z^2 \left[\left(\frac{\partial \overline{v}}{\partial z}\right)^2 - \frac{\alpha g}{T} \left(\frac{\partial \overline{T}}{\partial z} + \Gamma\right)\right]^{1/2} \left(\frac{\partial \overline{T}}{\partial z} + \Gamma\right)$$

and

$$\frac{F_E}{L} = -\rho k^2 z^2 \left[\left(\frac{\partial \overline{v}}{\partial z}\right)^2 - \frac{\alpha g}{T} \left(\frac{\partial \overline{T}}{\partial z} + \Gamma\right)\right]^{1/2} \frac{\partial \overline{q}}{\partial z}$$

These two equations may be transformed to finite difference form and then evaluated using simultaneous observations of wind speed, temperature and humidity at two heights. The flux densities of heat and water vapor then may be calculated from observations at z_1 and z_2 by the equations

$$F_h = -c_p\rho \frac{(\bar{v}_2 - \bar{v}_1)k^2[T_2 - T_1 + \Gamma(z_2 - z_1)]}{(\ln z_2/z_1)^2} (1 - \alpha \overline{\text{Ri}})^{1/2} \quad (5.34)$$

$$\frac{F_E}{L} = -\rho k^2 \frac{(\bar{v}_2 - \bar{v}_1)(\bar{q}_2 - \bar{q}_1)}{(\ln z_2/z_1)^2} (1 - \alpha \overline{\text{Ri}})^{1/2} \quad (5.35)$$

The bar over Ri denotes that the average value of this quantity should be taken between two heights. For small values of $z_2 - z_1$ the average value of Ri is accurately

$$\overline{\text{Ri}} = \frac{g}{T} \frac{T_2 - T_1 + \Gamma(z_2 - z_1)}{(\bar{v}_2 - \bar{v}_1)^2} (z_2 - z_1)$$

The aerodynamic method usually has been applied under the assumption that the neutral wind profile holds up to the heights where the temperature and humidity observations are taken. This is equivalent to setting Ri in Eqs. (5.34) and (5.35) equal to zero. The results obtained in this way often contain errors in the calculated fluxes which probably amount to at least 10 to 20 per cent. The errors may be ascribed to the assumption of equality of the transfer coefficients and to the assumption of the neutral profile. By comparing the results of calculations based on Eqs. (5.34) and (5.35) for the diabatic case with the results of direct measurement, the ratio of the transfer coefficients may be evaluated.

The *energy method* is based on conservation of energy as expressed in Eq. (5.2) and therefore requires measuring the upward heat flux density at the earth's surface (F_g) and the net radiative flux density from the surface (F_n). Equation (5.2) provides one relation connecting F_h and F_E; an independent relation is obtained by dividing Eq. (5.14) by Eq. (5.15) and assuming $K_h = K_E$. The result is

$$\frac{F_h}{F_E} \equiv \beta = \frac{c_p(\partial T/\partial z + \Gamma)}{L \, \partial \bar{q}/\partial z} \approx \frac{c_p}{L} \frac{T_2 - T_1 + \Gamma(z_2 - z_1)}{\bar{q}_2 - \bar{q}_1}$$

The ratio β is called the *Bowen ratio*; it can be calculated (under the above assumption) from temperature and humidity observations at two heights. By combining this equation with Eq. (5.2), the two fluxes may be expressed by

$$F_h = \frac{\beta}{1 + \beta} (F_g - F_n) \quad (5.36a)$$

and

$$F_E = \frac{1}{1 + \beta} (F_g - F_n) \quad (5.36b)$$

The accuracy of the energy method depends strongly on accurate determination of β. A constant value of $\frac{1}{10}$ is often used for β over the ocean, but this is almost equivalent to an open admission of defeat. When $F_E \ll F_h$, β is large and F_h can be determined accurately even if the assumption that $K_h = K_E$ is invalid. When $F_h \ll F_E$, β is small and F_E can be determined accurately. In the more general case β is of order one, and the energy method has some of the same limitations as the aerodynamic method.

Although subject to obvious deficiencies, the aerodynamic and energy methods are in common use. The energy method is simpler, and therefore is usually preferable as long as F_g can be determined. When this is not possible or where $F_g \sim F_n$, the aerodynamic method is preferable.

5.7   Vertical Aerosol Distribution

Turbulent eddies transport dust, smoke and other aerosol particles from regions of high concentration to regions of low concentration by the process discussed in Section 5.3. For the common case of maximum concentration at or near the earth's surface, the turbulent transport is upward. However, aerosol particles larger in diameter than about 10^2 microns have an appreciable settling speed. After a time interval of several hours approximate equilibrium between gravitational settling and upward turbulent transport is often achieved. Under this condition

$$ws = -K \frac{ds}{dz} \tag{5.37}$$

where w represents the settling rate and s the concentration of particles. This equation is valid in this simple form only for particles having the same settling rate. Under neutral stability and close to the ground K is proportional to z. Integration of Eq. (5.37) then yields

$$s = s_0 \left(\frac{z}{z_0}\right)^{-w/ku_*}$$

Equation (5.37) can also be integrated easily when K is constant with height, and with somewhat more effort for the case of the diabatic wind profile when K is determined by Eq. (5.30).

The considerations above may be applied to determination of the vertical distribution of blowing sand, blowing snow, salt particles from the sea spray, and other aerosols. It is necessary that the concentration of particles at height z_0 as well as the friction velocity and settling rate be known.

Particles suspended in the atmosphere near the earth's surface limit the visual range.* If the relation between particle size and number and ex-

---

* See Section 7.11 for quantitative discussion of visual range.

tinction coefficient can be determined, measurements of the extinction coefficient with height may specify the vertical particle distribution, and Eq. (5.37) may then be used to calculate the coefficient of eddy transfer. To design and carry out experiments for this purpose provides a stimulating challenge.

5.8 Nocturnal Cooling

In the absence of solar radiation the earth and atmosphere radiate to space, and the ground surface or the tops of clouds cool rapidly. The resulting nocturnal cooling may be represented as the sum of several readily understood phenomena: essentially black-body radiation of the earth's surface, radiation from the atmosphere, radiation from clouds and from hills, trees, and other obstacles, evaporation or condensation at the surface, heat conduction within the earth, and turbulent conduction within the atmosphere. These processes are expressed formally by Eq. (5.2). Determination of the surface temperature as a function of time is a particularly complex mathematical problem because important interactions between terms occur. For example, as the surface cools a temperature inversion develops in the air just above the surface, and in this layer turbulent transfer is severely limited. Or, as surface temperature falls the radiation emitted by the surface falls. Because interactions of these sorts are extensive and complicated, only a very simplified analysis is carried out below.

An hour before sunset on 8 August 1953 at O'Neill, Nebraska, the vertical distribution of temperature in the neighborhood of the ground surface was as shown in Fig. 5.6. The surface radiated strongly to the atmosphere and to space at the rate of 0.45×10^6 erg cm^{-2} sec^{-1} (0.64 cal cm^{-2} min^{-1}). The atmosphere supplied only 0.39×10^6 erg cm^{-2} sec^{-1} (0.56 cal cm^{-2} min^{-1}) so that the surface cooled. Two hours later the surface had cooled by almost 10°C, and the direction of heat conduction within the earth and the atmosphere had reversed from its earlier direction. Subsequent cooling proceeded only very slowly, and by midnight the total flux toward the surface had become equal to the total flux away from the surface. At this time, assuming evaporation negligible, Eq. (5.2) combined with Eqs. (4.39) and (5.14) may be written

$$\int_0^{u_\infty} \gamma \sigma T^4 \, du - \left[k_g \left(\frac{\partial T}{\partial z} \right) \right]_{-0} = \sigma T_g{}^4 - \left[\rho c_p K_h \left(\frac{\partial \overline{T}}{\partial z} + \Gamma \right) \right]_{+0} \quad (5.38)$$

where the earth's surface has been assumed to be black, the subscripts -0 and $+0$ refer, respectively, to values just below and just above the surface. In order to clarify the major physical processes a series of approxi-

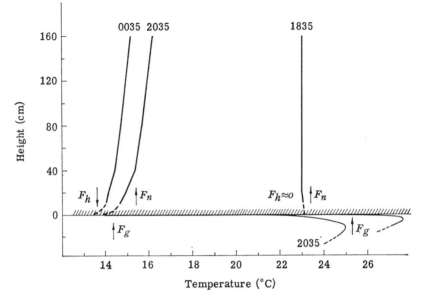

FIG. 5.6. Vertical temperature distribution at 1835, 2035, and 0035 CST on the night of 8–9 August 1953 at O'Neill, Nebraska. Directions of vertical fluxes are indicated by arrows (after H. Lettau and B. Davidson, "Exploring the Atmosphere's First Mile," p. 402. Pergamon, New York, 1957).

mations is useful. Under the inversion conditions which prevail at the time of minimum temperature turbulent motions are severely inhibited, so that K_h is small and the turbulent transfer of enthalpy may be neglected. Heat conduction within the upper layer of the earth may be represented by the linear function

$$F_g(0) = -k_g \frac{T_g - T(-z_1)}{z_1} \tag{5.39}$$

where z_1 represents the thickness of the surface layer of assumed constant conductivity. Where the surface layer is thin and of relatively small conductivity the temperature approaches a linear function of height after only a few hours of surface cooling. Such conditions are found with snow cover of a few inches to a foot or so, in ice overlying water and, with less accuracy, in dry cultivated soil overlying compact moist soil.

Atmospheric radiation may be represented by integrating the first term in Eq. (5.38). The result may be expressed by either side of the identity

$$\sigma \overline{T_a^4} \epsilon_f \equiv \sigma T_e^4 \tag{5.40}$$

where $\overline{T}_a$ represents the mean air temperature, ϵ_f represents the flux emissivity of the atmosphere, and T_e represents the effective temperature of the atmosphere.

Equation (5.38) may now be solved for T_g as a function of $T(-z_1)$ and k_g. An approximate solution is given by expressing T_g by $T_e - \Delta T$ and expanding in a binomial series. This procedure yields the minimum temperature at the ground surface in the form

$$T_{gm} = \left[\frac{4 + \kappa T(-z_1)/T_e}{4 + \kappa} \right] T_e \qquad (5.41)$$

where $\kappa \equiv k_g/\sigma T_e^3 z_1$. Upon introducing measured soil conductivity and atmospheric data obtained by radiosonde into Eq. (5.41), the minimum surface temperature can be calculated as is required in problem 6. For small conductivity or great thickness of the surface insulating layer (small κ) T_{gm} approaches the effective temperature. If a value of 0.70 is used for ϵ_f, Eqs. (5.40) and (5.41) yield, for no heat conduction from air or earth, $T_{gm} \approx 0.92 T_a$, which means that for a uniform air temperature of 0°C, the lowest possible temperature of the surface is about -22°C. The effect of conduction within the earth is to raise the lowest-possible minimum, and for very large conductivity or small thickness of the insulating layer (large κ) T_{gm} approaches $T(-z_1)$. The qualitative effects of cloud cover or solid obstructions, of change in the optical thickness of water vapor, of change in turbulent conduction, and of condensation and evaporation may be expressed by appropriate changes in Eq. (5.41).

Although we have assumed here that the air temperature remains constant as the surface cools, this cannot be true very close to the surface. Here, the large positive vertical temperature gradient results in convection toward the surface. The vertical temperature profile assumes a form qualitatively similar to that of the wind profile. This implies that the transfer coefficients for enthalpy and for momentum are qualitatively similar.

5.9 Fog Formation

In Chapter III it was emphasized that condensation almost always occurs in natural air whose dew point even slightly exceeds the temperature. Because the daytime dew point depression near the ground often amounts to only a few degrees, it appears that evaporation from the surface or cooling of the air at night by contact with the ground should lead to the formation of fog. Indeed, fog does form under certain conditions, but not nearly so readily as might be expected from the discussion so far. It is necessary to look in more detail at the processes which bring the temperature and the dew point together.

Two processes act to prevent supersaturation and fog formation. First, in the case of a cold surface the turbulent eddies which transfer heat downward also may transfer water vapor downward; the situation may be understood most easily by inspection of Fig. 5.7. Initial isothermal temper-

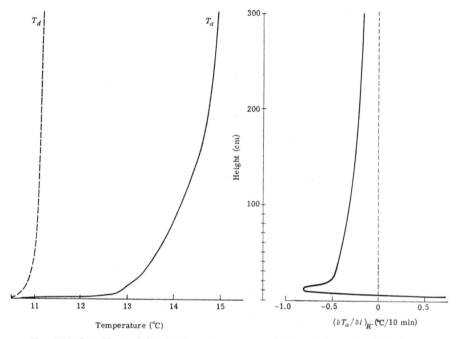

FIG. 5.7. *Left:* Vertical distribution of temperature (T_a) and dew point (T_d) observed on 26 August 1954 above a water surface at Friday Harbor, Washington.
    Right: Vertical distribution of radiative temperature change calculated from Eq. (4.45) for the curves shown on the left.

ature and dew point distributions are modified by transfer to the colder surface, so that they assume the forms shown at the left-hand side of the figure. Both heat and water vapor are transferred downward to the cold surface by the same turbulent eddies and with approximately equal effectiveness, and the dew point may remain below the air temperature. In the case of a warm wet surface, heat and water vapor are transferred upward, and in this case also the dew point may be less than the temperature. Second, the hygroscopic nature of much of the earth's surface (for example, sea water, dry soil, or dry vegetation) results in depression of the dew point below the temperature even at the surface, as indicated in Fig. 5.7. This depression amounts to 0.2°C at the sea surface.

In spite of these inhibiting factors the following processes bring about supersaturation and fog formation under favorable conditions:

(a) Mixing of parcels of saturated, or nearly saturated, air at different temperatures and vapor densities may produce a supersaturated mixture.
(b) Radiative flux divergence may result in cooling air below its dew point.
(c) Adiabatic cooling accompanying upward motion or falling pressure may result in supersaturation and condensation. "Upslope" fog is formed in this way.
(d) Condensation may occur on giant hygroscopic nuclei at relative humidities below 100%. This is especially effective in industrial regions, and the product is sometimes known as "smog."
(e) The molecular diffusion coefficient for water vapor in air exceeds the thermal diffusivity for heat so that water vapor may evaporate so rapidly from warm surfaces that supersaturation occurs within the laminar layer adjacent to the surface.

Processes (a) and (b) deserve extended discussion.

Mixing of two masses of different temperatures and vapor densities results in a mixture in which these properties are, respectively, linear averages of the initial properties, appropriately weighted by the masses. The dew point of the mixture is higher than the corresponding linear average of the individual dew points because of the exponential form of the Clausius-Clapeyron equation [Eq. (2.88)]. The dew point increment, the difference between the linear average of the dew points and the dew point corresponding to the average specific humidity, may then be expressed by

$$\delta T_d \approx \frac{d^2 T_d}{dq_s^2} \frac{\Delta q_s^2}{2}$$

where Δq_s represents the difference between the initial and the average specific humidity. Upon recalling that $dq_s/q_s \approx de_s/e_s$, the Clausius-Clapeyron equation may be substituted into the above equation with the result

$$\delta T_d \approx - \frac{R_w T_d^2}{2L q_s^2} \Delta q_s^2$$

For a mixture of equal masses Δq_s is half the difference between the individual specific humidities. For the range of atmospheric conditions in which fog is common the coefficient is slowly varying, and the right-hand side of the above equation has the approximate value $7(\Delta q_s/q_s)^2$ °C. For the case in which $\Delta q_s/q_s$ is 0.1, the dew point increment produced by mixing of equal masses is about 0.07°C. If the dew point depression of the two masses were initially within this increment, mixing may result in saturation and fog formation. The conditions favorable for fog formation by this process are: high relative humidity, large temperature gradient, and the presence of turbulent eddies.

Fog formation by radiative flux divergence will be discussed by reference to radiative processes in air above a cold surface. An air parcel at temperature T_g in contact with the surface also of temperature T_g exchanges radiation with the surface with no gain or loss of energy. It exchanges radiation with the atmosphere above with gain of energy because T_g is below the air temperature. An air parcel at a height of, say, 10 centimeters is warmer than the surface and so exchanges radiation with the surface with net loss. It exchanges with the warmer atmosphere with net gain, so that there may exist a certain height where the total net exchange vanishes. At a height of, say, one meter the air is much warmer than the surface, so loses energy by exchange with the surface while it gains only slightly by exchange with the atmosphere above. These calculations may be made using Eq. (4.45). The results are summarized at the right-hand side of Fig. 5.7. The effect of the radiational temperature change shown is to prevent saturation just above the surface and to drive the temperature curve strongly toward the dew point curve in the neighborhood of 10 to 20 centimeters. The modified temperature profile results in a compensating heat flux by turbulence so that net cooling and warming is likely to be very small. In the case shown in Fig. 5.7 the dew point was too low to be reached by radiational cooling, but it may be inferred that under more nearly saturated conditions fog should be expected to form first at a height of a few decimeters above a cold moist surface. The effect of radiational cooling is detectable at a height of 10 to 20 centimeters in the T_a curve of Fig. 5.7 and appears to amount to about 0.1°C. The conditions favorable for fog formation by this process are: high relative humidity and large temperature gradient. Turbulent eddies act to prevent fog formation by radiation.

Above a warm moist surface the temperature and humidity decreases with height; the results are modified to the extent that the sign of the radiational temperature change is reversed. Thus, there is strong cooling just above the surface with warming above 10 centimeters or so, and fog forms in contact with the warm surface. In this case processes (a), (b), and (e) above may all be effective.

After fog has formed, the physical processes are altered greatly. In the warm surface case the air is statically unstable, so buoyant eddies carry the fog upward, the depth of the layer increases and the top is uneven and turbulent. Fog which has formed in this way sometimes moves from the warm surface to a cold surface, e.g., warm water to cold land. The surface energy source is cut off, and the character of the upper surface of the fog is altered. A sharp temperature inversion may develop at the top of the fog as shown in Fig. 5.8 as a result of upward radiation from the fog. This stable stratification leads to a smooth, uniform top to the fog layer, and the unstable stratification within the fog leads to a well mixed homogeneous layer. The combined effect is to confine smoke or other pollutants

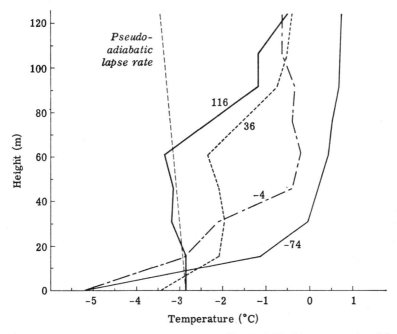

FIG. 5.8. Vertical temperature distribution at Hanford, Washington, on the night of 9 December 1947. The numbers indicate the number of minutes before (−) or after (+) fog was first observed [after R. G. Fleagle, W. H. Parrott, and M. L. Barad, *J. Meteorol.* **9**, 53–60 (1952)].

within the fog layer and to thoroughly mix them within this layer. This is why the air pollution disasters in the Meuse Valley of Belgium in December 1930, at Donora, Pennsylvania, in October 1948 and at London, England, in December 1952 each occurred within a persistent blanket of fog.

5.10 Air Modification

In previous sections the various forms of energy transfer that occur in the atmosphere have been described. In Sections 5.8 and 5.9 we sought to provide insight into two special examples of energy transfer, nocturnal cooling, and fog formation. Here the more general problem of air modification by energy transfer from the earth's surface will be considered. The objective is to gain an understanding of the factors which determine the vertical fluxes and of the resulting vertical distribution of temperature and water vapor. The vertical fluxes continually modify the lower atmosphere on a worldwide scale and thereby cause the development of storms and winds of all descriptions, in short, of the weather.

At the end of Section 5.5 it was pointed out that the fluxes near the ground are much smaller in stably stratified air than in unstably stratified air. The result is that upward fluxes of heat and water vapor are much more important than the downward fluxes and that the earth's surface functions as an energy source for the atmosphere. An example of very large upward fluxes and consequent strong air modification occurs with cold air flowing over the much warmer Great Lakes; the famous snowstorms of Buffalo are the direct result. The opposite effect of a strong downward heat flux accompanied by considerable cooling of the air is rare but may occur locally in chinook or foehn winds; however, the widespread and vigorous upward fluxes are much more effective in air modification.

One of the most important examples of large scale air modification occurs when cold air flows from a continent or ice covered surface over a warm ocean. The analysis of this problem will be simplified somewhat by assuming that the surface temperature is uniform and that a steady state has been achieved, that is, that temperature and humidity is constant in time at every point. It may be recognized that the depth of the layer which is modified by the warm ocean must increase with distance from the shore. Upon applying the principle of conservation of energy to this layer, the upward flux through the ocean surface must equal the individual rate of heating integrated over the column. Figure 5.9 provides a schematic diagram for the problem. After introducing the definition of potential temperature into the First Law of Thermodynamics, the upward flux is expressed by

$$F_h(0) = \int_0^{z_2} \rho c_p \frac{T}{\theta} \frac{d\theta}{dt} dz$$

where z_2 represents the height to which modification extends. In this and the following development all quantities represent averages but the bar notation has been omitted for aesthetic reasons. And, if for simplicity it is assumed that the wind is horizontal and that the wind direction is independent of height

$$F_h(0) = c_p \int_0^{z_2} \rho v \frac{T}{\theta} \frac{\partial \theta}{\partial x} dz \qquad (5.42)$$

where x represents the coordinate in the direction of the wind with origin at the coast, and v represents the wind speed.

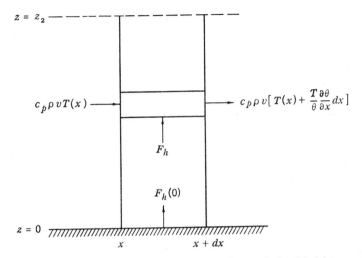

FIG. 5.9. Vertical heat flux density (F_h) into a column of air of height z_2 and the resulting horizontal inflow and outflow of enthalpy as expressed by Eq. (5.42).

Assume now that within the surface layer of depth z_1 the vertical heat flux is independent of height and that above z_1 the transfer coefficient is so large that the vertical potential temperature gradient approaches zero. Equation (5.34) then may be used to express the surface heat flux in the form

$$F_h(0) = -\frac{c_p\rho k^2 v_1(\theta_1 - \theta_0)(T/\theta)}{\left(\ln\frac{z_1}{z_0}\right)^2}(1 - \alpha\,\overline{\mathrm{Ri}})^{1/2} \qquad (5.43)$$

The height of the surface layer is not precisely defined. It is convenient to relate this height to the roughness length, so that the factor

$$\left(\frac{1}{k}\ln\frac{z_1}{z_0}\right)^{-1}$$

is a constant. Also, assume that $(1 - \alpha\,\overline{\mathrm{Ri}})^{1/2}$ is a constant and that in the surface layer $T \approx \theta$. Equation (5.43) now may be written in the form

$$F_h = -ac_p\rho v_1(\theta_1 - \theta_0)$$

where a is a dimensionless constant which replaces

$$\left(\frac{1}{k}\ln\frac{z_1}{z_0}\right)^{-2}(1 - \alpha\,\overline{\mathrm{Ri}})^{1/2}$$

Substitution of this equation into (5.42) yields

$$a\rho_1 v_1(\theta_0 - \theta_1) = \int_0^{z_2} \rho v\,\frac{T}{\theta}\,\frac{\partial\theta}{\partial x}\,dz$$

The right-hand side of this equation may be approximated by integrating from z_1 to z_2, recalling that in this layer θ is independent of height. Therefore

$$a\rho_1 v_1(\theta_0 - \theta_1) = \frac{d\theta_1}{dx} \int_{z_1}^{z_2} \frac{T}{\theta} \rho v \, dz \approx \frac{T}{\theta} \frac{d\theta_1}{dx} \overline{\rho v} \, z_2 \tag{5.44}$$

The height z_2 depends on the temperature distribution of the original cold air before it reaches the ocean and on the change in potential temperature which occurs during passage over the sea. This last difference is expressed by $\theta_1(x) - \theta_1(0)$ (which must be distinguished from $\theta_0 - \theta_1$ measured in the vertical). Because the potential temperature is continuous at z_2, this height may be expressed by

$$z_2 = \frac{\theta_1(x) - \theta_1(0)}{\Gamma - \gamma} \frac{T}{\theta} \tag{5.45}$$

where Γ represents the adiabatic lapse rate and γ the lapse rate of the original cold air.

Upon substituting Eq. (5.45) into (5.44), for simplicity assuming that $\overline{\rho v}$ is independent of x, introducing a normalized potential temperature (θ^*) and integrating

$$\theta^* + \ln(1 - \theta^*) = \xi \tag{5.46}$$

where

$$\theta^* \equiv \frac{\theta_1 - \theta_1(0)}{\theta_0 - \theta_1(0)} \quad \text{and} \quad \xi \equiv \frac{a(\Gamma - \gamma) \rho_1 v_1 x}{[\theta_1(0) - \theta_0] \overline{\rho v}(T/\theta)^2}$$

Equation (5.46) is plotted in Fig. 5.10. Observations have been used to establish the value of a as about 2×10^{-3}. This value corresponds under neutral conditions to a height of the surface layer (z_1) equal to 15 m and a roughness length (z_0) equal to 1.0 cm.

Because water vapor and heat are transferred by the same turbulent mechanism, modification of humidity may be treated by a method essentially similar to that used for modification of temperature. A difference arises because the height of the layer of turbulent mixing is determined by the potential temperature distribution according to Eq. (5.45), whereas the humidity distribution has no appreciable effect on this height. Consequently, turbulent mixing may produce a discontinuity in humidity at the top of the layer. Following a development analogous to that introduced for temperature modification, and introducing a discontinuity in specific humidity represented by $q_1 - q_2(0)$, the humidity modification may be expressed by

$$\frac{\rho_1 v_1}{\overline{\rho v}} a(q_0 - q_1) = \frac{dq_1}{dx} z_2 + (q_1 - q_2(0)) \frac{dz_2}{dx} \tag{5.47}$$

When the height h increases, dry undisturbed air with humidity $q(0)$ enters

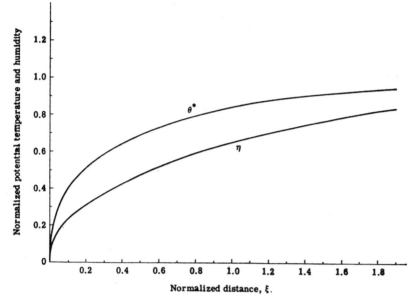

FIG. 5.10. Normalized potential temperature (θ^*) and normalized humidity (η) plotted as a function of normalized distance traveled over a warm ocean (ξ) as calculated from Eqs. (5.46) and (5.48) [after J. A. Businger, *Med. Verh. Kon. Ned. Met. Inst.* **61** (1954)].

the mixing layer and quite suddenly attains by turbulent mixing the humidity $q_1(x)$.

Equation (5.47) is solved by eliminating h with Eq. (5.45) and x with Eq. (5.46). This yields an equation relating the humidity and the temperature; upon assuming that $q_2(0)$ is constant with height, this equation takes the form

$$\frac{d\eta}{d\theta^*} = \frac{\theta^* - \eta}{\theta^*(1 - \theta^*)}$$

where η represents the normalized specific humidity $[q_1 - q_2(0)][q_0 - q_1(0)]^{-1}$. The boundary condition is $\eta = 0$ when $\theta^* = 0$. Integration of this equation using the series expansion, $\eta = a_0 + a_1\theta^* + a_2\theta^{*2} + \ldots$, yields

$$\eta = \frac{1 - \theta^*}{\theta^*} \ln(1 - \theta^*) + 1 \tag{5.48}$$

In Fig. 5.10 η is shown as a function of ξ.

The example of air modification treated here serves to illustrate the process in a simple but important case. In principle the analysis can be extended to a more general form, but this is an extremely complex task. We have glimpsed the gate to the jungle.

5.11   Global Summary of Energy Transfer

The physical processes of energy transfer by radiation, evaporation, and turbulent conduction vary greatly from place to place. Among the factors responsible for this variation the most important are the effect of latitude on solar flux density and the different surface properties of water and land. Each of us is familiar with geographical variations in energy transfer, and we make use of this familiarity in planning travel to the beach in mid-July, in buying tire chains in preparation for a winter trip to Minneapolis, or in deciding to plant corn in Iowa.

The insolation absorbed at the earth's surface was described in Chapter IV; this is the first step in the series of energy transformations which ultimately is responsible for winds and ocean currents as well as for all biological processes. In this chapter the mechanisms responsible for the intervening energy transformations have been discussed. The equations developed in the course of this discussion may be used to calculate evaporation and heat transfer at many points over the surface of the earth and at many times during the year. The distribution of the mean annual heat

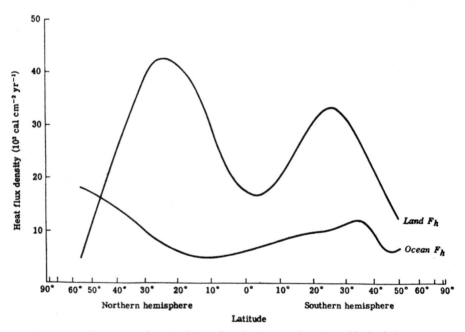

FIG. 5.11. Mean annual vertical heat flux density as a function of latitude for ocean areas and for land areas (after M. I. Budyko, "Gidrometeorologichleskoe iz datel 'stvo," 255 pp. Leningrad, 1956; translated by Nina A. Stepanova, "The Heat Balance of the Earth's Surface," U. S. Weather Bureau, Washington, D.C., 1958).

transfer with latitude is shown in Fig. 5.11 for land areas and for ocean areas; readers who are properly skeptical hardly need to be told that results of this sort reflect great perseverance and great courage, but accuracy is likely to be uncertain.

Figure 5.11 reveals that on the average over both land and sea heat is transferred upward from the earth to the atmosphere. The fact that this occurs in spite of the normal increase of potential temperature with height reflects the large values of eddy transfer coefficients which accompany superadiabatic lapse rates. Figure 5.11 also shows that between 45°N and 45°S upward transfer over land is three or four times as large as it is over the oceans.

Of more interest is the latitudinal distribution of evaporation over land and over the oceans. By comparing Fig. 5.11 and 5.12 the energy utilized in evaporation over the oceans may be recognized as amounting to roughly ten times the vertical heat flux; over land the two energy transfers are more nearly equal.

Evaporation from land areas reaches a strong maximum in the equatorial region, but at all latitudes precipitation exceeds evaporation. This implies, because water vapor stored in the atmosphere is constant over long periods, that there is a net transport of water by the atmosphere from ocean to

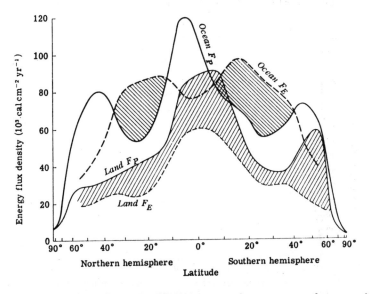

FIG. 5.12. Energy flux densities corresponding to the mean annual evaporation F_E and precipitation F_P for land areas and ocean areas (after M. I. Budyko, *op. cit.*, Fig. 5.11; W. Köppen, and R. Geiger, "Handbuch der Klimatologie," Vol. 1B, p. 492. Bornträger, Berlin, 1936).

land areas. Evaporation from ocean areas exhibits maxima amounting to about 150 cm year^{-1} in the subtropical regions and minima over the poles and the equator. Consequently, water must be transported by the wind systems from the subtropical oceans to middle latitudes (40° to 60°) and to the equatorial regions where the major precipitation systems of the world are concentrated. In studying Fig. 5.12 one should remember that land occupies only 30% of the surface area of the earth, so that the distributions of evaporation and precipitation for the whole earth are close to those shown for the oceans.

Aside from the slow accumulations of thermal energy called changes in climate, the annual insolation absorbed by the earth and its atmosphere must equal that given up in the form of radiation to space. Similarly, the total energy of all forms absorbed by earth and atmosphere separately must equal, respectively, that given up by the earth and atmosphere. Errors in calculation of individual terms are far larger than any possible small residual, but no calculations are needed to establish these propositions. However, the net radiation absorbed or emitted at a point or in particular regions may be quite large, as is shown in Fig. 5.13. The energy absorbed by the oceans between 15°S and 30°N as shown in Fig. 5.13 must be transported poleward by ocean currents where it is released from the surface into the atmosphere. Similarly, the net radiation absorbed by the

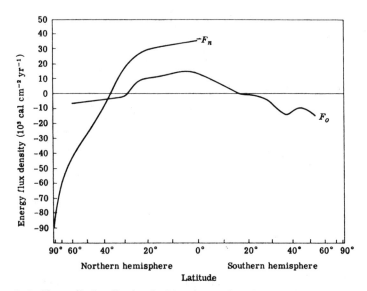

FIG. 5.13. Net radiation F_n absorbed by the earth and atmosphere as a function of latitude [after H. G. Houghton, *J. Meteorol.* **11**, 1 (1954)] and heat absorbed by the oceans (F_0) (adapted from M. I. Budyko, *op. cit.*, Fig. 5.11).

atmosphere and earth between the equator and 34°N must be transported to high latitudes by ocean currents and air currents. Except close to the equator, the energy absorbed in the oceans is less than half the total absorbed by earth and atmosphere. Therefore, we may conclude that although the mass of the oceans is more than 100 times the mass of the atmosphere, more heat is carried poleward by winds than by ocean currents, a result of considerable satisfaction to chauvinistic meteorologists.

The study of the wind systems which transport energy poleward is a part of atmospheric physics to be discussed in a subsequent volume. Here, we shall only point out that the ocean and the atmosphere are two thermodynamic engines which operate somewhat differently. For the atmospheric engine the primary heat source may be regarded as the tropical land and ocean surfaces, and the primary heat sink is the upper part of the water vapor atmosphere. In transporting thermal energy from the source to the sink potential energy of the system tends to decrease with corresponding increase in kinetic energy. For the ocean engine, on the other hand, the primary source is the tropical ocean surface, and the primary sink is the polar ocean surface. Because the source and sink are at nearly the same geopotential, only weak fluid motion is to be expected.

List of Symbols

First used
in Section

a Dimensionless constant, $\left(\dfrac{1}{k} \ln \dfrac{z_1}{z_0}\right)^{-2} (1 - \alpha \overline{R}_i)^{1/2}$ 5.10

c_g	Specific heat of ground	5.1
c_p	Specific heat at constant pressure	5.1
e	vapor pressure	5.9
F_E	Latent heat flux density	5.1
F_g	Heat flux density into the ground	5.1
F_h	Enthalpy (convective heat) flux density	5.1
F_m	Flux density of momentum	5.3
F_n	Net radiation flux density	5.1
F_s	Flux density of property s	5.3
g	Force of gravity per unit mass	5.5
h	Height	5.10
k	Thermal conductivity, von Karman constant	5.2, 5.4
K	Coefficient of turbulent transfer	5.3
l	Mixing length	5.3
L	Latent heat of vaporization	5.1
p	Pressure	5.3
q	Specific humidity	5.3
R_w	Specific gas constant for water vapor	5.9

List of Symbols (Continued)

Subscripts

Problems

1. Calculate by two methods the rate of heat conduction into the soil for each two-hour period shown in Fig. 5.1 for k_g equal to 3.1×10^4 erg cm^{-1} sec^{-1} °C^{-1} and the following values of $c_g \rho_g$.

Depth (cm)	$c_g \rho_g \times 10^{-7}$ (erg cm^{-3} °C^{-1})
0.5	1.2
1.5	
2.5	
3.5	
4.5	
5.5	
10	1.6
20	1.5
30	1.2
40	1.1
80	1.2

Which method is likely to be more accurate?

2. Show from the equation for turbulent enthalpy flux density, $F = c_p\,\overline{\rho w'T'}$, that Eq. (5.14) is the proper form of the transfer for heat.

3. Find the stress exerted by the wind on the ground surface and the roughness for 0630 9 August 1953 at O'Neill, Nebraska when the temperature near the ground was nearly independent of height and the following wind speeds were observed.

Height (meters)	Speed (m sec^{-1})
6.4	7.91
3.2	7.18
1.6	6.34
0.8	5.35
0.4	4.46

4. Compare the following wind profile observed at 1230 9 August 1953 at O'Neill, Nebraska with the corresponding profile computed from Eq. (5.29).

Height (meters)	Wind speed (m sec^{-1})	Temp. (°C)
6.4	9.99	27.64
3.2	9.10	28.12
1.6	8.06	28.44
0.8	6.91	29.07
0.4	5.72	29.52
0.2	—	30.12
0.1	—	30.67

To do this, first calculate the stress using the roughness obtained in problem 3 for the adiabatic profile and the observed wind speed at a height of 0.4 meter. Then, calculate ζ at 1.6 meters from Eq. (5.28) using smoothed profiles and obtain S at each height. Integrate Eq. (5.27) numerically to find $\bar{v}$ at each height.

5. Compute K_h/K_m from Eq. (5.31) using the data of problem 4 at heights of 0.8, 1.6, 3.2 meters. F_h was measured as 2.5×10^5 erg cm^{-2} sec^{-1}.

6. Calculate from Eq. (5.41) the minimum possible surface temperature for the data given in Fig. 4.19 for an ice layer of 10 cm thickness having a thermal conductivity of 2.1×10^5 erg cm^{-1} sec^{-1} °C^{-1}. For the sounding shown the specific humidities and temperatures are as follows:

Pressure (mb)	Specific humidity (g/kg)	Temperature (°C)
360	0.19	−38
400	0.40	−32
460	0.58	−23
500	1.0	−20
540	1.45	−18
600	1.4	−12
690	1.3	− 6
700	0.95	− 5
820	1.2	+ 1
850	1.9	+ 3
900	2.5	+ 4
930	3.7	+ 3
970	3.2	+ 1
990	2.3	− 2

General References

Brunt, *Physical and Dynamical Meteorology*, relates the energy transfer processes to the large scale atmospheric structure and to the dominant systems of atmospheric motions.

Geiger, *The Climate near the Ground*, provides a great quantity of data and discussion of the details of the temperature and humidity distributions.

Sutton, *Micrometeorology*, discusses turbulence and its effects in much greater detail than is given here, and places much emphasis on the theory of diffusion.

Pasquill, *Atmospheric Diffusion*, relates diffusion to the fundamentals of atmospheric turbulence, and extends substantially the theory presented in this chapter.

Priestley, *Turbulent Transfer in the Lower Atmosphere*, distinguishes clearly between mechanical turbulence and free convection, and discusses the evidence concerning the dependence of the eddy transfer coefficients on static stability.

Lettau and Davidson, *Exploring the Atmosphere's First Mile*, provides the most complete set of data relating to ground surface energy transfer. The tables are a gold mine, but the discussion is uneven and uncritical.

Budyko, *The Heat Balance of the Earth's Surface*, gives the most complete description of the global distribution of vertical energy transfer.

CHAPTER VI

Geomagnetic Phenomena

"The search for truth is in one way hard and in one way easy,
For it is evident that no one can master it fully nor miss it wholly." ARISTOTLE

MANY OF THE ATMOSPHERIC PHENOMENA discussed so far: those which depend on gravitation, on changes of state and phase, or on energy transformations, have a very great effect on the physical nature of the environment in which each of us lives. The effects of these phenomena are greatest in the denser portion of the atmosphere, and consequently they come directly to the attention of every observant individual. Phenomena associated with the geomagnetic field are more subtle and only occasionally penetrate our attention barriers. However, their effects are important to a wide range of geophysical problems. For example, the source of the geomagnetic field concerns the physical composition and state of the earth's deep interior; the geomagnetic field plays a central role in the analysis of motions in the thermosphere; and interaction of solar plasma with the geomagnetic field is responsible for auroral phenomena and the radiation zones or belts which surround the earth.

A few direct geomagnetic measurements in the outer portion of the atmosphere have been made, but the interior of the earth remains inaccessible. For these reasons much of our knowledge has come through deductive inference. Deduction has proved to be so powerful a method, for example, in discovery of the planet Neptune, or in predictions of the energy available through nuclear disintegration that physical science can scarcely be imagined apart from it. However, where verification by experiment or observation is not possible, deduction may lead to serious error, for it takes no account of facts which are not yet known. Examples may be recalled from the history of the ether concept or, most pertinently to this discussion, from Kelvin's 1864 estimate of the earth's cooling rate. In ignorance of radioactive sources of internal energy, Kelvin calculated that the maximum possible age of the solid earth is 400 million years, about 10% of the currently accepted minimum age. History, then, suggests that inferences concerning the geomagnetic field should be carefully examined and accepted only tentatively where direct verification is not available.

We shall proceed to develop fundamental electromagnetic theory in so far as it is necessary for our purposes, and then shall apply the theory to an account of the principal geomagnetic phenomena.

6.1 Magnetic Induction

In Section 3.14 the force acting on a test charge at rest was expressed by Coulomb's Law, and the electric field was introduced as a conceptual aid in understanding electrostatic phenomena. Here, two additional laws will be introduced and the magnetic field will be defined. The discussion of this section will be limited to particle speeds for which relativity corrections are not necessary; that is, speeds very close to the speed of light will not be considered.

Ampère's Law

Two electrical charges in motion exert forces on one another which depend on the individual velocities as well as on separation of the charges. A series of brilliant experiments carried out by Ampère between 1820 and 1825 have led to the following generalization. Two charges each moving with constant velocity experience a force in addition to the Coulomb force which for the test charge q is given by

$$\frac{\mathbf{F}_m}{q} = \frac{\mu Q \, \mathbf{v}_q \times (\mathbf{v}_Q \times \mathbf{r})}{4\pi r^3}$$

(6.1)

where the constant of proportionality μ is called the *permeability* of the medium. The force $\mathbf{F}_m$ may be called a *magnetic force*. Equation (6.1), which is a form of Ampère's Law, may be illuminated by considering Fig. 6.1 and the following special cases. If the two charges move parallel to one another along a line normal to the vector separating them, the force exerted on q is directed toward Q. On the other hand, if the charges move along a single straight line, the magnetic force vanishes. It is convenient to define the *magnetic flux density* or *magnetic induction* associated with the charge Q by

$$\mathbf{B} \equiv \frac{\mu Q}{4\pi r^2} \frac{\mathbf{v}_Q \times \mathbf{r}}{r}$$

(6.2)

Therefore, Eq. (6.1) may be written

$$\frac{\mathbf{F}_m}{q} = \mathbf{v}_q \times \mathbf{B}$$

(6.3)

An essential difference between the electric and magnetic forces is that the electric force may act in the direction of motion, whereas the magnetic force acts normal to the motion. Hence, the magnetic force can only change the direction of a moving charge but cannot do work on it.

In fact, the magnetic force may be considered an apparent force which arises from the relative motion of test charge and observer. The magnetic

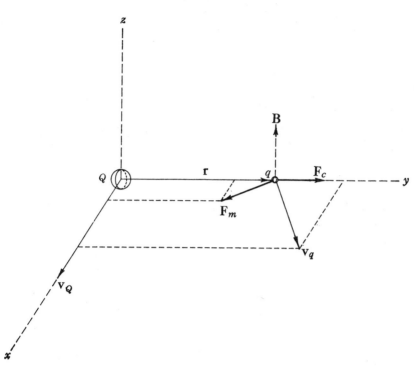

FIG. 6.1. The Coulomb force $\mathbf{F}_c$, the magnetic force $\mathbf{F}_m$, and the magnetic flux density $\mathbf{B}$ associated with charges q and Q of the same sign moving with constant velocities in the xy plane.

induction, on the other hand, is a property of space independent of the relative motion of test charge and observer.

Equation (6.2) may be cast into a particularly pregnant form which expresses the flux density produced by an electric current. Let $d\mathbf{B}$ represent the differential magnetic flux density produced by a differential charge dQ moving at velocity $\mathbf{v}$. The current I is defined by dQ/dt, and $\mathbf{v}$ is defined by $d\mathbf{s}/dt$. Equation (6.2) then takes the form

$$d\mathbf{B} = \mu I \, \frac{d\mathbf{s} \times \mathbf{r}}{4\pi r^3} \tag{6.4}$$

The magnetic induction at a distance a from an infinite straight conductor carrying the current I is found, by integrating over the length of the conductor, to be

$$B = \frac{\mu I}{2\pi a}$$

and is circular in the plane normal to I. We recognize that this is equivalent to the line integral of the magnetic induction around a circle enclosing the current. Thus

$$\oint \frac{\mathbf{B}}{\mu} \cdot d\mathbf{l} = I = \iint \mathbf{J} \cdot d\mathbf{A} \tag{6.5}$$

where $\mathbf{J}$ represents the current density and $\mathbf{A}$ area enclosed by the line integral. Although derived here for the case of a straight conductor, Eq. (6.5) may be developed directly from Eq. (6.4) and therefore holds for the line integral of the magnetic induction around any closed circuit enclosing a current.

Faraday's Law

Soon after Ampère's discovery of the magnetic force Michael Faraday and Joseph Henry about 1831 found independently that a charge at rest experiences a force if the magnetic flux density changes locally. The results of Faraday's and Henry's experiments are expressed by Faraday's Law in the form

$$\oint \mathbf{E} \cdot d\mathbf{l} = -\iint \frac{\partial \mathbf{B}}{\partial t} \cdot d\mathbf{A} \tag{6.6}$$

where $\mathbf{E}$ represents the *induced electric field*. Equation (6.6) states that the work done per unit charge in a closed circuit equals the negative integral of the local rate of change of magnetic flux taken over the area of the circuit.

The induced electric field which appears in Faraday's Law has an origin different from the Coulomb electric field and yet each is referred to as an electric field and the same symbol ($\mathbf{E}$) is used to represent each. Can this be justified? Imagine an isolated test charge at rest in an unchanging magnetic field; no force acts on it. If another charge approaches, the test charge experiences a Coulomb force and work is done in accelerating the charge. On the other hand, if no charge approaches but the magnetic flux density changes, the test charge experiences a force which the test charge cannot distinguish from the Coulomb force. Therefore, there is no need to treat the two forces separately; and the electric field is recognized as arising both from static charges and from changing magnetic fields.

One important difference does exist. The work done in moving a charge in an electrostatic field always vanishes upon closing the circuit, whereas, as shown in Eq. (6.6), the work done in a closed circuit in an induced electric field is, in general, finite.

The differential equation which governs the motion of charged particles results from introducing the electric and magnetic forces into Newton's

Second Law. The equation of motion for a particle of mass m and charge q then is given by

$$\frac{d\mathbf{v}}{dt} = \frac{q}{m} [\mathbf{E} + \mathbf{v} \times \mathbf{B}] \equiv \frac{q}{m} \mathbf{E}' \tag{6.7}$$

The vector $\mathbf{E}$ represents the sum of the electrostatic field and induced electric field. This equation is fundamental to the study of ionized gases (hydromagnetics) as well as to the behavior of charges in cyclotrons and other particle accelerators.

Ampère's and Faraday's Laws are independent statements of fundamental physics which apply to free space as well as to conductors. They may, however, be combined for application to closed conductors. Consider a conductor moving in a magnetic field. Equation (6.3) shows that separation of the positive and negative charges occurs, and this establishes an electrostatic field within the moving conductor. If the circuit is closed, a current flows in the circuit and work is done per unit charge which is expressible by

$$\oint \mathbf{v} \times \mathbf{B} \cdot d\mathbf{l}$$

The energy is supplied by the kinetic energy of the moving conductor. The line integral may be transformed by Stokes' theorem (Appendix I.F) to

$$\iint \nabla \times (\mathbf{v} \times \mathbf{B}) \cdot d\mathbf{A}$$

If the work per unit charge (the induced voltage*) expressed by Eqs. (6.3) and (6.6) are added, the sum may be expressed by

$$\mathcal{E} \equiv - \iint \left[\frac{\partial \mathbf{B}}{\partial t} + \mathbf{v} \cdot \nabla \mathbf{B} + \mathbf{B} (\nabla \cdot \mathbf{v}) - \mathbf{v} (\nabla \cdot \mathbf{B}) - \mathbf{B} \cdot \nabla \mathbf{v} \right] \cdot d\mathbf{A}$$

The fourth term on the right vanishes because the magnetic field has no sources or sinks. The remaining terms combine so that the work done in the circuit may be expressed by

$$\mathcal{E} = - \frac{d\phi}{dt} \tag{6.8}$$

where

$$\phi \equiv \iint \mathbf{B} \cdot d\mathbf{A} \tag{6.9}$$

---

* The term "electromotive force," which is often used for $\mathcal{E}$, has been avoided here because of the confusion which it introduces when the word "force" is taken literally.

The tedium of the expansion and reduction has been obscured between the lines.*

Equation (6.8) expresses the fact that the work done in a closed circuit (the induced voltage) is equal to the individual rate of change of normal magnetic flux passing through the circuit. The word *individual* here means that the derivative is taken following the circuit, so that the effects of local change in flux density, motion through the magnetic field and change in area normal to the magnetic flux enclosed by the conductor are combined.

The equations of this section account for the major electromagnetic phenomena associated with simple electrical currents. Electromagnetic phenomena also occur in the absence of currents in dielectrics like the atmosphere or interplanetary space, and to account for these phenomena extension of this discussion is required.

6.2   Maxwell's Equations

Electromagnetic theory rests on a formulation of Ampère's and Faraday's Laws due to Maxwell which remains as one of the greatest scientific achievements. Maxwell considered a circuit containing a capacitor, and he defined the scalar *displacement* within the capacitor by the identity

$$D \equiv \frac{Q}{A} \tag{6.10}$$

where Q represents the charge and A the area of one plate of the capacitor. More generally, Eq. (6.10) applies to the charge on any surface A, say the surface of a sphere. For the sphere we may easily calculate from Eqs. (3.24) and (6.10) that the vector displacement is related to the electric field by

$$\mathbf{D} = \epsilon\mathbf{E} \tag{6.11}$$

and this may be considered the definition of the vector displacement. Further, the *displacement current* per unit area may be defined by

$$\mathbf{J}_D \equiv \frac{\partial \mathbf{D}}{\partial t} \tag{6.12}$$

and if ϵ is independent of time

$$\mathbf{J}_D = \epsilon \frac{\partial \mathbf{E}}{\partial t}$$

Now, recall that within a dielectric, positive and negative charges are impelled to move in opposite directions and, in consequence, electrical dipoles are created within the dielectric. The product of charge and separa-

---

* Panofsky and Phillips (see Bibliography) p. 144, accomplish this in an elegant and simple way.

tion of positive and negative charges, the dipole moment per unit volume, is called the polarization (**P**). Evidently, **P** and **D** have identical units, and the displacement within a dielectric may be represented by the sum of the displacement in vacuum and the polarization. Thus

$$\mathbf{D} = \epsilon\mathbf{E} = \epsilon_0\mathbf{E} + \mathbf{P}$$

It is convenient to express the dielectric coefficient by

$$\epsilon = \epsilon_0\left(1 + \frac{\mathbf{P}\cdot\mathbf{E}}{\epsilon_0 E^2}\right) \equiv \epsilon_0(1 + \chi) \tag{6.13}$$

where χ represents the property of the dielectric known as the *susceptibility*. Susceptibility may be positive, negative, or zero and, indeed, may be complex. The latter case is considered in a discussion of the index of refraction in Section 7.6.

It is convenient to define the *magnetic field intensity* by the identity

$$\mathbf{H} \equiv \frac{\mathbf{B}}{\mu} \tag{6.14}$$

The relation between **H** and **B** is analogous to the relation between **D** and **E**. **H** and **D** are dependent only on the source of the respective fields, whereas **B** and **E** depend also on the local properties of the medium.

Now, Ampère's Law in the form of Eq. (6.5) should hold for a dielectric if the displacement current is added to the current I. Thus, with the aid of Eqs. (6.12) and (6.14), Eq. (6.5) becomes

$$\oint \mathbf{H}\cdot d\mathbf{l} = \iint (\mathbf{J}_D + \mathbf{J})\cdot d\mathbf{A}$$

This equation together with Eq. (6.6) may be taken as fundamental, for they contain Ampère's and Faraday's Laws. However, these equations apply to closed curves of finite and arbitrary form, whereas we should like to have relations between the field vectors at a point. This can be achieved by applying Stokes' theorem with the result

$$\nabla \times \mathbf{H} = \frac{\partial \mathbf{D}}{\partial t} + \mathbf{J} \tag{6.15}$$

$$\nabla \times \mathbf{E} = -\frac{\partial \mathbf{B}}{\partial t} \tag{6.16}$$

These are the first two of the four fundamental differential equations of electromagnetic theory known as Maxwell's equations. For the important case of no current flow and for μ and ϵ independent of time they may be written

$$\nabla \times \mathbf{H} = \epsilon \frac{\partial \mathbf{E}}{\partial t} \tag{6.17}$$

$$\nabla \times \mathbf{E} = -\mu \frac{\partial \mathbf{H}}{\partial t} \tag{6.18}$$

These equations reveal the symmetry in the roles played by $\mathbf{E}$ and $\mathbf{H}$ in electromagnetic waves.

The third Maxwell equation expresses Coulomb's Law in the more general differential form

$$\nabla \cdot \mathbf{D} = \rho$$

where ρ represents the space charge density. Integrating this equation over a sphere containing in the center the finite charge Q results in Eq. (3.24). The fourth equation has already been used in the development of Eq. (6.8). It expresses the fact that the magnetic field has no sources or sinks and is written in the form

$$\nabla \cdot \mathbf{B} = 0$$

6.3   Field of a Magnetic Dipole*

In many important cases magnetic fields are produced by electrical charges flowing in circular paths; such a field is called a magnetic dipole field because of its close correspondence to the electrostatic dipole. To determine the components of the dipole field, first note that Eq. (6.5) for uniform permittivity shows that for any closed circuit not enclosing a current the closed line integral of the magnetic induction taken around the circuit vanishes. Under this condition there must exist a scalar potential function Φ_m such that $\nabla \Phi_m = \mathbf{B}$. This is a familiar theorem in theoretical physics which can be easily proved. A direct analogy is provided by the relation between the geopotential and the field of gravity discussed in Chapter I. The condition requires that we consider the field only outside the path of the current, and, in fact, we shall consider only the region far from the current for which the distance is large compared to the radius of the current flow.

Because the field is uniform in the direction parallel to the current flow, the problem is two-dimensional. We may recognize from Fig. 6.2 that the fields produced by the current flowing in the right and left halves of the circle are opposite in direction, and the field produced by the right side is

---

* The material of this section is not essential to understanding succeeding sections.

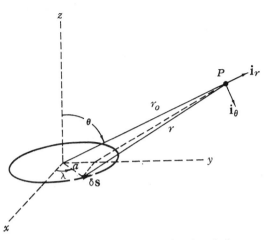

FIG. 6.2. The geometry of the magnetic dipole. An electrical current flowing in the segment of circle ds produces a magnetic induction field at P.

the greater because on the right r is smaller than it is on the left. The contributions to the potential made by the components of the current flow in the y direction are compensated by equal and opposite contributions, so that only the component in the x direction need be considered. From the discussion of Eq. (6.2) it follows that the magnetic fields of the x components of the two halves of the circuit lie in the yz plane and form a small angle at P. For $r \gg a$ this angle is negligible, and the direction of the magnetic field is in the direction of the unit vector $\mathbf{i}_\theta$, that is, normal to $\mathbf{r}_0$. Therefore, the gradient of the potential is in the θ direction, and the potential may be calculated by integration of the equation

$$d\Phi_m = \frac{\partial \Phi_m}{\partial \theta} d\theta = \mathbf{i}_\theta \cdot \left[\int r \, \delta\mathbf{B} \right] d\theta \qquad (6.19)$$

where $\delta\mathbf{B}$ represents the differential increment of field produced by the x component of the current flow in δs. The integral in Eq. (6.19) may be expressed from Eq. (6.4) in the form

$$\int r \, \delta\mathbf{B} = \left(\frac{\mu}{4\pi} \right) I a \int_0^{2\pi} \frac{\mathbf{i} \times \mathbf{i}_r}{r} \sin \lambda \, \delta\lambda \qquad (6.20)$$

where $\mathbf{i}$ and $\mathbf{i}_r$ represent unit vectors in the x and r_0 directions, respectively, and the component of δs in the x direction has been expressed by $a \sin \lambda \, \delta\lambda$. With the aid of Fig. 6.2 it is easily shown that r is represented by

$$r^2 = r_0{}^2 + a^2 - 2r_0 a \sin \theta \sin \lambda$$

Therefore, for $r \gg a$ the first two terms of a series expansion gives

$$\frac{1}{r} = \frac{1}{r_0}\left(1 + \frac{a \sin \theta \sin \lambda}{r_0}\right) \tag{6.21}$$

and upon substituting Eq. (6.21) into (6.20)

$$\int r \, \delta\mathbf{B} = \mathbf{i}_\theta \frac{\mu I a}{4\pi r_0} \int_0^{2\pi} \left(1 + \frac{a \sin \theta \sin \lambda}{r_0}\right) \sin \lambda \, \delta\lambda$$

If the right-hand side is integrated and substituted into Eq. (6.19), the result is

$$d\Phi_m = \left(\frac{\mu}{4\pi}\right)\frac{I\pi a^2}{r_0{}^2} \sin \theta \, d\theta$$

One further integration gives for the magnetic scalar potential

$$\Phi_m = -\frac{M}{r_0{}^2} \cos \theta \tag{6.22}$$

where the *magnetic moment* of the dipole is defined by

$$M \equiv \left(\frac{\mu}{4\pi}\right) I\pi a^2$$

and the potential is chosen as zero where $\theta = \pi/2$.

Having found the magnetic scalar potential, the magnetic induction is formed by taking the gradient of the potential. In spherical coordinates

$$\mathbf{B} = \mathbf{i}_r \frac{2M \cos \theta}{r^3} + \mathbf{i}_\theta \frac{M \sin \theta}{r^3} \tag{6.23}$$

Equation (6.23) shows that the magnetic induction field is radial along the z or polar axis and is circular in the xy plane or the plane of the equator. The field is illustrated in Fig. 6.3.

Equation (6.23) also may be developed from Eq. (6.2) without deriving the magnetic scalar potential. The resolution into components and the integration is left as problem 1.

6.4   Behavior of Plasma in Electromagnetic Fields

The fundamental principles discussed in the preceding section may be applied readily to isolated charged particles, but application to a large number of charged particles (a *plasma*) is very much more difficult because interactions occur between particles and between the plasma and the external field. Plasma theory is incomplete, and the relevant quantitative data for the atmosphere are still too inaccurate to permit definitive calculations. For these reasons we shall discuss only qualitatively the major processes which are believed to control plasma behavior in the atmosphere.

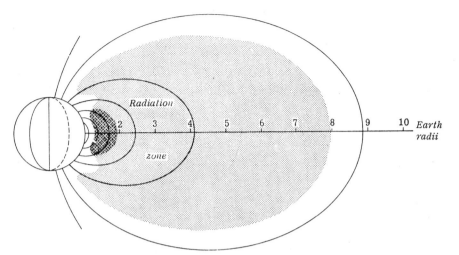

FIG. 6.3. The dipole field of the earth calculated from Eq. (6.22) using $M = 8 \times 10^{15}$ weber meters and the location of the radiation zone.

The references given at the end of the chapter provide much quantitative detail.

Consider a number of electrons and positive ions moving with equal speeds in the xz plane through a magnetic field which is directed along the y axis as shown in Fig. 6.4a. According to equation (6.3) the positive ions are deflected in a counterclockwise sense as seen in looking toward the north and perform circles in the xz plane. The electrons are deflected in a clockwise sense as seen in looking north. The radii and the angular frequency of these motions is found to be, respectively,

$$r_+ = \frac{m_+ v}{e_+ B} \qquad r_- = \frac{m_- v}{e_- B} \tag{6.24}$$

$$\nu_+ = \frac{e_+ B}{2\pi m_+} \qquad \nu_- = \frac{e_- B}{2\pi m_-} \tag{6.25}$$

where m represents mass, v the speed, ν the frequency, and the signs of the charges are indicated by subscripts. Equations (6.25) express the *cyclotron-* or *gyro-frequency* for the positive ion and the electron. At a height of one earth radius above the equator where the geomagnetic field is about $0{:}04 \times 10^{-4}$ weber m^{-2} an electron with speed of 2×10^3 km sec^{-1} (corresponding to travel from the sun in one day) has a radius of about 2.5 meters. The corresponding radius of the orbit of a proton is about 4.5 kilometers. Equations (6.25) show that the gyro-frequency of the electron is much greater than that of the proton. For these reasons the magnetic field within the orbit induced by the electron is much greater than that induced by the

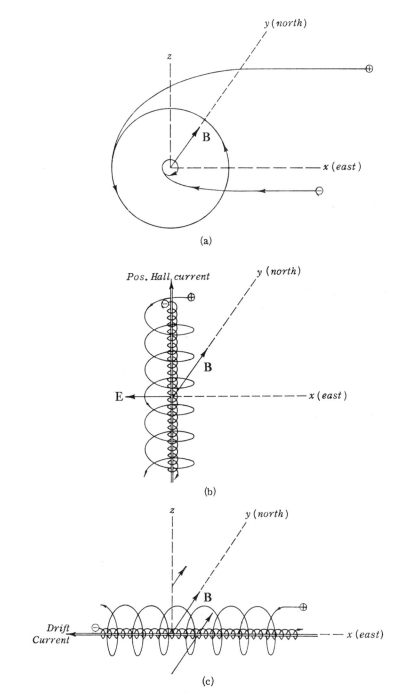

(a)

(b)

(c)

proton as is illustrated in problem 2. In each case the induced field opposes the external field. Outside the orbit the induced field is in the direction of the external field, so that the rotating charge has no net effect on the intensity of the geomagnetic field.

Next, consider the effect on the orbiting charges of an electric field directed along the x axis as shown in Fig. 6.4b. Each time the proton moves in the direction of the E vector ($-x$ axis) it is accelerated and each time it moves in the x direction it is decelerated; as a result the radius of curvature is increased in the $-x$ direction and decreased in the x direction, and the proton "drifts" in the $-z$ direction. At the same time the electron experiences acceleration and deceleration in the opposite phase with the result that it also drifts in the $-z$ direction. So long as the numbers and velocities of the ions and electrons are equal, there is no resultant current, but departure from these conditions may occur in either of two ways. If the electric field is created suddenly, the electrons respond more rapidly than the protons so that a net positive current (Hall current) flows in the z direction. And, if a charged particle collides with a neutral particle and gives up all or most of its momentum, it is then accelerated by the electric field. The electrons respond more rapidly than the protons so that in this case too a Hall current flows at right angles to E and B. A current also flows in the E direction (called the Pedersen current). The Pedersen current is favored by collision frequency which is close to the gyro-frequency, and the Hall current is favored by difference in mobility between positive and negative ions. Very large variations in current are possible depending on density and distribution of charge and on density of the neutral particles.

The Hall current is particularly important when it produces polarization within the medium. Suppose that in Fig. 6.4b the Hall conductivity decreases above and below the xy plane. Then the Hall current results in polarizing the conducting layer with the upper surface charged negatively and the lower surface charged positively. The electric field associated with the polarization may become strong enough to permit a strong current in the direction of the primary E field. This effect is important near the geomagnetic equator where the horizontal B field leads to polarization of the horizontal E layer. At other latitudes the Hall current is directed

---

FIG. 6.4. Paths followed by charged particles in electric and magnetic fields.

(a) An electron ($-$) and a proton ($+$) deflected into circular orbits normal to the field B.

(b) The Hall current associated with a B field in the y direction (north) and an E field in the $-x$ direction (west).

(c) The drift current associated with a B field in the y direction (north) which decreases in magnitude in the z direction.

at an angle to the vertical and therefore is less effective in polarizing the
E layer. Ionospheric currents are further discussed in Section 6.8.

The gravitational field produces an effect qualitatively similar to Hall
current. In this case both positive and negative charges are accelerated
downward with the result that they drift in opposite directions, the positives to the east, the negatives to the west. Both contribute to a drift
current toward the east, and this current flows whether or not there are
collisions with neutral particles. The apparent centrifugal force associated
with motion along the geomagnetic field lines is directed outward from the
earth, so that the spiraling charges are accelerated by centrifugal force as
they move outward, and vice versa. The result is that radius of curvature
of the orbiting charges is greater on the outer than on the inner side, and
a drift current flows toward the west. Because the drift currents produced
by the gravitational and centrifugal forces are in opposite directions, their
net effect depends upon whether $g_0{}^*(R_0/R)^2$ is greater or less than v^2/R.
Here, v represents the velocity along the geomagnetic lines and R, R_0, and
$g_0{}^*$ are defined in Chapter I. At five earth radii, the centrifugal effect
dominates, and the drift current is toward the west if the velocity along
the geomagnetic lines exceeds about 3.5 km sec^{-1}. In the magnetosphere this
speed is very greatly exceeded, so that a drift current toward the west is
to be expected. In the ionosphere the effects of gravity and of centrifugal
force in producing drift currents are probably very much less than the
effects of electric fields.

Finally, consider a magnetic field in which flux density decreases in the
z direction as shown in Fig. 6.4c. Positive and negative charges experience
larger radius of curvature in the weaker field than in the stronger, so they
drift in opposite directions forming a positive current in the $-x$ direction.
In this way ring currents toward the west may be formed in the very high
atmosphere where collisions are rare.

The effects of gravitation and of the radial decrease of magnetic field,
which produce ring currents in opposite directions, may be compared by
differentiating Eq. (6.24), first allowing only v to vary and then allowing
only B to vary. The ratio of the differentials is

$$\frac{dr_m}{dr_g} = -\frac{v\,dB}{B\,dv}$$

where the operator d represents the difference between top and bottom
of the path and the subscripts m and g refer, respectively, to the magnetic
and gravitational effects. But from Eq. (6.23) where R represents distance
from the center of the earth

$$\frac{dB}{B} = -\frac{3dR}{R}$$

Also, from the discussion of geopotential in Chapter I

$$v \, dv = -g^* \, dR \approx -g_0^* \left(\frac{R_0}{R}\right)^2 dR$$

Therefore

$$\left|\frac{dr_m}{dr_g}\right| = \frac{3v^2 R}{g_0^* R_0^2} \tag{6.26}$$

If the ratio given by Eq. (6.26) exceeds unity the magnetic effect predominates and the resultant ring current is toward the west. At a height of four earth radii the ratio is unity for v equal to 2 km sec^{-1}, which corresponds to a temperature for protons of 170°K. We must conclude that the magnetic effect dominates and the ring current should flow toward the west.

6.5   The Geomagnetic Field

It has been known for hundreds of years that a suspended bar magnet tends to orient itself in approximately the north-south plane and that it tends to dip from the horizontal to a degree which depends upon latitude. The pole of the magnet which points toward the north is called a *north magnetic pole* and the pole which points toward the south is called a *south magnetic pole*. From these observations it is possible to visualize the earth's magnetic field as similar to the field in the neighborhood of a magnetic dipole situated near the earth's center with magnetic moment of 8×10^{25} gauss cm^3 (8×10^{15} weber meters) as shown in Fig. 6.3. From Eq. (6.23) it follows that this field could be produced by a current of 2.6×10^9 amperes flowing around the axis at a radius of 10^3 kilometers. On the plane of the equator the dipole field decreases from 0.32 gauss (0.32×10^{-4} weber m^{-2}) at the earth's surface to 0.04 gauss at a height of one earth radius. The axis of the dipole field (geomagnetic axis) is inclined to the axis of rotation by about 11.5°. The intersections of the earth's surface with the geomagnetic axis are called the *geomagnetic poles;* the north geomagnetic pole is located in central Greenland and the south geomagnetic pole in Antarctica, south of Australia. The actual field at the surface of the earth, which differs from the dipole field for several reasons, is represented in Fig. 6.5. The magnitude of the field at the surface varies from 0.32 to 0.6 gauss between geomagnetic equator and pole.

In the 18th century it was observed that the geomagnetic field undergoes slow changes extending over a century or more called *secular* variations. During the hundred years ending about 1936 the geomagnetic dipole moment decreased by about 7%. Precise measurement at many points on the earth's surface also reveals that at any time the earth's magnetic field differs in an interesting manner from that of a dipole. If the dipole-defect

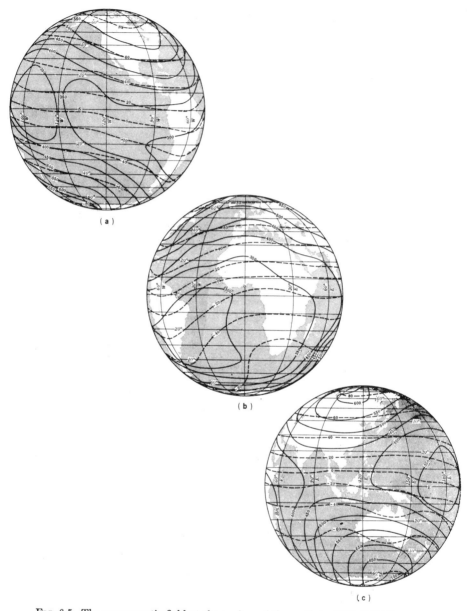

FIG. 6.5. The geomagnetic field at the surface of the earth in 1955. Solid lines represent the absolute value of the magnetic induction and dashed lines represent the dip angle [adapted from E. H. Vestine, *Trans. Am. Geophys. Union* **41**, 4 (1960), and attributed to U. S. Hydrographic Office and "Encyclopaedia Britannica"].

is plotted and lines of constant value of the defect are drawn, the resulting patterns suggest the appearance of an atmospheric pressure chart. If dipole-defects are plotted at intervals of several years, it may be noted that the isopleths move over the map in a manner analogous to the movement of isobars—that is, the identities of the patterns are maintained, but their movements and changes are predictable only within wide limits.

In addition to the secular changes discussed above, the earth's magnetic field undergoes short period fluctuations of relatively great magnitude called *transient* variations. These variations may be classified under "quiet conditions" or "disturbed conditions," depending upon whether or not the magnetic field undergoes large and frequent fluctuations. During "quiet conditions" there is observed a solar diurnal variation and a lunar diurnal variation. These variations increase in amplitude from winter to summer. During "disturbed conditions," in addition to the normal diurnal variation, there is a diurnal variation of the disturbance as well as irregular variations. The frequency of "disturbed conditions" has a period of approximately eleven years corresponding to half the sunspot cycle and a period of approximately 27 days corresponding to the period of the sun's rotation.

6.6 Dynamo Theory of Geomagnetism

Although there still does not exist a complete and verified explanation of the earth's magnetic field, the dynamo theory which has been developed by Bullard* and by Elsasser† is plausible and consistent with the known facts. Ignorance of the structure and behavior of the earth's interior stand in the way of verification.

The dynamo theory may be summarized as follows: radioactive decay in the earth's interior heats the fluid which surrounds the core; the resulting expansion of the fluid leads to convection. Now, if a fluid current descends toward the earth's interior, its absolute angular frequency tends to increase in accordance with conservation of angular momentum. Consequently, the central core of the earth may have an angular frequency greater than that of the outer shell or of the crust. This implies that changes in the intensity of convection or of heat generated should be accompanied by changes in the magnetic field. Observations have revealed changes in the rotation rate with a period of the order of 25 years, which is consistent with the observations of secular change in geomagnetic field referred to earlier.

Now, consider the operation of the disk dynamo, which serves for this purpose as an analog of the earth. The disk is driven by a central axle and is surrounded by a circular coil of wire connected as shown in Fig. 6.6. If

---

* E. C. Bullard, *Proc. Roy. Soc.* A197, 433 (1949).

† W. M. Elsasser, *Revs. Modern Phys.* 22, 1 (1950).

no magnetic field is present, the disk may be rotated without notable result. However, if there exists a small magnetic field normal to the disk, then each radial element in the disk cuts magnetic lines of force as the disk rotates. In this case Eq. (6.8) gives for the work done per unit charge (the induced voltage)

$$\mathcal{E} = \int_0^{r_0} \mathbf{v} \times \mathbf{B} \cdot d\mathbf{l} \qquad (6.27)$$

The vectors $\mathbf{v}$, $\mathbf{B}$, and $d\mathbf{l}$ are mutually perpendicular, so that Eq. (6.27) becomes

$$\mathcal{E} = \int_0^{r_0} B\omega r \, dr$$

where ω represents angular frequency of the disk and r_0 the radius of the disk. An exercise is provided in problem 3. All quantities are expressed

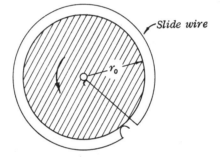

FIG. 6.6. The disk dynamo used to illustrate the essential elements of the earth's dynamo.

in MKS units; therefore $\mathbf{B}$ is measured in webers m^{-2} and $\mathcal{E}$ is expressed in volts. Integration from the center of the disk to the rim gives

$$\mathcal{E} = \tfrac{1}{2}B\omega r_0^2$$

The resulting current in the circular coil is limited by the resistance of the complete circuit. The current in the circular coil produces an enhancement of the magnetic field normal to the coil. By integrating Eq. (6.4) the enhancement of the field is given by

$$\Delta B = \frac{\mu I}{2r_0}$$

The strengthened magnetic field produces an increased potential difference and increased current which in turn results in increase in magnetic field strength. Analogous effects are often used in electronics to bring about large changes from a small stimulus and are referred to as "positive feedback." In the case of the disk dynamo the magnetic field increases until the energy dissipated as heat in the circuit equals the energy supplied to rotate the shaft. An exercise is provided in problem 4.

If the earth's magnetic field is to be attributed to a mechanism analogous to that described for the disk dynamo, there must be horizontal currents flowing in the earth. Horizontal currents are observed on the earth's surface, but their interpretation is difficult because they arise from ion currents in the ionosphere as well as from sources within the earth.

Rotation of the earth also induces voltage directed from equator (negative) to pole (positive) which is derivable from Eq. (6.27). Because the magnetic force depends upon the velocity of the electric charges, the magnetic force appears only when measured in a fixed nonrotating coordinate system. Therefore, observations made on the earth cannot detect the equator to pole voltage difference, although for a fixed coordinate system the measured voltage would be large, as is illustrated by problem 5.

6.7 Ion Distribution

Observations of the earth's magnetic field reveal a diurnal variation amounting over most of the earth to about 0.1% of the permanent field; the variation is tuned to local time. This fact led Balfour Stewart in 1882 to postulate the existence of free electrical charges and an electrical current in the upper atmosphere. The ionized layer, called the *ionosphere*, was first observed in 1924 by Appleton and Barnett[*] in England and Breit and Tuve[†] in the U. S. using reflection of radio waves.

Existence of the ionosphere depends on ion production in the upper atmosphere by absorption of ultraviolet radiation as discussed in Section 4.19. At a height of 80 kilometers the air density is roughly 2×10^{-8} g cm^{-3} or 4×10^{14} air molecules per cubic centimeter. There are only 10^{-2} g cm^{-2} between the air at this height and interplanetary space, whereas at sea level there are 10^3 grams of air above each square centimeter of surface. This mass of 10^{-2} g cm^{-2} is just about sufficient under normal conditions to absorb completely the very short-wave radiation from the sun which carries sufficient energy to produce ionization. Above this height the intensity of ionizing radiation increases with height as suggested in Fig. 4.22. The ratio of neutral particles to ions decreases step by step as the minimum ionizing intensity is reached successively for each constituent of the atmosphere. At the same time the number of molecules per unit volume decreases as indicated in Fig. 6.6. As we have seen in Section 4.19, the rate of ion production per unit volume may be expressed by

[*] E. V. Appleton and M. A. F. Barnett, *Proc. Roy. Soc.* **A109**, 621 (1925).

[†] G. Breit and M. Tuve, *Phys. Rev.* **28**, 554 (1926).

$$p = \int_0^\infty \beta_\lambda k_\lambda F_\lambda \rho \, d\lambda \qquad (6.28)$$

where β_λ represents a constant of proportionality indicating the ionizing efficiency of the absorbed radiation (F_λ), k_λ represents the absorption coefficient and ρ the air density. The vertical distribution of ion production, as expressed in Eq. (6.28), is illustrated schematically in Fig. 4.22. For a particular band of wavelengths maximum rate of ion production occurs at a height which is characteristic of the absorbing molecules. We should expect, therefore, that there might be a layer of maximum ion concentration somewhere above a height of 80 kilometers. There are, in fact, at least three and sometimes four such layers below 400 kilometers which, together, are called the ionosphere. Each of these layers must be formed by predominant absorption of a particular wavelength or band of wavelengths by a particular constituent of the atmosphere. As we have seen, the details still present many problems. The major ionization phenomena responsible for existence of the ionosphere are given in Table 4.1.

We shall now express the rate of change of ion concentration as the difference between rate of electron production and recombination. From Eq. (4.62) it follows that

$$\frac{dn}{dt} = p - \alpha n^2 \qquad (6.29)$$

Observations using radio waves reflected from the ionosphere (described in Chapter VII) provide measurements of n and dn/dt. If observations are made when p is zero (say, during an eclipse of the sun), Eq. (6.29) can be used to calculate the recombination coefficient. Then, if the recombination coefficient is assumed to remain constant, the rate of electron production may be calculated at any time. In the E layers during much of the day recombination approximately balances ionization $(dn/dt \approx 0)$, so that $p = \alpha n^2$. And because radiation flux density is proportional to the cosine of the zenith angle of the sun, it is possible to predict ion concentration for any time of day within limits imposed by the accuracy of observations and the assumptions referred to above. These calculations are required in problem 6.

Recombination of isolated positive and negative ions is a very rare occurrence due to the fact that potential energy of the Coulomb field between oppositely charged ions is converted into kinetic energy upon close approach of the two ions. The energy of the pair usually exceeds ionizing potential. However, if neutral particles are present, the excess energy may be transferred to these particles during collision with the result that recombination is greatly facilitated. Recombination also may occur readily if one of the ions is polyatomic. In the case of a diatomic ion the excess energy may be utilized in dissociating the diatomic ion into two neutral monatomic molecules. The dissociation process, which is analogous to reaction (4.49), probably accounts for much of the recombination occurring in the E layer. Because the density of neutral particles decreases with height and because temperature increases with height in the ionosphere, recombination decreases markedly with increasing height. As a result the D and E layers are distinct only during the day and virtually disappear at night, whereas the F₂ layer is essentially permanent. However, the latter is subject to large and sometimes erratic fluctuations in ion concentration. The position and ion concentration of each of the layers are indicated in Figs. 6.7 and 7.19.

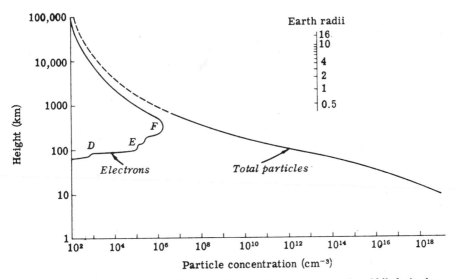

FIG. 6.7. Electron density as a function of height for summer noon in middle latitudes [after J. E. Jackson, CSAGI Meeting, Moscow, 1958: published by H. Newell, *in* "Physics of the Upper Atmosphere" (J. A. Ratcliffe, ed.), p. 108, Academic Press, New York, 1960] and total particle concentration as a function of height (Appendix II.E).

Even in the ionosphere neutral molecules far outnumber ions. The ratio in the E layer is roughly $1/10^8$ and in the F₂ layer $1/10^4$. Each of the three component conductivities is proportional to charge density and is inversely related to the frequency of collision with neutral particles. As a result, all three conductivities exhibit maxima in the E layer and decrease markedly above 150 kilometers. For this reason the ionospheric current flows most readily in the region of the E layer.

With increasing height above the F₂ layer (300 kilometers) the ratio of ions to neutral particles increases, and time variations in ion density of several orders of magnitude probably occur. Particularly at low geomagnetic latitude, satellite observations have shown increase with height of particulate radiation flux density to a maximum of 10^{15} electron volts cm^{-2} sec^{-1} ster^{-1} at a height of three or four thousand kilometers (one half the earth radius). This is the lower region of the radiation zone discovered by Van Allen* and his colleagues and by Vernov† and his colleagues. This region consists of electrons of energies up to one million electron volts (1 Mev) and protons of energy of the order of 10^2 Mev. These energies represent speeds very close to the speed of light. The density of particles is uncertain. Between heights of $\frac{1}{2}$ to 1 earth radius the energy flux may decrease, but in some cases increases again and reaches a maximum between 2 and 3 earth radii. This outer region consists almost entirely of electrons with energies between 10^3 and 10^6 electron volts. The radiation of the inner region is harder than that of the outer region, presumably as a result of decay of high energy neutrons by collision with air molecules. The outer zone extends to higher latitudes than does the inner zone, but is absent above the regions of the geomagnetic poles.

Measured concentrations of high-energy charged particles have changed radically between observations, apparently in response to solar emission of plasma, and the distinction between the two radiation zones is not always clear. Nuclear explosions at heights of several hundred kilometers have resulted in the formation of radiation shells between the two quasi-permanent zones which conform to the geomagnetic field. Following the explosions, particulate energy flux along the shell increased by an order of magnitude.

Beyond the radiation zone the earth's atmosphere merges with the solar atmosphere. The behavior of comet tails when approaching the sun, as well as other evidence, shows that a permanent "solar wind" of nearly equal numbers of protons and electrons flows outward from the sun. This has been explained as resulting from a pressure gradient associated with the coronal temperature gradient. Streams of plasma of higher density also are ejected by the sun for periods up to weeks or months. These streams are often associated with the appearance of sunspots (cool regions in the photosphere) and solar flares (hot regions), and with the occurrence of magnetic storms (fluctuations in the geomagnetic field). The physical characteristics of the solar wind are still uncertain, but measurements have been made which indicate that at 20 to 30 earth radii there are 10 protons cm^{-3} moving radially from the sun at 3×10^2 km sec^{-1} and with temperature of 10^5 °K.

On the other hand, observations of a delay of 24 to 48 hours between

---

* J. A. Van Allen, *J. Geophys. Research* **64**, 1683 (1959).

† S. N. Vernov *et al.*, *Doklady Akad. Nauk S.S.S.R.* **125**, 304 (1959); S. N. Vernov and A. E. Chudakov, *in* "Space Research," Proc. First Intern. Space Sci. Symposium (H. Kahllmann Bijl, ed.); p. 751. North-Holland, Amsterdam, 1960.

solar eruption and onset of a geomagnetic storm suggest a solar wind velocity in excess of 10^3 km sec^{-1}. However, energy transfer from the sun may occur by propagation of hydromagnetic waves rather than by mass transport so that solar wind velocity may be considerably less than 10^3 km sec^{-1}. Suggestions of much greater particle speed in interplanetary space have met respectful attention because the solar wind may carry with it a magnetic field which imprisons high-energy ionized particles.

6.8 The Ionospheric Current

The ionosphere is routinely observed by radio means, and direct measurements of ion concentration and other properties have been made by rockets. The ionospheric electrical current, which was detected prior to discovery of the ionosphere by inference based on observations of fluctuations of the geomagnetic field at the ground, has been verified by rocket observations. Here, we shall outline qualitatively the theory of the current which is now widely accepted as fundamentally valid, the *dynamo theory*. The details are too complex to be treated here, and not all the difficulties have yet been overcome.

The ionosphere is driven by the atmospheric tides. We recall from Chapter I that the gravitational fields of the sun and moon produce organized horizontal and vertical motions which can be visualized by imagining "bulges" in the atmosphere immediately under the tide generating body and on the side opposite this body as shown in Fig. 1.5b. In addition, diurnal heating produces a bulge on the sunlit side of the earth; and the atmosphere appears to respond by forming a similar bulge on the dark side. The latter, however, is not evident above the mesosphere, so that the diurnal heating wave proves to be dominant in the region of the E layer. The air flows radially outward from the region somewhat east of the sub-solar point, with speed increasing with distance as shown in Fig. 6.8a. Ions carried in these currents experience a force normal to the plane of the velocity and magnetic induction vectors as given by Eq. 6.3. Because the motions are essentially horizontal, the ion displacements are primarily horizontal at high latitude and vertical at low latitude. Therefore, at high latitude positive ions are impelled to the left of the tidal motion as is indicated in Fig. 6.8b. Now, if we consider the region of tidal expansion shown on the left of Fig. 6.8a, at high latitude the field of the magnetic force per unit charge is directed toward lower latitudes and toward the west. Positive charge is therefore concentrated at low latitudes in the west; and an electric field is directed along the equator toward the east. The horizontal component of the sum of the magnetic force field and electrostatic field is shown in Fig. 6.8b.

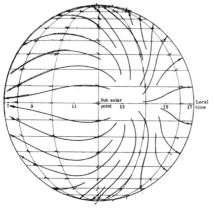

(a1)

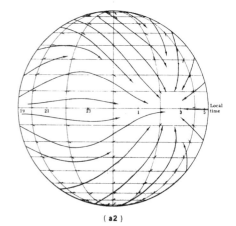

(a2)

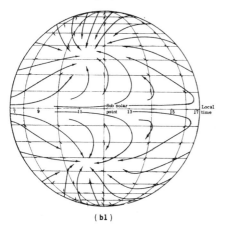

(b1)

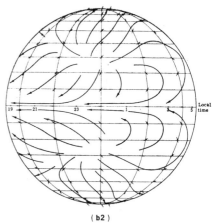

(b2)

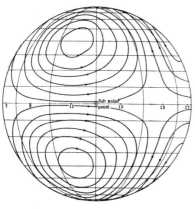

(c1)

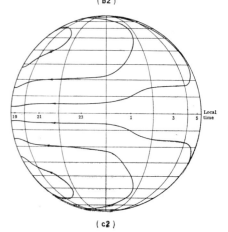

(c2)

The mechanism just described may be imagined as analogous to a dynamo with an external connection between its poles. But because the electrical conductivity of the ionosphere is anisotropic, current flow is not simply related to Fig. 6.8b. It is convenient to distinguish three conductivities: σ_0 in the direction of the component of the electric field parallel to $\mathbf{B}$, σ_1 in the direction of the component of the electric field normal to $\mathbf{B}$, and σ_2 in the direction of the Hall current (normal to both electric and magnetic fields). The quantitative theory of the σ's and of the ionospheric dynamo has been given by Baker and Martyn,* but the pleasure of that exploration will be diluted very little here. Each of the conductivities is directly proportional to the ion concentration and to mobility. Conductivity σ_0, which always exceeds σ_1 and σ_2, is inversely proportional to frequency of collision between ions and neutral particles. Conductivity σ_1 depends on the relation between collision frequency and gyro-frequency; it vanishes for very small and very large collision frequency and reaches its maximum when the collision and gyro-frequencies are equal. Conductivity σ_2 is directly related to the difference in mobilities of the positive and negative ions.

Polarization of the ionosphere by the Hall current, as noted in Section 6.4, adds a vertical electrostatic field to the horizontal field, and this produces a horizontal Hall current. The polarization is a function of the horizontal field and the conductivities, and this makes it possible to eliminate the polarization field by introducing "effective" conductivities which are functions of the σ's and the angle between the geomagnetic field and the plane of the ionosphere. The effective conductivities may be much greater than σ_1 and may even approach σ_0.

Hall polarization is greatest at the geomagnetic equator where the geomagnetic field is horizontal. As a result the current flow along the equator

* W. G. Baker and D. F. Martyn, *Phil. Trans. Roy. Soc. London* **A246**, 281 (1954). W. G. Baker, *Phil. Trans. Roy. Soc. London* **A246**, 295 (1954); see also S. Chapman, *Nuovo Cimento* (*Suppl.*) **4**, 1385 (1956).

FIG. 6.8. (a) Distribution of the solar tidal wind velocity in the E layer at the time of the equinoxes.
(b) Distribution of the horizontal component of the sum of the induced field and the electrostatic field associated with the solar tidal motion in the E layer for the equinoxes [from E. H. Vestine, *in* "Physics of the Upper Atmosphere" (J. A. Ratcliffe, ed.), p. 487, Academic Press, New York, 1960, after J. Maeda, *J. Geomag. Geoelect.* **7**, 121 (1955)].
(c) Distribution of horizontal electric current in the ionosphere associated with the solar diurnal tide for the equinoxes. Each line represents a current flow of 10^4 amperes (after S. Chapman and J. Bartels, "Geomagnetism," p. 229. Oxford Univ. Press, London and New York, 1948).

is concentrated into a narrow dense *electrojet*. The equatorial electrojet accounts for the daily enhancement of the geomagnetic field at, for instance, Huancayo, Peru where the field around local noon exceeds the average field by about 1%. An estimate of the strength of the electrojet is required in problem 8. The total current flow in one of the daytime circuits shown in Fig. 6.8c under normal conditions is about 62×10^3 amperes, most of which flows on the sunlit side of the earth. The equatorial electrojet is not sufficiently emphasized in Fig. 6.8c.

Current flow in the ionosphere and the resulting geomagnetic variations are further complicated by the fact that current in the E layer induces current flow in the F layer. It is supposed that, because of the high conductivity along the sloping **B** lines, the horizontal **E'** field of the E layer is transmitted more or less unchanged along the **B** lines to the F layer. Here the **E'** field induces current flow which probably differs from that of the E layer, but which is almost entirely unknown. In summary, the E layer acts as a dynamo driven by the mechanical tidal motions, while the F layer acts as a motor driven by the E layer dynamo.

The phenomena and processes discussed so far in this section represent the average condition or the typical "undisturbed" condition at time of equinox. Analyses have been carried out for other times of the year. The greatest interest, however, attaches to "disturbed" periods. At such times the local geomagnetic field may change suddenly and erratically, and maps of the changes may be interpreted as reflecting current flow in the region of the ionosphere. In high latitudes, in the auroral zone, there is often an *auroral electrojet* which flows along lines of geomagnetic latitude from east to west on the morning side and from west to east on the evening side of the earth. Such electrojets may carry several hundred thousand amperes, may last for hours, and may recur on successive nights at about the same time. Another phenomenon of great fascination is the *sudden commencement*, a magnetic disturbance which occurs simultaneously all over the earth. In each of the phenomena a connection with the behavior of the very high atmosphere is evident.

6.9   Particle Motion in the Magnetosphere

Above heights of 500–1000 kilometers, the ionized particles experience few collisions, and they are strongly bound by the geomagnetic field. This was the region which was described in Section 2.21 as the magnetosphere. The characteristic motion of ions in this region, as pointed out in Section 6.4, consist of a helical spinning about the geomagnetic lines of magnetic induction, with particles oscillating from north mirror point to south mirror point with periods of a few seconds (for electrons) to a few tens of seconds (for protons). As a result of the decrease of field strength with height a ring current flows toward the west, and this is strong enough to result in decreasing the geomagnetic field within the ring current by as much as 5 to 20%.

Particular interest attaches to an explanation of the source of the high energy particles in the radiation zone, because calculations by Birkeland and Störmer in the early years of the 20th century had indicated that charged particles from the cosmos should be excluded from the very zone in which greatest energy flux was found. On the other hand, charges once in the region should be trapped by the geomagnetic field. One mechanism which is undoubtedly responsible for some of the particles depends upon neutron decay into a proton-electron pair within the zone of trapped particles. Some neutrons arrive in the solar wind, and others are probably formed in the upper part of the nonionized atmosphere by collision of very high-energy cosmic rays with neutral air molecules. The neutrons are unaffected by the geomagnetic field, and so may enter the radiation zone and decay into ions which are subsequently trapped by the geomagnetic field. Neutron decay is presumed to account for most of the radiation of at least the inner part of the radiation zone.

Particle supply to the outer part of the radiation zone (2 to 3 earth radii) probably is related to dramatic increase in the density of plasma flowing from the sun. As the plasma approaches the earth the geomagnetic field is "compressed," permitting the charges to penetrate much further toward the earth than would be possible in the undisturbed geomagnetic field. However, as the plasma comes closer to the earth, the geomagnetic field resists more and more strongly, and finally at a distance from the earth which fluctuates between 5 and 14 earth radii the field "carves" a cavity in the solar wind. Due to the charge separation which results from the different deflections of protons and electrons the surface of the cavity acquires a charge. For a plasma stream approaching the earth from the direction of the sun, negative charge is developed on the morning side, positive charge is developed on the evening side; and a current tends to flow toward the east. The circuit may be completed in the "lee" of the earth where the two streams come together again. In the plane of the equator a ring current flows around the earth from west to east, and this enhances the geomagnetic field inside the ring current.

The approaching plasma stream may have two characteristic effects on the magnetosphere. First, the geomagnetic field is modified so that particles from the outer radiation zone are "dumped" into the lower atmosphere at the mirror points and collide with neutral particles. This may occur because the mirror points are depressed to heights where sufficient neutral particles are present. Second, the arrival of the plasma recharges the outer radiation zone with high energy electrons. The mechanism of the energy transfer is still unknown; it may occur through injection of high energy electrons or through energy transfer by hydromagnetic waves from the plasma to electrons already present in the zone.

6.10   The Aurora

The high-energy electrons which reach the region below one thousand kilometers at the mirror points of the outer radiation zone collide with neutral particles, give up their kinetic energy, and raise the neutral atom to a higher energy state. The excited atom, after a short time interval, may radiate its newly acquired energy in the luminescent form known as the aurora. The light of the aurora is, of course, characteristic of the spectra of the emitting gas.

Systematic auroral observations have been made since the eighteenth century, and much of our knowledge of upper atmospheric properties has come from these observations and the auroral theory developed by Störmer* and others during the early years of the twentieth century. Recent rocket observations have confirmed the earlier deductions while at the same time providing unique quantitative data. Aurorae are observed frequently at high geomagnetic latitude in both hemispheres and with decreasing frequency as the geomagnetic equator is approached as indicated in Fig. 6.9. Maximum frequency is observed along a circle of radius about 18° latitude which presumably marks the poleward limit of the geomagnetic lines which pass through the zones of trapped radiation. Aurorae are visible on occasion at heights ranging from about 80 kilometers to above 650 kilometers with maximum frequency of the lower limit being just above 100 kilometers. These data refer only to frequency of observation; no map of the frequency of occurrence yet exists.

The aurora as seen from College, Alaska is illustrated in Fig. 6.10 and has been dramatically described by Chapman† as follows:

"After darkness has fallen, a faint arc of light may sooner or later be seen low on the north horizon, or centered somewhat to the east of north. Gradually it rises in the sky and grows in brightness. . . . As it mounts in the sky, its ends, on the horizon, advance to the east and west. Its light is a transparent white when faint, and commonly pale yellow-green when bright—rather like the tender color of a young plant that germinates in the dark. The breadth of the arc is perhaps thrice that of a rainbow. The lower edge is generally more definite than the upper. The motion upward toward

---

* C. Störmer, *Arch. sci. phys. et nat.*, [4] Genève **24**, 5 (1907), and many subsequent papers.

† S. Chapman, *Am. Scientist* **49**, 249 (1961).

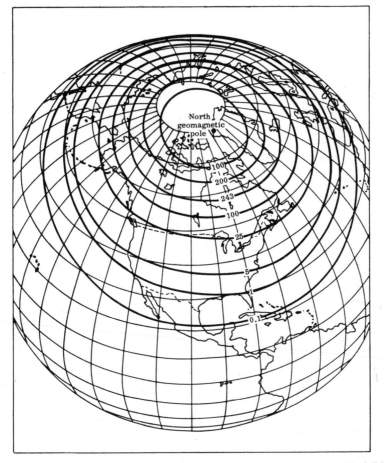

FIG. 6.9. Average distribution of annual frequency of auroral sightings if visibility were unimpaired, northern hemisphere [after S. Chapman, *Am. Scientist* **49**, 254 (1961)].

the zenith may be so slow that the scene is one of repose. As the arc rises, another may appear beyond it, and follow its rise. At times four, five, or even more arcs may thus appear. They rise together, and some of them may cross the zenith and pass onwards into the southern half of the sky. . . . Though it seems almost in repose, a closer look may discern small changes going on, changes of form, or faint waves of light progressing along the band or bands.

"This may be all that appears on some nights. But on others the aurora enters after a while on a new and distinctly different phase, much more active and varied. The transition from the quiet to the active phase may

FIG. 6.10. The aurora as seen from College, Alaska (photograph by V. P. Hessler).

be speedy, even sudden. The band becomes thinner, rays appear in it, it begins to fold and also to become corrugated in finer pleats. It becomes a rayed band of irregular changing form, like a great curtain of drapery in the sky. . . . Its color may remain yellow-green, but often a purplish-red border appears along the lower edge, perhaps intermittently. . . . Vivid green or violet or blue colors sometimes appear. Sometimes there seems to be an upward motion along the rays, or motion to the east or west along the band. The curtains may sweep rapidly across the sky as if they were the sport of breezes in the high air; or they may vanish and reappear, in the same place or elsewhere. This grand display may continue for many minutes or even hours, incessantly changing in form, location, color and intensity; or intermissions may occur, when the sky has little or no aurora.

"At times the observer may look into a great auroral fold nearly overhead, when the rays in its different parts will seem to converge, forming what is called a corona or crown. Often such a corona rapidly fluctuates in form, and its rays may flash and flare on all sides, or roll around the center. "At the end of an outstanding display the aurorae may assume fantastic forms, no longer in connected curtains or bands. There may be a widespread collection of small curtains, stretching over a large part of the sky, which brighten and fade, or, as it is said, pulsate. Finally, the sky may be covered by soft billowy clouds, not unlike a mackerel sky with rather large 'scales'; but these 'scales' and patches appear and disappear, with periods of not many seconds. At last the sky becomes altogether clear, with no more aurora. But later the whole sequence may begin anew, and continue till dawn pales the soft auroral light."

The auroral arcs and bands described by Chapman tend to follow the lines of geomagnetic latitude; they may be thousands of kilometers in length and only a few kilometers thick.* Most aurorae occur on the night-time side of the earth, indicating that the incidence of the solar wind influences the injection of particles into the auroral zone, and there is a more or less systematic dependence of auroral characteristics on time. Occurrence of the aurora is clearly typical and normal along the auroral zone, its absence for a whole night is rare.

The characteristics of aurorae are subtle and not easily observed in their entirety, but their regularities have led Akasofu and Chapman† to suggest the following model of the onset of a typical aurora. At 5 or 6 earth radii the centrifugal force of the flight of the electrons back and forth along the geomagnetic lines accounts for development of a westward ring current in the plane of the geomagnetic equator which may be strong enough to cancel or even reverse the local geomagnetic field. In such case the region of field reversal must be bounded on its outer and inner sides by "neutral" lines, lines of zero geomagnetic field. Electrons which come close to the neutral lines are no longer required to orbit around lines of force, and so the current along the line of force can be enhanced at the expense of the orbital energy. This enables the electrons within a sharply defined layer to descend farther along the magnetic field than their fellows who did not come close to the neutral line. Thus observations made in polar regions may reveal the properties of the magnetosphere in the equatorial region; auroral arcs are regarded by Akasofu and Chapman as transcriptions of the neutral lines.

Auroral rays mark the direction of the geomagnetic field; in high latitudes the rays are within 10° or 15° of vertical and at lower latitudes they are more nearly horizontal. Typical rays may extend between 250 and 90

---

* Personal communication from J. S. Kim.

† S.-I. Akasofu and S. Chapman, *Phil. Trans. Roy. Soc. London* **A253**, 359 (1961).

kilometers above the earth, but much higher rays are also observed. These originate in the sunlit atmosphere 500 to 1000 kilometers above the earth, and are visible to the observer in the earth's shadow. Alfvén* has explained the occurrence of folds in the bands or draperies as resulting from localized space charge. The resulting radial acceleration of charge induces a vertical motion of the charges which distorts the band into the typical folded pattern.

Spectroscopic analysis of the aurora shows that nitrogen and atomic oxygen in both ionized and neutral forms are present; the spectral lines are narrow showing that the emitting gases are at relatively low temperature (hundreds of degrees Kelvin). There are also lines in the spectrum emitted by neutral hydrogen, and these exhibit a remarkable quality. When viewed at right angles to the auroral rays, Doppler broadening of the spectral lines indicate random speeds of order 400 km sec^{-1}, much greater than the speeds of oxygen and nitrogen atoms. When viewed along the rays from below, a similar broadening is observed; but in addition there is a Doppler shift of central wavelength toward shorter wavelengths indicating that the hydrogen is streaming through the atmosphere toward the earth at speeds up to 3×10^3 km sec^{-1}. The Doppler shift is discussed in Section 7.14.

We observe the geomagnetic field and its effects through glasses that permit seeing a very small fraction of the whole. Nevertheless, this area of atmospheric physics affords unambiguous evidence of direct solar-atmosphere effects. On the other hand, despite the persistent effort of dedicated individuals, solar-weather relationships continue to elude detection.

List of Symbols

		First used in Section
a	Distance from wire carrying current, radius of ring current	6.1, 6.3
A	Area vector	6.1
B	Magnetic induction	6.1
D	Displacement vector	6.2
e	Charge of electron	6.4
E	Electric field vector	6.1
E'	Sum of electric field and magnetic force per unit charge	6.1
F	Radiation flux density	6.7
F_m	Magnetic force vector	6.1
g	Acceleration of gravity	6.4
H	Magnetic field vector	6.2
i_r, i_θ	Unit vectors in r and θ directions, respectively	6.3
I	Current	6.1

* H. Alfven, "Cosmical Electrodynamics," p. 206. Oxford Univ. Press, London and New York, 1953.

J	Current per unit area	6.2
k	Absorption coefficient	6.7
l	Length vector	6.1
m	Mass	6.1
M	Magnetic moment	6.3
n	Number of ions per unit volume	6.7
p	Rate of ion production per unit volume	6.7
P	Polarization vector	6.2
q	Charge of test charge	6.1
Q	Charge	6.1
r	Radius vector	6.1
R	Distance from center of earth	6.4
s	Distance vector	6.1
t	Time	6.1
v	Velocity	6.1
α	Recombination coefficient	6.7
β	Ionizing efficiency coefficient	6.7
ϵ	Permittivity	6.1
$\mathcal{E}$	Work per unit charge in closed circuit	6.1
θ	Colatitude, zenith angle	6.3, 6.5
λ	Azimuth angle	6.3
μ	Permeability	6.1
ϕ	Magnetic flux	6.1
Φ_m	Magnetic scalar potential	6.3
χ	Susceptibility	6.2
ν	Frequency	6.4
ρ	Charge density, air density	6.2, 6.7
ω	Angular frequency	6.6

Subscripts

D	Displacement current
g	Gravitational field
m	Magnetic
0	Sea level
λ	Monochromatic

Problems

1. Calculate the r and θ components of the magnetic induction field of a circular ring of current directly from Eq. (6.2) by resolving the field into components parallel to the z axis and perpendicular to the projection of the **r** vector on the yz plane as shown in Fig. 6.2.

2. Find the modification to the geomagnetic field of 4×10^{-5} weber m^{-2} produced at the center of their orbits by a proton and an electron which travel at 10^4 km sec^{-1} normal to the magnetic field.

3. Find the magnetic force induced in a conductor of one meter length which rotates at one revolution per second in a horizontal plane about one of its ends if the vertical component of the earth's magnetic field is 4×10^{-5} weber m^{-2}.

4. Find the magnetic flux density produced at the center of a circle of one meter radius by a current of one ampere.

5. Imagine a wire extended from pole to equator which is motionless as seen from a nonrotating coordinate system outside the earth. Estimate the voltage difference which would be generated between pole and equator if the vertical component of the earth's magnetic field can be approximated by 0.6 gauss sin ϕ where ϕ represents latitude.

6. During the eclipse of 31 August 1932 observations of ion density in the E layer at the time of the eclipse maximum indicated a decrease in concentration of electrons from 63×10^3 cm^{-3} to 50×10^3 cm^{-3} in 10 minutes. If we assume that there was no ion production during this 10 minute period, what is the effective electron recombination coefficient? Find the rate of ion production following the eclipse when observations indicated a maximum concentration of 90×10^3 electrons cm^{-3}.

7. Estimate the magnitude of the electrojet over Huancayo, Peru if it is assumed that the current flows at a height of 100 kilometers and if the geomagnetic field is observed to increase by 4×10^{-7} weber m^{-2}.

General References

Halliday and Resnick, *Physics for Students of Science and Engineering*, and Sears, *Electricity and Magnetism*, Vol. 2, give good elementary accounts of electromagnetic principles.

Panofsky and Phillips, *Electricity and Magnetism*, presents electromagnetic theory on an advanced level. The discussion is clear and the physical bases are carefully explained.

Alfvén, *Cosmical Electrodynamics*, is a well-organized account of theory and relevant presatellite observations. Hydromagnetic waves, solar physics, aurorae and cosmic rays are discussed.

Runcorn, *The Magnetism of the Earth's Body*, in "Handbuch der Physik," v. 47 gives a summary of observations and of several theories of the geomagnetic magnetic field.

Chapman and Bartels, *Geomagnetism*, is a two-volume definitive account of observations and theory of geomagnetism prior to the use of satellites and space probes.

Mitra, *The Upper Atmosphere*, prior to the IGY was the definitive reference covering all topics concerned with the upper atmosphere. It contains detailed summaries of observations and theories of geomagnetic phenomena which are still of great value.

Massey and Boyd, *The Upper Atmosphere*, provides a coherent interpretation in elementary terms of geomagnetic and other upper atmosphere phenomena.

Ratcliffe, *Physics of the Upper Atmosphere*, gives an up to date, comprehensive, clearly written account of upper air phenomena. It omits consideration of the permanent geomagnetic field, but otherwise discusses each of the geomagnetic phenomena included in this chapter much more thoroughly than has been attempted here.

Proceedings of the I.R.E., *Special Issue on the Nature of the Ionosphere*, provides data on ionospheric currents from the IGY and interpretation.

Chamberlain, *Physics of the Aurora and Airglow*, is the definitive monograph on the aurora. It emphasizes spectroscopic methods of observation and analysis.

Kallmann Bijl (ed.), *Space Research*, is a voluminous collection of papers on satellite and spaceprobe measurements from the beginning to 1959.

CHAPTER VII

Atmospheric Signal Phenomena

"To observations which ourselve we make,
We grow more partial for th' observer's sake." ALEXANDER POPE

IN EARLIER CHAPTERS the properties of cloud particles and the vertical distribution of density, pressure, temperature, humidity, and ion concentration have been set forth. In this chapter the effects of these atmospheric properties on the transmission of electromagnetic and acoustical waves will be considered. These effects we shall call *signal phenomena*. Signal phenomena are sometimes deliberately sought for in the observational study of certain properties of the atmosphere; in other cases they are nuisances disturbing communication or introducing errors, for example, in the work of the surveyor. They appear, also, as curious, or remarkable, or beautiful sights whose wonder is adequate reason for trying to understand them. Everyone is familiar with the most common and spectacular phenomena such as the rainbow and the halo, but to the keen observer there is a wealth of more subtle but no less fascinating phenomena including supernumerary bows, focusing of sound, and the green flash. The observer may be of the genus bird watcher, who makes careful notes of what he observes, he may be a poet inspired by the striking beauty of a visual phenomenon, or he may be a scientist who is challenged to achieve physical understanding.

The scientist's interest is to interpret specific phenomena as applications of as few general principles as possible. Only a few of the many signal phenomena will be treated in this chapter, but these examples should establish a basis in understanding such that the individual reader may provide his own interpretations of many phenomena which must go unmentioned here.

PART I: GENERAL PROPERTIES OF WAVES

7.1 Nature of Waves

Most solid substances are so constituted that when a particle is displaced from its equilibrium position by an external force, the displaced particle exerts forces on adjacent particles. This property is called *elasticity;* it is defined as the ratio of the applied force to the displacement, or the ratio of stress to strain. Elasticity varies widely from one substance to another.

For example, steel has a very high elasticity because when a particle which forms part of a steel object is displaced from its equilibrium position, relatively large forces act on adjacent particles which in turn result in their displacement. The reaction force results in the return of the first particle to its equilibrium position; a periodic oscillation then occurs as a result of the inertia of the moving particle. On the other hand, putty or even rubber has a low elasticity since the forces exerted by a displaced particle are small. As a result of elasticity, individual particles perform an oscillatory motion and the oscillation is transmitted from particle to particle. The elastic material is said to perform *wave motion*, and it is easy to recognize that transmission of energy is associated with wave motion.

In solid bodies the forces which are transmitted by wave motion from one particle to an adjacent particle are intermolecular forces. For small displacements, these forces and the reactive forces are large; so it may be concluded intuitively that the resulting wave motion is rapid. Other forces also result in wave motion. For example, the tension in a stretched string produces a restoring force in a portion of the string displaced from the equilibrium position. The reaction to this force acting on adjacent portions of the string results in their displacement, and the displacement moves as a wave along the string. Or, the gravitational attraction acting on water displaced from its equilibrium position results in the familiar motion known as surface water waves. In this last example, the wave motion is not dependent on the elasticity of the medium, but rather on the gravitational field of force acting on water displaced from its equilibrium position. Electromagnetic forces acting on the ionized plasma of the magnetosphere induce *hydromagnetic* waves in an analogous manner. Another field of force, due to the earth's rotation, acts on moving air; when displacement of a portion of the atmosphere occurs, wave motions are produced which are closely related to large scale pressure systems and their attendant weather. This rather complex type of wave motion occupies an important place in the study of atmospheric motions. It is not too early, however, to realize that simple types of wave motion are in certain fundamental respects similar to more complex types of wave motion.

Waves are commonly classified as *longitudinal* if the oscillation of the particles of the medium is predominantly parallel to the direction of wave propagation and as *transverse* if the oscillation of the particles is predominantly perpendicular to the direction of wave propagation. Sound waves and tidal waves in both ocean and atmosphere are of the longitudinal type, whereas waves in a stretched string and electromagnetic waves are of the transverse type. Large scale atmospheric oscillations associated with pressure systems also are of the transverse type.

For simple harmonic oscillation along the z axis the displacement may be expressed as a function of y and t by

$$z = A \cos \frac{2\pi}{\lambda} (y - ct) \qquad (7.1)$$

where A represents the amplitude of the wave, λ the wavelength and c the wave speed (more rigorously, the *phase* speed) in the y direction. The quantity $(2\pi/\lambda)(y - ct)$ is called the phase of the wave.

More complicated forms of waves may be represented by the sum of a series of sine or cosine terms with different values of λ and c.

7.2 Phase Speed

The speed of waves depends upon the characteristics of the transmitting medium. This dependence will be illustrated by developing the speed of several types of waves.

Wave on a Stretched String

Consider a transverse wave traveling along a stretched string with a velocity c. If the coordinate system is permitted to move at the speed c in the direction of the wave, the principles of ordinary mechanics must hold in the moving system as well as in the stationary system. In the moving system the wave is stationary and the string moves with velocity $-c$. The forces acting on a short portion of the deformed string are the centrifugal force resulting from motion of the string in a curved path and the centripetal force resulting from tension in the curved portion of the string. The two forces must be equal since the wave is stationary. The centrifugal force of a length element, Δs, may be expressed by $m \Delta s\, c^2/r$, where m represents mass per unit length, and r radius of curvature of the segment Δs. The centripetal force may be developed from Fig. 7.1. Upon isolating the segment Δs, it may be recognized from the force diagram that in the limit as $\Delta s \rightarrow 0$

$$\frac{F}{T} = \frac{\Delta s}{r}$$

Upon equating centrifugal and centripetal forces

$$T \frac{\Delta s}{r} = m \Delta s \frac{c^2}{r}$$

and

$$c = \sqrt{\frac{T}{m}}$$

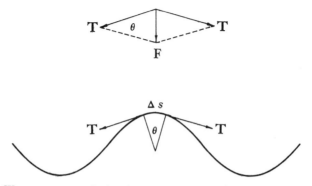

Fɪɢ. 7.1. Waves on a stretched string under tension T and corresponding force diagram for segment Δs.

The speed of the wave depends only on the tension and mass per unit length of the string. It should be noted that it has been assumed implicitly that the amplitude of the waves is sufficiently small that the tension may be considered constant.

The Acoustic Wave

Consider a longitudinal wave traveling with a velocity c through a fluid enclosed in a cylinder. If the coordinate system is permitted to move at the velocity c, the wave is motionless with respect to the moving coordinate system. The wave system is composed of alternate stationary regions of compression and rarefaction, and the fluid moves through these regions with a velocity $-c$.

In the half-wavelength between adjacent planes of maximum and minimum pressure the fluid particles must undergo a change in speed dc; they slow down in approaching a plane of compression and speed up in approaching a plane of rarefaction. The difference in pressure between these two planes may be computed from Newton's Second Law. Thus, for unit cross-sectional area

$$dp = -\rho \, ds \, \frac{dc}{dt} = -\rho c \, dc$$

and therefore

$$\frac{1}{c}\frac{dp}{dc} = -\rho \tag{7.2}$$

In order to eliminate dc the volume of fluid flowing through unit cross section in time dt may be expressed by $V = c \, dt$. Now, if the change in c is given by dc, the corresponding change in V is $dV/V = dc/c$. And, since V is proportional to specific volume (α) Eq. (7.2) may be rewritten in the form

$$\frac{dp}{d\alpha} = -\frac{c^2}{\alpha^2} \qquad (7.3)$$

From Chapter II, one should expect that the ratio of pressure change to volume change depends upon the flux of heat between adjacent regions of compression and rarefaction. Consider the following two extreme cases: (1) heat flux is sufficiently rapid that no temperature gradient exists within the fluid; (2) heat flux is negligible. Express the ratio of pressure change to volume change in the first case by writing the equation of state (2.10) for constant temperature. Thus

$$p\, d\alpha + \alpha\, dp = 0$$

Substitution of this equation in (7.3) gives

$$c^2 = p\alpha$$

and, by again substituting the equation of state,

$$c = \sqrt{R_m T}$$

This is the velocity of sound for isothermal expansion, often called the "Newtonian" velocity of sound.

In the second case, adiabatic conditions prevail. From Eq. (2.46c) the ratio of pressure change to volume change may be written

$$\frac{dp}{d\alpha} = -\frac{c_p}{c_v}\frac{p}{\alpha}$$

Substitution of this equation in (7.3) gives

$$c = \sqrt{\frac{c_p}{c_v} R_m T} \qquad (7.4)$$

This is the velocity of sound for adiabatic conditions. The fact that observations of the velocity of sound almost exactly agree with Eq. (7.4) indicates that the compression and expansion which occurs in acoustic waves is very nearly adiabatic.

The Surface Water Wave

Imagine the cylinder of the preceding discussion to be lying in a horizontal position and to be partly filled with water. As surface waves move through the cylinder, pressure changes occur within the water. At the bottom of the cylinder the particles must move longitudinally, back and forth, as the surface waves pass. These particles, therefore, cannot distinguish between the passage of sound and surface waves. In this case, however, the fluid is incompressible, and dV/V is expressible by dh/h or dc/c and dp may be expressed by $\rho g\, dh$, where h represents water depth. Equation (7.2) now yields for the *shallow water wave speed*

$$c = \sqrt{gh}$$

In deep water waves, the water particles move very nearly in circles; consequently, the pressure at any reference depth depends both on the depth and on the vertical acceleration of the water particles above the reference. Vertical acceleration depends on wavelength, so the wave speed in deep water turns out to depend on wavelength. Similar dependence of wave speed on wavelength is found in electromagnetic waves in air. Media in which wave speed depends on wavelength are called *dispersive media*.

7.3 Electromagnetic Waves

Electromagnetic waves exhibit many of the features which are characteristic of the waves already mentioned, but they are different in that they may move through vacuum. No motion and no medium are needed for their transmission, although the medium does influence the phase speed.

In Section 6.2 Maxwell's first two equations have been expressed as Eqs. (6.17) and (6.18) for no current flow and for permittivity and permeability independent of time. If an electromagnetic field is considered which varies

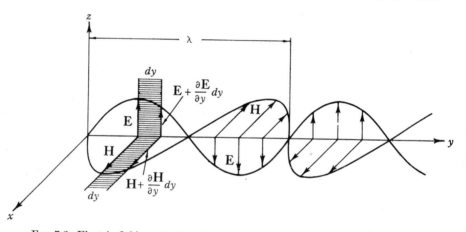

FIG. 7.2. Electric field vector **E** and magnetic intensity vector **H** associated with an electromagnetic wave of length λ traveling in the y direction.

only in one space dimension, as is illustrated in Fig. 7.2, Eqs. (6.17) and (6.18) reduce to

$$\frac{\partial H}{\partial y} = -\epsilon \frac{\partial E}{\partial t}$$

$$\frac{\partial E}{\partial y} = -\mu \frac{\partial H}{\partial t}$$

These equations may be combined by differentiating the first with respect to y and the second with respect to t with the result

$$\frac{\partial^2 H}{\partial y^2} = \mu\epsilon \frac{\partial^2 H}{\partial t^2}$$

Cross differentiation in the opposite sense to eliminate H yields

$$\frac{\partial^2 E}{\partial y^2} = \mu\epsilon \frac{\partial^2 E}{\partial t^2}$$

These equations are classic forms of the linear wave equation in one dimension. The reader is asked to show in problem 1 that the waves of the stretched string are governed by a similar differential equation. Solutions may be written immediately in the form

$$H = H_m \sin \frac{2\pi}{\lambda} (y - ct)$$

$$E = E_m \sin \frac{2\pi}{\lambda} (y - ct)$$

Upon substituting the solutions into the wave equations each equation yields

$$c^2 = \frac{1}{\mu\epsilon}$$

showing that the magnetic and electric fields maintain their identical phase; together they constitute an electromagnetic wave. The speed is given by

$$c = \sqrt{\frac{1}{\mu\epsilon}} \tag{7.5}$$

For vacuum the speed of electromagnetic waves is

$$c_0 = \sqrt{\frac{1}{\mu_0\epsilon_0}} = 2.99793 \times 10^{10} \text{ cm sec}^{-1}$$

For air μ and ϵ differ only slightly from their respective values in vacuum, but the small differences play the central role in some of the applications which follow.

7.4 Dispersion and Group Velocity

If one looks carefully at a series of water waves, he may observe that an "individual" wave experiences changes in shape and amplitude as it is overtaken by faster (longer) waves or as it overtakes slower (shorter) waves. In fact, a group of waves may move into undisturbed water as a coherent group at a speed less than the speed or phase velocity of the individual waves. Waves may be readily observed overtaking the group and increasing in amplitude, then moving ahead of the group and decreasing in amplitude. And, because any wave signal is a group of waves of finite length, it must be recognized that it is the *group velocity*, not the phase velocity, which is measured.

Imagine two waves of slightly different wavelength moving through a medium in which phase speed increases with wavelength. The two waves alternately reinforce and interfere with each other with the result shown in Fig. 7.3. The displacement of the surface from its equilibrium position may be expressed by

$$z = A \sin (kx - \omega t) + A \sin (k'x - \omega' t)$$

where k represents wave number and ω angular frequency. This may be transformed to

$$z = 2A \cos \left(\frac{k - k'}{2} x - \frac{\omega - \omega'}{2} t \right) \sin \left(\frac{k + k'}{2} x - \frac{\omega + \omega'}{2} t \right)$$

Because, as already specified, the waves are of nearly equal length, k and ω are nearly equal to k' and ω', respectively. Therefore

$$z = 2A \cos \left(\frac{x}{2} \delta k - \frac{t}{2} \delta \omega \right) \sin (kx - \omega t) \qquad (7.6)$$

Equation (7.6) describes an envelope with amplitude $2A$ within which there are frequent oscillations as illustrated in Fig. 7.3. To find the separation between successive maxima (separate groups), hold t constant and let $(x/2)\delta k$ increase from 0 to π. The corresponding increment in x is given by

$$\frac{\Delta x}{2} \delta k = \pi$$

or

$$\Delta x = \frac{2\pi}{\delta k}$$

Similarly, if x is held constant and $(t/2)\, \delta \omega$ increases from 0 to π, the corresponding increment in t is given by

$$\Delta t = \frac{2\pi}{\delta \omega}$$

These increments represent, respectively, the wavelength and the period of the group; the group velocity is therefore defined as

$$c_g \equiv \frac{\Delta x}{\Delta t} \equiv \frac{\delta \omega}{\delta k} \approx \frac{d\omega}{dk}$$

This equation may be expressed in terms of index of refraction by substituting $n = c_0/c$ and $c = \omega/k$, hence the group velocity becomes

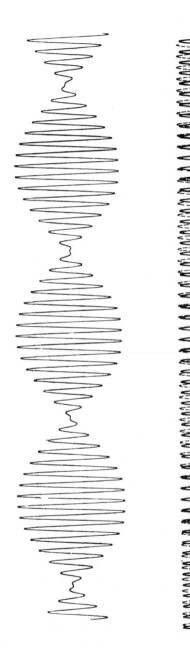

FIG. 7.3. Superposition of two sine waves of equal amplitude and of slightly different wavelength.

$$c_g = \frac{c_0/n}{1 + \dfrac{\omega}{n}\dfrac{dn}{d\omega}} \qquad (7.7)$$

or, in terms of wavelength λ and phase speed c

$$c_g = c - \lambda \frac{dc}{d\lambda}$$

PART II: SCATTERING OF RADIATION

7.5   The Physical Concept

Waves interact with matter on which they are incident in complex and wonderful ways. In trying to understand these processes it will be helpful to imagine a simple and familiar example. Consider a surface water wave incident on a floating block of wood. The block is set into vertical oscillation by the passage of the wave, and the vertical oscillation produces a circular wave which travels outward from the block in concentric circles. We say that some of the incident energy has been *scattered* in all directions by the block. Now, if the block is very small or the wave very long, the block rises and falls with the surface of the water, and little energy is scattered. On the other hand, if the block is very large or the waves very short, the block is nearly motionless, and we say that the wave is *reflected*. However, reflection is clearly a special case of the general phenomenon of scattering. For a particular block, the energy scattered may be studied as a function of incident wavelength. One might expect to find that the proportion of energy scattered in a particular direction varies with wavelength, and that the total energy scattered in all directions reaches a maximum for a particular "resonant" wavelength.

Electromagnetic waves are scattered in an analogous manner. Passage of the wave induces oscillation of the atomic electrons with the result that waves are emitted from the atom. The proportion scattered in a particular direction depends on incident wavelength, and on the size and the *permittivity* of the scattering particle.

The scattered energy may be calculated by imagining that at a particular instant a plane polarized electromagnetic wave exerts forces on the electrical charges of an atom such that the positive charge is displaced in one direction (the direction of the **E** vector) and the negative charge is displaced in the opposite direction. As the wave passes, the phase reverses and the charges are displaced toward each other; hence the electromagnetic wave produces oscillation of charges at the frequency of the wave. The pair of opposite charges is called an electric *dipole*, and the product of the positive charge and the maximum separation of the charges is defined as the *dipole moment*. This form of polarization is called *dielectric* polarization. There exists a second form, *parelectric* polarization, which arises from the fact that individual molecules may possess an electric moment. Although the moments are randomly oriented in the absence of an electric field, the moments may be aligned by an impressed field. In general, molecular orientation responds only rather slowly to an electric field, so that parelectric polarization is most important for low-frequency waves. Polarization plays a crucial role in scattering of radar waves by raindrops and in many other applications.

Development of the theory of scattering by electromagnetic waves requires consideration of the complete range of wavelengths and a range of size of scattering particles from electrons to raindrops or hailstones. A theory of this generality has been developed by Mie,* who integrated Maxwell's equations and expressed the energy scattered in a particular direction by a sphere by an infinite series of terms representing products of associated Legendre polynomials and spherical Bessel functions. The results reveal that the phenomena known as scattering, reflection, absorption, diffraction, and refraction are all contained in the same solution, the distinctions being determined by the nature of the scattering elements and the various scale parameters. However, in the quest for clarity and simplicity, we shall sacrifice the elegance of Mie's solution. After considering in mathematical detail some of the interactions of electromagnetic waves and matter, the phenomena of small particle scattering, diffraction, refraction, and dispersion will be treated separately by approximate methods.

7.6 Complex Index of Refraction

In air or water the permeability μ is nearly equal to μ_0 and the directions of the electric field and the polarization are the same. Under these conditions substitution of Eq. (6.13) into (7.5) gives for the *index of refraction*

$$n' \equiv \frac{c_0}{c} = \sqrt{1 + \frac{\mathbf{P} \cdot \mathbf{E}}{\epsilon_0 E^2}} = \sqrt{1 + \chi} \qquad (7.8)$$

where χ is referred to as the *susceptibility* of the medium. The susceptibility depends strongly on frequency, so that the index of refraction varies with frequency or wavelength. For this reason electrostatic measurements of susceptibility may not be used to calculate index of refraction for light or other high-frequency waves.

---

* G. Mie, *Ann. Physik* **25**, 377 (1908).

Because, as mentioned in Section 6.2, the susceptibility may be complex, the index of refraction also may be complex. In this case

$$n' \equiv n + i\kappa \tag{7.9}$$

where n represents the real index of refraction and κ the *absorption index*. The relation of κ to absorption of radiation can be recognized readily if the solution to the wave equation is written in the form

$$\mathbf{E} = \mathbf{E}_m \exp \left\{ i \frac{2\pi}{\lambda} (y - ct) \right\}$$

After combining with Eqs. (4.5), (7.8), and (7.9) for $t = 0$

$$\mathbf{E} = \mathbf{E}_m \exp \left(i2\pi\nu \frac{ny}{c_0} \right) \exp \left(-2\pi\nu \frac{\kappa y}{c_0} \right) \tag{7.10}$$

The first exponent describes an oscillation, whereas the second exponent describes an exponential decay of the amplitude with y which must be related to Beer's law [Eq. (4.8)]. Later in this section the flux density will be shown to be proportional to the square of the amplitude of the electromagnetic wave; therefore the monochromatic absorption coefficient is expressed by

$$k_\nu = \frac{4\pi\nu\kappa}{\rho c_0} \tag{7.11}$$

In order to investigate the dependence of index of refraction on frequency, consider the polarization of a dielectric subjected to an oscillating electric field. For a single dipole the dipole moment may be expressed as proportional to electric field, and therefore the polarization may be expressed by

$$\mathbf{P} = N\alpha\mathbf{E} \tag{7.12}$$

where N represents the number of dipoles (atoms or molecules) per unit volume and α is called the *polarizability*.

The polarization of a dipole has been defined in Section 6.2 as the product of the charge e of the dipole and the displacement $\mathbf{s}$; therefore the polarization of the dielectric is also expressed by

$$\mathbf{P} = N e\mathbf{s}$$

and upon combining this with Eq. (7.12)

$$\alpha = \frac{es}{E} \tag{7.13}$$

The polarizability may be calculated by considering a model of a dipole consisting of an electron of charge e and mass m which is displaced a distance z from its equilibrium position by a periodic force eE and is subject to a restoring force mkz and a damping force $m\gamma(dz/dt)$. For this system Newton's Second Law may be written

$$\frac{d^2z}{dt^2} + \gamma\frac{dz}{dt} + kz = \frac{eE}{m} \qquad (7.14)$$

The homogeneous solution of this differential equation may be written $z = z_0 \exp(-2\pi i\nu t)$, and this requires that

$$(-4\pi^2\nu^2 + k - 2\pi i\nu\gamma)z = \frac{eE}{m} \qquad (7.15)$$

This equation shows that the displacement z is proportional to the electric field, and therefore the work done on the dipole is proportional to E^2. Therefore, the flux density of the electromagnetic wave, the energy passing unit area per unit time per unit solid angle, is also proportional to E_m^2.

Equation (7.15) has maximum amplitude when the real terms on the left cancel, therefore the natural or *resonant* frequency is defined by

$$\nu_0 \equiv \frac{k^{1/2}}{2\pi}$$

In this case Eq. (7.13) can be written in the form

$$\alpha = \frac{e^2}{m}\frac{1}{4\pi^2(\nu_0^2 - \nu^2) - 2\pi i\gamma\nu}$$

or, upon rationalizing,

$$\alpha = \frac{e^2}{m}\left\{\frac{(\nu_0^2 - \nu^2)}{4\pi^2(\nu_0^2 - \nu^2)^2 + \nu^2\gamma^2} + \frac{i}{2\pi}\frac{\nu\gamma}{4\pi^2(\nu_0^2 - \nu^2)^2 + \nu^2\gamma^2}\right\}$$

Substitution of this result into Eq. (7.12) and then into Eq. (7.8) yields

$$n'^2 = 1 + \frac{Ne^2}{\epsilon_0 m}\left\{\frac{(\nu_0^2 - \nu^2)}{4\pi^2(\nu_0^2 - \nu^2)^2 + \nu^2\gamma^2} + \frac{i}{2\pi}\frac{\nu\gamma}{4\pi^2(\nu_0^2 - \nu^2)^2 + \nu^2\gamma^2}\right\}$$

and because, according to Eq. (7.9),

$$n'^2 = n^2 - \kappa^2 + 2in\kappa$$

it follows that

$$n^2 - \kappa^2 = 1 + \frac{Ne^2}{\epsilon_0 m}\frac{(\nu_0^2 - \nu^2)}{4\pi^2(\nu_0^2 - \nu^2)^2 + \nu^2\gamma^2} \qquad (7.16)$$

and

$$2n\kappa = \frac{Ne^2}{2\pi\epsilon_0 m}\frac{\nu\gamma}{4\pi^2(\nu_0^2 - \nu^2)^2 + \nu^2\gamma^2} \qquad (7.17)$$

For the atmosphere $n \approx 1$ and $\kappa \ll n - 1$. For frequencies in the neighborhood of the resonant frequency, $\nu^2 - \nu_0^2 \simeq 2\nu_0\,\Delta\nu$, where $\Delta\nu$ represents $\nu - \nu_0$. Under these conditions the index of refraction and the absorption coefficient may be expressed from Eqs. (7.16), (7.17), and (7.11) in the forms

$$n \approx 1 - \frac{Ne^2}{\epsilon_0 m \nu_0} \frac{\Delta\nu}{16\pi^2(\Delta\nu)^2 + \gamma^2} \qquad (7.18)$$

and

$$k_\nu = \frac{4\pi\kappa\nu}{\rho c_0} \approx \frac{Ne^2}{\rho c_0 m \epsilon_0} \frac{\gamma}{16\pi^2(\Delta\nu)^2 + \gamma^2} \qquad (7.19)$$

Equations (7.18) and (7.19) describe, respectively, the frequency dependence of the index of refraction and of the absorption coefficient in the neighborhood of the resonant frequency of an oscillating dipole. The functions are illustrated in Fig. 7.4. The frequency dependence of the index of re-

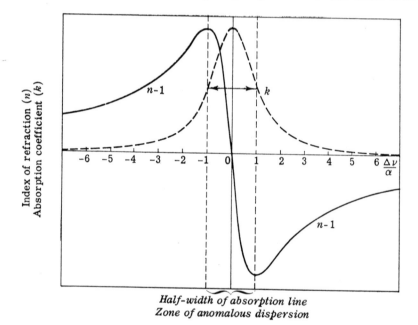

Half-width of absorption line
Zone of anomalous dispersion

FIG. 7.4. The real and imaginary parts of the complex index of refraction computed from Eqs. (7.18) and (7.19).

fraction determines the mode of *dispersion* and, according to Eq. (7.7), the group velocity. For frequencies such that $\nu_0 - (1/4\pi)\gamma > \nu$ the index of refraction is greater than unity and the wave speed is less than c_0. As frequency increases index of refraction increases. This mode is referred to as *normal dispersion,* and under these conditions the separate waves of light are *dispersed* by a prism into the component colors with short wavelengths (blue) refracted most and long wavelengths (red) refracted least. For frequencies in the range defined by

$$\nu_0 + \frac{1}{4\pi}\gamma > \nu > \nu_0 - \frac{1}{4\pi}\gamma$$

the index of refraction decreases with increase in frequency; this is referred to as *anomalous dispersion*. For frequencies in the range $\nu > \nu_0 + (1/4\pi)\gamma$ normal dispersion again occurs, but the index of refraction is smaller than unity and the phase speed is greater than c_0. A particular case of importance in this range occurs when electromagnetic waves are incident on free electrons or ions. In this case the individual charges are not bound to a nucleus and so do not experience the restoring force expressed in Eq. (7.14) by kz. Consequently ν_0 vanishes, and when at the same time the damping constant is negligible, Eq. (7.16) may be written in the form

$$n = \left(1 - \frac{Ne^2}{4\pi^2 m \epsilon_0 \nu^2}\right)^{1/2} = \left(1 - \frac{Ne^2}{\pi m \nu^2}\right)^{1/2} \tag{7.20}$$

Application of this equation to refraction of radiowaves by the ionosphere is discussed in Section 7.15.

Equation (7.19) is identical to Eq. (4.21) if

$$k_l = \frac{Ne^2}{4\rho c_0 m \epsilon_0} = \frac{\pi Ne^2}{\rho c_0 m} \quad \text{and} \quad \alpha = \frac{1}{4\pi}\gamma$$

This may be interpreted to mean that damping generates an absorption line centered at the resonant frequency; the characteristic line shape prescribed by Eq. (7.19) is called the Lorentz line shape.

Damping of the oscillating dipole may occur by the following mechanisms. Electromagnetic waves may be radiated by the dipole, and loss of energy |associated with the emission of radiation results in reduced amplitude of the dipole. A second mechanism of damping results from collisions of molecules. If it is assumed that an oscillating dipole loses its energy during a collision, then the decay time is equal to the average time τ between two collisions. The damping constant is inversely proportional to the decay time as may be recognized from the solution for the damped linear oscillator [Eq. (7.14) with $E = 0$]. For this case Eq. (7.15) becomes

$$-4\pi^2(\nu^2 - \nu_0^2) - 2\pi i \nu \gamma = 0$$

or, because $\nu \approx \nu_0$,

$$2\pi\nu = 2\pi\nu_0 - \frac{i\gamma}{2}$$

and because $z = z_0 \exp(-2\pi i \nu t)$, the solution of the differential equation takes the form

$$z = z_0 \exp\left(-\tfrac{1}{2}\gamma t\right) \exp\left(-2\pi i\nu_0 t\right)$$

This shows that the decay time is equal to $2/\gamma$. Now, the time between collisions is inversely proportional to the density and the velocity of the molecules. Therefore, for constant temperature the damping constant is proportional to the pressure, and the width of absorption lines increases with pressure even though the total absorption is independent of pressure.

7.7   Radiation Emitted by an Oscillating Dipole

Oscillation of an electric dipole in a fixed direction produces a plane polarized electromagnetic wave whose amplitude at a particular point depends on its position with respect to the dipole, on the frequency of the incident electromagnetic wave, on the size, and on the electrical properties of the scattering particles. These parameters are related to the flux density of the radiation, as the reader will see in the following discussion if his stamina is equal to the task. Conservation of energy requires that the flux density be inversely proportional to r^2. Because the electric vector oscillates in a fixed plane, the amplitude of the radiated electric vector is a maximum at right angles to the dipole axis ($\theta = \pi/2$) and falls to zero along the axis ($\theta = 0, \pi$) as shown in Fig. 7.5. Therefore the amplitude is proportional to sin θ, and the flux density is proportional to sin^2 θ. Be-

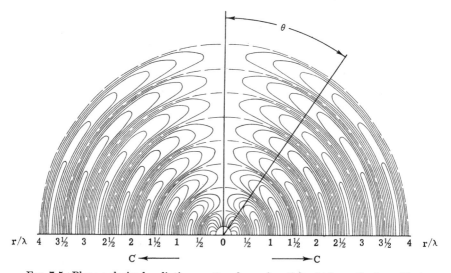

FIG. 7.5. Plane polarized radiation scattered as a function of θ by a dipole oscillating along the vertical axis. The lines represent the distribution of absolute magnitude of the electric field at an instant [after G. Joos, *Theoretical Physics*. Hafner, New York, 1950, p. 327].

cause electromagnetic waves arise from acceleration of charges, the amplitude is proportional to acceleration of the charge and to magnitude of the charge. The displacement of charge in the dipole is proportional to $\sin (2\pi\nu)(t - r/c)$, so that the acceleration is proportional to $4\pi^2\nu^2$ or to $1/\lambda^2$. The amplitude also must be proportional to the number of dipoles per unit volume or (assuming constant density) to the volume of the oscillating particle. Therefore, upon combining all these considerations the flux density may be represented by

$$F_m \propto \frac{a^6 \sin^2 \theta}{\lambda^4 r^2}$$

where a represents the radius of the volume of the oscillating particle.

In addition to the factors mentioned above, the amplitude must be proportional to the polarizability of the dielectric of which the scattering particles are composed. For a material in which the separation of dipoles is large, interaction between dipoles is negligible; and the polarizability is expressed by Eq. (7.12). But for liquids or solids interactions between dipoles make the electric field dependent on the dipoles, and a further step is necessary. In this case a dipole embedded in a dielectric and surrounded by identical dipoles experiences electric field strength which may be represented by the sum of the external field ($\mathbf{E}$) and the field ($\mathbf{E'}$) arising from the surrounding electric charges. Its polarization ($\mathbf{P}$) therefore is given by the right side of Eq. (7.12) plus an additional term which must be proportional to $\mathbf{P}$.

Imagine a minute spherical cavity in the material with a single dipole at the center. The interior surface of the cavity has a certain charge density which can be expressed by $P \cos \theta$ where θ is the angle between the direction of P and the normal to the spherical surface. From Coulomb's Law an increment of field strength contributed by a differential element of surface area $(d\sigma)$ is

$$dE' = \frac{P \cos \theta \, d\sigma}{4\pi\epsilon_0 a^2}$$

where a represents the radius of the cavity and ϵ_0 is used because the cavity is considered to be empty. Upon setting $d\sigma = a^2 \sin \theta \, d\theta \, d\phi$, replacing ϵ_0 by $1/4\pi$, taking the component in the direction of $\mathbf{P}$, and integrating with respect to ϕ

$$dE' = 2\pi P \cos^2 \theta \sin \theta \, d\theta$$

Finally, integrating from $\theta = 0$ to $\theta = \pi$ gives

$$\mathbf{E'} = \frac{4\pi\mathbf{P}}{3}$$

Instead of Eq. (7.12) the polarization now may be expressed by

$$P = N\alpha \left[E + \frac{4\pi P}{3} \right]$$

Substitution of this equation into (7.8) leads to

$$\frac{n^2 - 1}{n^2 + 2} = \frac{4\pi N\alpha}{3}$$

This shows that the polarizability is proportional to $(n^2 - 1)/(n^2 + 2)$. Therefore, the flux density of the electric field is represented by

$$F_m \propto \frac{a^6}{r^2 \lambda^4} \left(\frac{n^2 - 1}{n^2 + 2} \right)^2 \sin^2 \theta \tag{7.21}$$

7.8 Small Particles

For particles which are small in diameter compared to the wavelength of the incident radiation Mie scattering theory yields results which are understandable on the basis of the oscillating dipole of the previous section. Consider a wave which sets the electrical charges of a dipole into oscillation. The oscillating charges themselves produce an electromagnetic field which propagates outward away from the dipole with the speed of light. At a distance r from the dipole the phase lags behind the dipole phase by $2\pi r/\lambda$. Each time the oscillating charges reverse direction, the electromagnetic field in the neighborhood of the dipole acts on them in such a way as to oppose the acceleration of the charges, and the energy of the field tends to return to the dipole. However, return of the energy also occurs at the speed of light, so that some energy will not have time to return to the dipole before the phase changes once again. If the frequency is high, much of the energy of the electromagnetic wave may in this way fail to be conserved by the dipole; it is radiated away.

A hydraulic analogy may help to clarify the phenomenon of radiation. Imagine a vertical cylinder which may be alternately pushed downward into and pulled upward out of a body of water. If the frequency is low, the work done in depressing the cylinder is regained in raising it. But, if frequency is high, waves radiate from the cylinder, and mechanical energy must be continuously supplied at the cylinder. Proportionality (7.21) shows that the distribution of radiated energy is proportional to $\sin^2 \theta$, so that energy is scattered in the forward and backward directions with equal flux density and drops to zero along the dipole axis as shown in Fig. 7.5. For a particular size of scattering particle proportionality (7.21) shows that flux density is inversely proportional to the fourth power of the wavelength of the incident radiation. A similar qualitative result must be expected in the case of randomly oriented dipoles; maximum radiation is emitted in the forward and backward directions, but flux density in the normal directions falls only one-half due to contributions from obliquely oriented dipoles. The dependence of scattered intensity on wavelength was used by Lord Rayleigh* to explain the blueness of the sky and the redness of the sunset. He recognized that light scattered by air molecules is richer in blue light (short wavelengths) than in red light (long wavelengths), and the residual direct beam is correspondingly deficient in blue and therefore predominantly red.

Proportionality (7.21) also is useful in explaining scattering of radar waves by droplets and ice crystals. This application will be discussed in Section 7.14.

7.9 Diffraction

Scattering by particles whose diameter is comparable to the wavelength of incident radiation cannot be represented adequately by the dipole model. Within certain ranges the Mie theory gives accurate results, but there are ranges for which the Mie series converges so slowly that calculations are not feasible. However, an approximate method may be used to represent phenomena occurring within this size range; these are known as *diffraction* phenomena.

Two intuitively attractive "principles" are used. In each case, the empirical statement long antedates the scattering theory which now provides its theoretical basis. *Huygen's principle* states that the shape of a wave may be predicted by considering that every point of equal phase simultaneously acts as a source of a spherical wave which diverges from each of these points. For a wave surface of infinite extent all phases propagated obliquely to the wave front are cancelled by interference from phases from other sources. Only the energy propagated in directions normal to the wave surface remains; the wave propagated in the backward direction must be eliminated by a separate specification. Scattering theory shows that for particles of the size discussed here the flux density contributed by each scattering element is a maximum in the forward direction and reaches zero in the backward direction.

Now apply Huygen's principle to a plane wave incident on an obstructing edge as shown in Fig. 7.6. Construct lines BP, CP, etc., such that each line is $\lambda/2$ longer than the preceding line. All waves originating at O, C, etc., arrive at P in phase and therefore reinforce each other, whereas waves originating at B, D, etc., are out of phase with those from O and C. If the amplitude which is observed at P due to the sector between O and B is

---

* Lord Rayleigh, *Phil. Mag.* 41, 107, 274, 447 (1871).

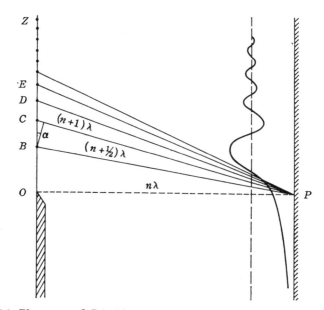

F<small>IG.</small> 7.6. Plane wave O-Z incident on an obstructing edge. The point P represents the geometrical shadow, and the curve represents the one-dimensional diffracted wave amplitude.

called A_1, the amplitude observed at P due to the sector between B and C, A_2, etc., the total amplitude due to all sectors may be written

$$A_0 = A_1 - A_2 + A_3 - A_4 + \cdots = \sum_{i=1}^{n} - A_i(-1)^i$$

But

$$A_2 \approx \frac{A_1 + A_3}{2} \qquad A_1 \approx \frac{A_{i-1} + A_{i+1}}{2}$$

and therefore

$$A_0 = \frac{A_1}{2} + \frac{A_n}{2} \approx \frac{A_1}{2} \qquad \text{for large } n$$

If the obstruction is removed, a like contribution from the lower portion of the wave surface results in an amplitude at P equal to A_1. It follows also that a greatly increased energy could be produced at P if alternate sectors were obstructed all along the wave; in this way Fresnel zones may be used to focus light much as a lens does.

Now consider the energy at a point below P. It is more convenient to imagine the obstruction moved upward, say to B. Then

$$A_B = A_2 - A_3 + A_4 - A_5 + \cdots$$

By the process demonstrated above $A_B \approx \frac{1}{2}A_2$. Upon moving the obstruction up to C, $A_C \approx \frac{1}{2}A_3$. Since the energy reaching P due to the individual sectors decreases gradually with increasing distance from P, the energy falls off gradually within the geometrical shadow.

The energy above the geometrical shadow may be investigated by moving the obstruction downward in steps. First consider it moved to $-B$. Then

$$A_{-B} \approx A_1 + \frac{A_1}{2} = \frac{3A_1}{2}$$

At $-C$

$$A_{-C} \approx A_1 - A_2 + \frac{A_1}{2} = \frac{3A_1}{2} - A_2$$

Thus, as shown in Fig. 7.6, above the geometrical shadow there are alternate maxima and minima (the first minimum in the geometrical shadow of point C), and below the geometrical shadow the amplitude falls off monotonically with distance.

If, now, an obstruction is introduced above point B, the diffraction pattern for the slit OB can be determined by analogous steps. The result is shown in Fig. 7.7. In a similar manner, an opaque obstacle produces the diffraction pattern shown in Fig. 7.8.

The second of the empirical principles to be used in this discussion is called *Babinet's principle*; it states that the disturbances produced by *complementary* screens are opposite in phase and identical in flux density.

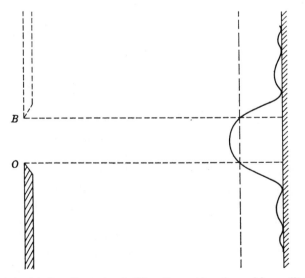

Fig. 7.7. (a) One-dimensional diffraction pattern formed by a slit OB.

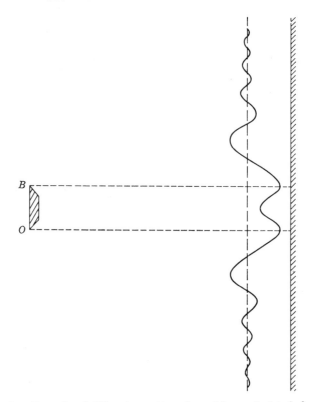

Fɪɢ. 7.8. One-dimensional diffraction pattern formed by an isolated obstacle *OB*.

Imagine a transparent screen containing one or more obstructing areas to be placed in the path of a series of wave fronts. Diffraction from the obstructions produces a disturbance in the wave amplitude reaching a particular point. Now, if the original screen is replaced by a complementary screen in which transmission occurs only in the areas which in the original screen were opaque, the diffracted wave amplitude which reaches a particular point must be disturbed in just the opposite way from that produced by the original screen. The sum of these two amplitude disturbances must be zero as would be the case with a completely transparent screen; and because flux density is proportional to the square of the amplitude, the two distributions of flux density must be identical. Babinet's principle is useful in understanding the diffraction produced by small particles illuminated by the sun. The observer sees the same diffraction pattern which would be produced by a screen with corresponding small openings.

Figure 7.6 may be used to develop a provocative relation between size of obstacle and angular radius of the diffraction fringes. When the obstructing screen was moved through one interval (from $-B$ to $-C$), the resulting intensity of P changed from a maximum to a minimum. Upon constructing a right triangle by dropping a perpendicular from B to CP, $\sin \alpha = \lambda/2BC$, and the relevant property of the distance BC is that it represents the dimension of an obstructing body. It follows that the smaller the size the larger α must be in order that the lines BP and CP differ by half a wavelength. Therefore, α represents the angular radius of the minimum flux density band, and the angular radius of the first maximum is given approximately by

$$\sin 2\alpha \equiv \sin \theta = \frac{\lambda}{a}$$

where a represents the width of the obstructing body. The series of minima called the diffraction pattern may be represented approximately by

$$\sin \theta = n\frac{\lambda}{a} \qquad n = 1, 2, 3, \ldots \qquad (7.22)$$

The analysis which has been carried out for one dimensional diffraction may be extended to diffraction in two dimensions (by circular cross section, for example). The problem is mathematically somewhat formidable, but the results differ from those developed here only in that in Eq. (7.22), n is replaced by $(n + 0.22)$.

Equation (7.22) shows that diffraction effects may occur for particles whose diameter lies between one and ten or twenty times the wavelength. For smaller particles no diffraction fringes can develop and for larger particles the angular radius of the diffraction fringes (for small n) is so small that they are very difficult to observe. In this case also the theoretical analysis becomes inaccurate. In order that diffraction fringes are well developed it is also necessary that the particle size be fairly uniform because otherwise there are no fixed phase relations among the waves emitted from the various particles.

Equation (7.22) provides an immediate and simple explanation of the *corona*, the series of colored rings of light which sometimes appear to an observer to surround the sun or moon. In these cases there is between the observer and the sun or moon a thin cloud consisting of small droplets of uniform size. Diffraction fringes of characteristic radius are formed by each of the component colors of white light with blue on the inside and red on the outside. Upon measuring the angular diameter of the corona of a particular wavelength, it is a simple matter to compute the mean diameter of the diffracting particles as is required in problem 2.

7.10   Refraction

In a fundamental sense refraction depends upon the fact that the individual scattering elements within certain media like air, ice, water, and glass scatter with an effective phase change. In media which contain bound charges (air, water, glass, etc.) the scattered wave and the incident wave are superimposed to form a wave slightly retarded over the original incident wave. The phase speed in these media is less than in vacuum. On the other hand, free electrical charges scatter with an advance in phase, with the result that the phase speed is greater than the phase speed in vacuum.

Let us look at the problem of phase change in more detail. Consider an electron in a conductor which oscillates with the incident electromagnetic wave. If it is not bound to a positive atomic nucleus, it experiences no restoring force and therefore oscillates in phase with the incident electric vector as shown in Fig. 7.9. The electric vector radiated by the oscillating

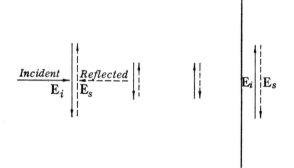

FIG. 7.9. Electric vectors (E_i) of an electromagnetic wave incident on a surface containing free electrons and the resulting electric vectors of the reflected wave (E_s).

electron, however, is opposite in direction to the electric vector of the incident wave; for there must be no net electric field in the surface of the conductor. There is, therefore, a change of phase of 180° between incident and scattered wave; this occurs both in reflection by conductors and in scattering by free charges. The effect of this phase reversal will be considered presently. First, however, it must be recognized that bound electrons may scatter without this phase change. Such electrons experience a restoring force, and, as shown in Section 7.6, they exhibit a natural frequency which increases as restoring force (or "stiffness" of the bond) increases. One may imagine an incident wave of constant frequency incident on an electron with adjustable natural frequency. As the natural frequency increases and approaches the incident frequency, the phase difference be-

tween incident electric force and acceleration of the electron decreases, the
amplitude increases, and at the resonant frequency the amplitude becomes
so large that absorption occurs. For still greater natural frequencies the
phase relations are reversed and the direction of displacement of the electron may be opposed to the direction of the incident electric vector. A
similar result occurs in a mechanical oscillator driven by a periodic force.

Now consider the effects of these two phase relations on the phase speed.
Imagine a large number of scattering particles randomly distributed, each
one of which scatters incident radiation. The electric vector at any point
is the resultant of the electric vector of the incident wave and the electric
vector of the scattered waves. The effect of the two scattering particles is
illustrated in Fig. 7.10. If scattering occurs in phase with the incident wave,
the incident wave reaches the two scattering particles P_1 and P_2 simulta-

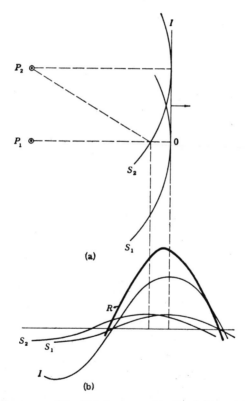

FIG. 7.10. (a) Plane wave (I) which has moved to the right past two scattering particles (P_1 and P_2) and scattered waves (S_1 and S_2) in the neighborhood of the point 0 for
scattering in phase with incident wave.

(b) Phase and amplitude relations of the incident and scattered waves
and the resultant wave (R) in the neighborhood of 0.

neously and passes on to the right with slightly diminished amplitude. Scattered waves S_1 and S_2 originate in phase with the incident wave; S_1 reaches 0 in phase with I, but S_2 is slightly retarded due to the greater distance from the point of origin. Superposition of the scattered waves and the incident wave yields the heavy curve which is retarded in phase over the curve representing the incident wave. The decrease in phase speed may be expected to be directly related to the number of scattering particles per unit volume, but there is not necessarily any decrease in amplitude of the wave.

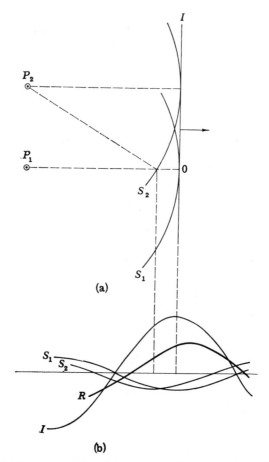

FIG. 7.11. (a) Plane wave (I) which has moved to the right past two scattering particles (P_1 and P_2) and scattered waves (S_1 and S_2) in the neighborhood of the point 0 for scattering out of phase with incident wave.

(b) Phase and amplitude relations of the incident and scattered waves and the resultant wave (R) in the neighborhood of 0.

Next, consider the case of scattering 180° out of phase as illustrated in Fig. 7.11. The incident wave reaches point 0 with slightly reduced amplitude due to out of phase scattering from P_1. At the moment that the incident wave reaches 0 with phase as shown, the scattered wave from P_2 is slightly retarded. Superposition of this scattered wave and the incident wave yields the heavy curve which is advanced in phase over the incident wave. Phase reversal also decreases the amplitude of the resultant wave as the scattered wave from P_1 illustrates, so that an electromagnetic wave can penetrate into a region of free electrical charges only a certain critical distance. At the same time each scattering event contributes radiation in the reverse direction; this is called *reflection*. In metals, charge density is so great that penetration is negligible, and we say that reflection occurs at the metallic surface.

Advance in phase by scattering results in increase in phase speed over the speed in vacuum. The phase speed, however, is a geometrical abstraction, and this result does not imply that energy can be transmitted at speeds greater than the speed in vacuum.

The difference in phase speed of a wave when passing from one medium to another leads to a geometrical relation known as Snell's law. The series of lines in Fig. 7.12 represent the successive positions of a single wave

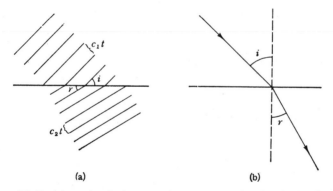

(a) (b)

FIG. 7.12. (a) Positions of a single wave phase at successive time intervals separated by t. The phase speed in the upper medium is c_1, in the lower medium c_2.

(b) The geometry of refraction of a ray drawn normal to successive positions of the wave phase.

phase at a series of equal time intervals. If the speed of the wave phase changes from c_1 to c_2 in passing the interface between the two media, then it is obvious from the diagram that

$$\frac{c_1}{c_2} = \frac{\sin i}{\sin r} \qquad (7.23)$$

where i represents the angle of incidence and r the angle of refraction. The bending of the wave train or "beam" which occurs in passing from one medium to another in which phase speed is different is called *refraction*. The apparent bending of a stick thrust into water at an oblique angle is a familiar example of refraction.

PART III: ATMOSPHERIC PROBING

7.11 Visual Range

The distance an object can be seen is called the *visual range;* the horizontal visual range near the surface of the earth is commonly called the *visibility*. Discussion of these atmospheric properties involves the characteristics of the human eye as well as the optical properties of the atmosphere.

The human eye is sensitive to light, the range of electromagnetic radiation between 0.4 and 0.7 micron. Within this range the sensitivity is strongly dependent on wavelength, and Fig. 7.13 indicates that green

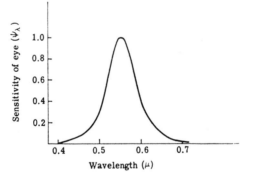

FIG. 7.13. The sensitivity curve of the human eye.

light of a certain intensity appears much brighter to the eye than does violet or red light of the same intensity. The sensitivity (ψ_λ) of the eye at wavelength λ is defined by the ratio of intensity at 0.555 micron, where the eye has maximum sensitivity, to intensity at wavelength λ required to yield equal impressions of brightness. The *monochromatic brightness* is defined by

$$B_\lambda = \psi_\lambda I_\lambda$$

where I_λ represents the monochromatic intensity emitted or reflected by the object. The *brightness* is defined by $B \equiv \int_0^\infty B_\lambda \, d\lambda$.

In order for an object to be visible there must be *contrast* between it and its surroundings. If B represents the brightness of the object and B_0 the brightness of the background, then the contrast of the object against its surroundings is defined by

$$C \equiv \frac{B - B_0}{B_0} \tag{7.24}$$

It is clear that this definition is insufficient to distinguish between two objects with different colors but the same brightness. However, this definition is adequate so long as only the contrast between "black" and "white" is considered.

In order to determine visual range a black object may be chosen which is large enough that the angle under which it appears can easily be resolved by eye. This requires at least 1 minute of arc. At short range B is very small and the contrast is close to minus one. At a greater distance the black object appears to have a certain brightness because light is scattered from all directions into the direction from object to observer. The apparent brightness increases with the distance to the object; the contrast becomes so small that the human eye is no longer able to perceive it at the visual range. For the normal eye the threshold contrast has a value of ± 0.02.

In general the intensity of the background varies along the line from object to observer as a result of the sum of attenuation and enhancement due to scattering of radiation from other directions. Under fairly uniform conditions the gains and losses are equal. However, the apparent brightness of a black object increases with distance, and according to Beer's law may be expressed by

$$B_\lambda = B_{s\lambda}(1 - e^{-\sigma_\lambda x})$$

where $B_{s\lambda}$ is the monochromatic brightness of the sky. The total apparent brightness of the black object is given by

$$B = \int_0^\infty B_{s\lambda}(1 - e^{-\sigma_\lambda x}) \, d\lambda$$

and the contrast is given by Eq. (7.24) in the form

$$C = -\frac{\int_0^\infty B_{s\lambda} e^{-\sigma_\lambda x} \, d\lambda}{B_s}$$

Assuming σ_λ to be independent of λ, x may be expressed by

$$x = \frac{1}{\sigma} \ln |C|$$

And taking $C = -0.02$ the visual range becomes

$$x_v = \frac{3.912}{\sigma} \tag{7.25}$$

Equation (7.25) states an explicit relation which permits calculating the extinction coefficient if the visual range is known. The assumption that σ_λ is independent of wavelength is not quite valid because, as pointed out in Section 7.8, air molecules scatter the blue light more strongly than the red; for this reason distant mountains appear to be blue and distant snowfields yellow. When the scattering is caused by larger particles, dust and small droplets with a size of the order of the wavelength of light, the assumption is much better. So in most cases with poor visibility the visual range is independent of wavelength.

At night it is possible to use artificial light sources for the determination of the extinction coefficient; and this value used in Eq. (7.25) leads to a prediction of the visual range for daytime. The concept of contrast loses its usefulness at night because the light source is many times brighter than the background sky, and the contrast becomes a very large but uncertain quantity. However, the extinction coefficient may be found by applying Beer's law to two identical light sources at two different distances, x_1 and x_2. The ratio of the flux densities is then

$$\frac{F_1}{F_2} = \frac{x_2{}^2}{x_1{}^2} \exp \left[\sigma(x_2 - x_1) \right]$$

or

$$\sigma = \frac{1}{x_2 - x_1} \left(\ln \frac{x_1{}^2}{x_2{}^2} + \ln \frac{F_1}{F_2} \right)$$

An exercise is given in problem 3.

7.12   Determination of Absorbing Constituents

Comparison of absorptivities of atmospheric constituents with corresponding absorptivities measured in the laboratory provides a measure of the total mass of these constituents provided the absorption bands are selected carefully. Inaccuracy is introduced because the absorption coefficient is dependent on temperature and pressure. If these measurements are made over short paths, the concentration of absorbing gas may be determined. This method has been used for determination of the concentration of water vapor and carbon dioxide.

It is also possible to compare the absorptivities at two different wavelengths for which the absorbing constituent has widely different absorption coefficients. This technique has been developed by R. C. Wood[*] for the measurement of water vapor. Using interference filters and a single radiation source, the intensities of wavelengths λ_1 and λ_2 are observed at a fixed distance represented by x. Beer's law then yields

[*] R. C. Wood, *Rev. Sci. Instr.* **29**, 36 (1958).

$$q = \frac{\rho x}{k_2 - k_1} \left(\ln \frac{I_1}{I_2} - \ln \frac{I_{10}}{I_{20}} \right)$$

where k_1 and k_2 are the absorption coefficients at λ_1 and λ_2, respectively, and the 0 subscripts indicate intensities without attenuation by water vapor. For accurate results two wavelengths of very different absorptivities should be chosen. The method is particularly useful in measurement of very low humidities. Why? The method is direct in that it depends only on properties of the air rather than on properties of a probe. It may be applied to measurement of space and/or time averages or to measurements of rapid fluctuations in humidity.

7.13   Optical Measurement of Lapse Rate

It may be recognized from Eq. (7.16) that because the index of refraction depends on the number of scattering particles per unit volume, the normal decrease of atmospheric density with height results in refraction of light. In order to discuss this effect quantitatively or even to determine in which direction refraction occurs, the relation of the frequency of light (ν) to the natural frequency of the scattering dipoles (ν_0) must be known. The natural frequencies of oscillation of the electrical dipoles created from atmospheric gases lie in the ultraviolet range. Therefore, upon recognizing that $k \ll n - 1$ and $\nu_0^2 \ll \nu^2$, Eq. (7.16) gives for the index of refraction for a particular constituent of the atmosphere

$$n - 1 = \frac{Ne^2}{\pi m \nu_0^2} \tag{7.26}$$

Among the atmospheric constituents only water vapor undergoes appreciable variation, so that Eq. (7.26) may be generalized by writing two terms of the form shown at the right-hand side, the first representing the contribution of dry air and the second the contribution of water vapor. Each is proportional to the corresponding density, hence

$$n - 1 = a\rho + b\rho_w/T \tag{7.27}$$

The temperature enters the second term because the water molecule is polar, and therefore the contribution of water vapor to the index of refraction is temperature dependent. For low temperature the permanent dipoles are readily aligned by an incident electric field, and the contribution to polarization and to index of refraction is large, whereas at high temperatures the probability of a molecule being polarized is small. The coefficients a and b are characteristic of dry air and water vapor, respectively, and their values depend on wavelength. For visible light a is approximately 0.234 g^{-1} cm^3 at standard pressure and temperature, whereas b is negligible.

For standard pressure and temperature the index of refraction is about 1.000293. For a wavelength of two centimeters a again is approximately 0.234 g^{-1} cm^3, and b is about 1.75×10^3 g^{-1} cm^3 °C. For infrared wavelengths the natural rotational frequency of the water vapor molecule is comparable to the incident frequency, and the index of refraction cannot be represented by Eq. (7.26). This is the region in which anomalous dispersion and absorption occur.

If, for visible light, the second term in Eq. (7.27) is neglected, the equation states that for the normal case of decrease of density with height the upper portion of a wave moves faster than the lower portion, and refraction toward the earth occurs, as shown in Fig. 7.14. It may be concluded

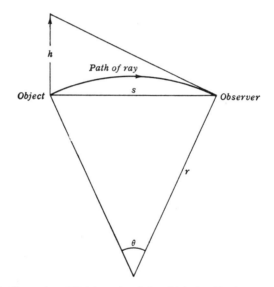

FIG. 7.14. Geometry of light ray in air in which density decreases upward.

that, in general, our friends are not so tall as they appear to be, although this depends on atmospheric conditions. Measurement of the radius of curvature of the light ray provides a measure of the gradient of density normal to the light path. In order to express the radius of curvature in terms of the gradient of density, note that the ray is everywhere normal to the wave phase which travels from the object to the observer in time expressed by s/c. This time must be independent of r. Thus, $\theta r/c$ is constant, and

$$\frac{1}{c}\frac{dc}{dr} = \frac{1}{r} \tag{7.28}$$

Now, $\cos \theta$ is expanded in a Taylor series and only the first two terms are retained with the result

$$\cos \theta = \frac{r}{r + h} = 1 - \frac{s^2}{2r^2}$$

or, approximately

$$\frac{1}{r} = \frac{2h}{s^2} \tag{7.29}$$

Notice that r is positive for refraction toward the earth, negative for refraction away from the earth.

For a nearly horizontal ray the r and z directions are nearly the same so that Eqs. (7.28) and (7.29) may be combined to give

$$\frac{dc}{dz} = \frac{2hc}{s^2} \tag{7.30}$$

Equation (7.30) shows that h, the difference between true and apparent height, is proportional to the rate of change with height of the speed of light in the atmosphere. Equations (7.27) and (7.8) together yield

$$\frac{dc}{dz} = -\frac{c(c_0 - c)}{c_0 \rho} \frac{d\rho}{dz}$$

Therefore, upon substituting the equation of state and the hydrostatic equation

$$\frac{dc}{dz} = \frac{c(c_0 - c)}{c_0} \left\{ \frac{\rho g}{p} + \frac{1}{T} \frac{\partial T}{\partial z} \right\} \tag{7.31}$$

Equations (7.31) and (7.30) may be combined, giving

$$\gamma \equiv -\frac{\partial T}{\partial z} = -\frac{2nTh}{s^2(n - 1)} - \frac{g}{R_m} \tag{7.32}$$

Thus the temperature lapse rate depends directly on the difference between the actual height of an object and its apparent height. An application is given in problem 4. A solidly mounted telescope focused on a fixed target provides a satisfactory instrument for measuring changes in lapse rate with time. It is important that the telescope and mount be shielded from the sun to prevent temperature changes from introducing large errors.

If two telescopes separated by a convenient vertical distance, say one meter, are focused on targets separated by the same distance, the difference in lapse rates between the two light rays is easily measured. An artillery range finder is a convenient instrument for this purpose. The instrument may be moved vertically in steps to provide a vertical profile of the lapse rate very near a uniform surface. Detailed temperature profiles determined over relatively cold water and over hot land by this method are shown in

Fig. 7.15. The great virtue of these observations is (a) that they are direct measures of average atmospheric properties rather than of properties of a thermometer and (b) that measurement does not influence the air temperature.

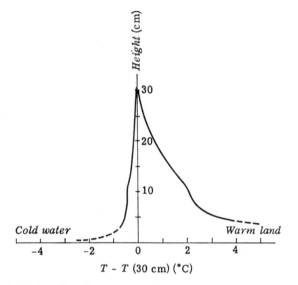

FIG. 7.15. Vertical profiles of temperature determined by optical measurements over cold water and over warm land [after R. G. Fleagle, *J. Meteorol.* **13**, 160 (1956); *Geophys. Research Papers* **No. 59**, Vol. II, 128 (1959)].

Due to the dependence of index of refraction on water vapor the gradient of water vapor might be measured in a similar way. To do this a filter might be used to admit a narrow range of wavelengths for which the second term on the right-hand side of Eq. (7.27) is important, say the infrared wavelengths near 2.5 microns. Simultaneous observation of light refraction and infrared refraction would enable one to solve two simultaneous equations for the lapse rates of temperature and vapor density.

7.14 Radar Investigation

Scattering of Radar Waves

For the case of back scatter ($\theta = \pi/2$) of electromagnetic radiation from particles small compared to the wavelength of the incident radiation, proportionality (7.21) may be written

$$F_m \propto \frac{a^6}{r^2 \lambda^4} \left(\frac{n^2 - 1}{n^2 + 2} \right)^2$$

Strong dependence on size of scattering particle and strong dependence on wavelength are dominant characteristics of this proportionality. Suppose the wavelength is held constant at roughly the centimeter range, and imagine the size of scattering particle to change. For very small particles the energy scattered is extremely small, but as size increases energy increases markedly. For millimeter particles (falling rain or ice crystals) enough energy is scattered that sensitive radars may be used to detect and study water distribution in clouds. In order to detect smaller cloud particles (10–100 μ) millimeter radar may be used. A limitation is introduced by the great difficulty which exists in building radar equipment capable of emitting sufficient energy at wavelengths much shorter than one centimeter and also by the increasing attenuation which occurs as wavelength is decreased. Consequently, radar investigations are rather sharply limited to the larger cloud particles, and most observations have been made with wavelengths which detect only precipitating particles. Clouds which do not contain precipitation-size droplets are nearly transparent to centimeter radar, so that attenuation of the radar beam is important only in proportion to the density of the strongly scattering (large) particles.

The dependence of scattered energy on index of refraction introduces a new and somewhat surprising phenomenon. Although for the wavelengths of visible light, water and ice have almost identical indices of refraction, for wavelengths of the order of centimeters the index of refraction for water is much the larger, and the ratio of the two increases from about two to about five as wavelength increases from one to five centimeters. It follows that water drops are about five times as effective as ice crystals in scattering radar waves.

The large index of refraction for water at radar wavelengths results from the fact that each individual water molecule possesses a dipole moment and that these dipoles can be lined up by electric fields, thereby adding to the polarization of the drop as a whole. This parelectric contribution to polarization is important only for frequencies low enough that the dipoles can be lined up by the electric field before reversal of the field occurs. Liquid and gas molecules are free to move, so that this lining up of dipoles can occur readily, whereas crystal molecules are fixed in position. Therefore, ice has a lower index of refraction than water. At the frequency of light waves the molecular dipoles do not respond to the electric field, and so at these frequencies the optical indices of refraction of ice and water are practically identical. Several interesting results follow. Radar "sees" liquid droplets of millimeter size clearly but does not see ice crystals of comparable size. When ice crystals fall into air above the zero isotherm, melting occurs and radar detects the melting as a fairly sharp bright band as shown in Fig. 7.16. The band is brighter than the region below the zero

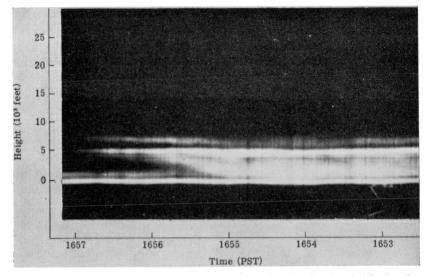

FIG. 7.16. Vertical distribution of echo intensity of radar showing bright band at University of Washington, 8 April 1960.

isotherm because the raindrops fall faster than the original ice crystals; these are therefore farther apart in the water phase than in the ice phase. Coalescence and change in radar cross section of the particles may enhance the conspicuousness of the bright band.

Curiously, for ratio of diameter of ice particle to wavelength between 1 and 2 back scattering from ice exceeds that from water of the same volume by as much as several orders of magnitude. This unexpected result which applies only to fairly large hailstones, was observed* and calculated† independently. As the ice acquires a coating of water in melting, back scattering may drop sharply.

The Use of Radar

Radar waves are emitted as pulses of electromagnetic waves of about a microsecond duration interrupted by an "off period" of about a millisecond. During this interval the preceding pulse travels to the target and back to the receiver; because the receiver usually utilizes the transmitting antenna, the maximum range is limited to the product of the speed of light and half the period between pulses. The emitted pulse is of high energy (up to 10^2 kilowatts), whereas the signal received after reflection from the

---

* D. Atlas, W. G. Harper, F. H. Ludlam, and W. C. Macklin, *Quart. J. Roy. Meteorol. Soc.* **86**, 468 (1960).

† B. M. Herman, and L. J. Battan, *Quart. J. Roy. Meteorol. Soc.* **87**, 223 (1961).

target may be only 10^{-12} to 10^{-13} watts; therefore, the transmitter must be off when the receiver is on and the receiver off when the transmitter is on. Radar waves are emitted in beams usually several degrees in width; for some purposes wider beams are used and for others very narrow beams may be used.

The radar receiver consists of a cathode ray tube in which a beam of electrons scans the tube and produces a visible spot when reflected energy is received from the target. The coordinates of the scope may be designed to provide information relating to distance (proportional to time between emitted pulse and received pulse) and direction (orientation of the trans- mitter). The following types of receivers are in wide use:

1. The A or R scope scans horizontally or vertically (but not both) across the tube and produces spots indicating distance of the target from the transmitter. The antenna is fixed in position. It is particularly useful in monitoring the height and thickness of clouds of precipitation droplets directly overhead.
2. The plan-position indicator (PPI) provides a plan view of the area around the transmitter on a polar coordinate grid. The antenna ro- tates in a horizontal plane, and the screen should retain its fluorescence long enough for the antenna to complete a full revolution. The PPI is particularly useful in describing the horizontal distribution of liquid water in storms ranging in size from one to a thousand kilometers.
3. The range-height indicator (RHI) provides a view of a vertical cross section with the transmitter at the lower corner. The antenna scans in the vertical plane. The RHI describes the vertical distribution of liquid water and is therefore particularly useful in cloud physics investigation.
4. Doppler radar provides direct measurements of velocity of targets along the line of sight of the radar beam based on the frequency differ- ence between emitted and reflected beam resulting from movement of the target.

Radar observations may provide valuable information in many different investigations in addition to direct description of the boundaries of regions containing liquid water or ice in large concentration. The reflected signal can be delicately amplified to provide quantitative measurements of liquid water content of clouds, or of rainfall, and to determine with high-resolu- tion changes in these concentrations with time. Observations of this sort have been useful in evaluating the importance of the ice crystal (Wegener- Bergeron) process and the coalescence process in precipitation.

The detailed structure and development of convective clouds has been described by radar observations. Not only have changes in cloud structure been observed, but by inference much has been learned about the wind velocity distribution including turbulence within the cloud. The development and movements of hurricanes, thunderstorms, squall lines, and tornadoes have been observed. Lightning can be detected by radar, and careful evaluations of the observations have provided unique data relating to the mechanism of lightning. Many hours have gone into radar monitoring of cloud seeding trials. Productive research in these difficult areas would be virtually impossible without radar.

Finally, radar is used to measure wind velocity by tracking targets carried by the wind (balloons with radar targets, strips of metal foil called "chaff," or cloud elements). Velocities may be determined from observations of target position on the scope at successive times or by measuring the Doppler shift in frequency associated with movement toward or away from the transmitter.

The Doppler shift may be developed from the following considerations. Imagine that the target is moving toward the train of waves and is reflecting the train back toward the transmitter. Because frequency is related to speed and wavelength by $\nu = c/\lambda$, the target intercepts waves with frequency given by

$$\nu' = \frac{c + v}{\lambda}$$

where v represents the speed of the target toward the transmitter. Thus, the change in frequency observed by the target is v/λ. An additional and identical change in frequency occurs at the receiver because successive reflected waves are "emitted" at successively closer positions, so that the total change in frequency is given by

$$\Delta\nu = \frac{2v}{\lambda} \tag{7.33}$$

In spite of the great usefulness of radar as an atmospheric probe, there are severe difficulties in its application. The basic difficulty is that the reflected signal depends on so many uncontrolled and unobserved quantities: concentration, size distribution, shape, temperature and phase of reflecting particles, distance and attenuation of signal between transmitter and target. In addition, as emphasized earlier in this chapter, there are complex problems in the detailed theoretical treatment of scattering as a function of wavelength and size of the scattering particles which have not yet been completely solved. Finally, there are special atmospheric properties which result in refraction of radar waves and which may greatly modify the capabilities of the instrument.

Refraction of Radar Waves

To determine the radius of curvature of a horizontal radar ray, express Eq. (7.28) in the form

$$\frac{1}{n}\frac{dn}{dr} = -\frac{1}{r} \tag{7.34}$$

Then, upon differentiating Eq. (7.27), introducing the equation of state and the hydrostatic equation, and replacing ρ_w by ρq

$$\frac{1}{r} = \frac{n-1}{n}\left(\frac{g}{R_m} + \frac{\partial T}{\partial z}\right) + \frac{b\rho q}{nT}\left(\frac{1}{T}\frac{\partial T}{\partial z} - \frac{1}{q}\frac{\partial q}{\partial z}\right) \tag{7.35}$$

For radar waves, as stated in Section 7.13, b is equal to about 1.75×10^3 g^{-1} cm^3 °C. The first term on the right is nearly identical in the cases of light and radar waves; the second term is proportional to the specific humidity and may be more important than the first. An exercise is given in problem 5. Under normal conditions of decrease of temperature and humidity with height the radius of the ray is about 4/3 the radius of the earth; consequently, radar usually "sees" beyond the horizon, but its range is limited.

For a rather modest humidity lapse or temperature inversion a horizontal ray may be bent with curvature equal to the curvature of the earth, and the range of the radar is then limited only by its power and sensitivity. Under stronger humidity lapse or temperature inversion even rays emitted with an upward component may be refracted back toward the earth; they are said to be trapped in a radar *duct*, as illustrated in Fig. 7.17. Ducts may

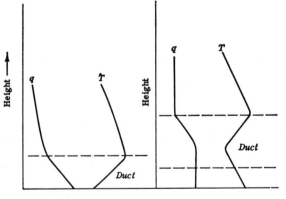

Specific humidity, temperature

Fig. 7.17. Vertical distributions of temperature (T) and specific humidity (q) which act as radar ducts.

develop just above the earth, for instance, when warm dry air moves over cool water or when moist ground cools by nocturnal radiation. Or elevated ducts may form under dry air which is subsiding or when turbulence creates a lapse layer in warm air moving over cold water. Usually, ducts at heights of a kilometer or more are only important in interpretation of radar observations made from aircraft, because signals emitted from the ground are incident on the elevated ducts at angles such that refraction is ineffective.

Within and in the vicinity of clouds strong gradients of temperature and humidity may develop in the horizontal direction as well as in the vertical. It is obvious that anomalous radar observations must be expected in these regions and that a general quantitative account is hardly feasible.

7.15   Radio Investigation of the Ionosphere

In Chapter IV the production of ions by absorption of ultraviolet radiation from the sun was discussed, and in Chapter VI some of the properties of the ionosphere were discussed. Here a brief account of the method of observation of the ionosphere will be given.

Free electrical charges respond to electromagnetic waves without experiencing the restoring force of the electrical dipole. They consequently have no natural frequency, and the index of refraction is given by Eq. (7.20). Upon introducing Snell's law

$$\sin^2 i = \sin^2 i' \left(1 - \frac{Ne^2}{\pi m \nu^2} \right)$$

where i represents the angle of incidence at the earth's surface and i' represents the angle of incidence where the concentration of ions is N. As illustrated in Fig. 7.18, the beam is refracted so that i' becomes $\pi/2$ at the highest point of the ray. Therefore, for vertical incidence, $\sin i = 0$ and

$$N = \frac{\pi m \nu^2}{e^2} \tag{7.36}$$

Equation (7.36) states that if vertically directed radio waves of frequency ν encounter N ions per cubic centimeter having a charge to mass ratio e/m, they will be refracted or reflected back toward the earth. The height of the refracting layer is determined by emitting short pulses of energy and measuring the time required for the pulse to make the round trip.

Although the original detection of the ionosphere was made using emitter and receiver separated as suggested in Fig. 7.18, it is more convenient to locate them together, and this is now the normal procedure. The incident waves penetrate the ionized region with steadily decreasing amplitude in the forward direction and steadily increasing amplitude in the backward

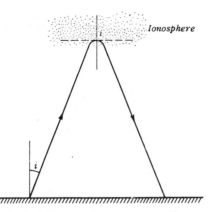

FIG. 7.18. Idealized path of a radio ray reflected by the ionosphere.

direction until the wave is said to have been reflected. So it may be recognized that in this case reflection and refraction are alternative concepts of equal validity. However, as stated in Section 7.5 the ray treatment of refraction is only an approximate description of the general problem of scattering from distributed particles.

Equation (7.36) permits calculation of the vertical distribution of ions if the charge and mass of the ions responsible for reflection are known and if the concentration of ions increases with height. Notice that since the mass of an electron is only $1/1837$ the mass of a proton, electrons are very much more effective in reflecting radio waves than are positive ions or heavy negative ions, and ion concentrations are usually reported as equivalent electron concentrations. The distribution of electrons is found by varying the radio signal frequency from a low to a high value. At low frequency small values of N result in reflection, so that the wave is reflected from a relatively small height. As frequency is increased, the beam penetrates to greater heights, and each frequency represents a unique value of N. Finally, beyond a critical frequency the beam does not encounter the concentration required for reflection and therefore passes through the ionosphere and escapes to space. A typical record on which distinct layers are evident is shown in Fig. 7.19.

Some sixty ionospheric monitoring stations distributed over the world keep continuous watch on the ion distribution in the atmosphere. In addition to providing the data needed for the description of the ionosphere, which is essential for the maintenance of radio communication, these observations have made possible complex calculations of temperature, of composition of the air, and of properties of the earth's magnetic field.

Communication over great distances is possible because reflection from

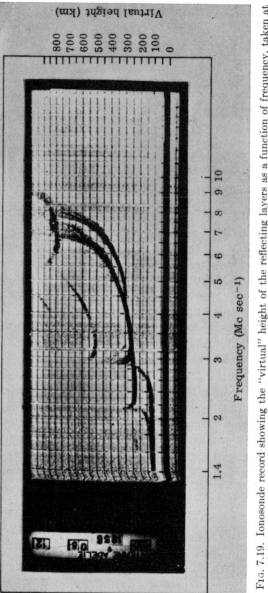

Fig. 7.19. Ionosonde record showing the "virtual" height of the reflecting layers as a function of frequency, taken at Terre Adelie, 8 Dec. 1956, 0900 GMT, lat. S 66° 49', long. E 141° 24' [after J. W. Wright and R. W. Knecht, Atlas of Ionograms, Nat. Bur. Standards, Boulder, Colorado (1957)].

the ionosphere can be utilized as shown in Fig. 7.20. At a small angle of incidence the beams of radio energy may penetrate the ionosphere, whereas at a larger angle they are reflected. Thus, communication is possible far beyond line of sight, but there may be a "skip distance" within which

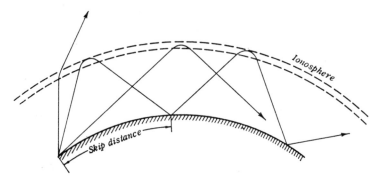

FIG. 7.20. Vertical cross section through the ionosphere showing paths of reflected and refracted radio beams.

signals cannot be received. Multiple reflections between the conducting earth and the ionosphere may make possible communication half-way or even all the way around the earth. Optimum frequency is chosen for the relevant distance and ionospheric condition; and if ion concentration changes in response to change in the ultraviolet emission from the sun, communication may be interrupted until the frequency of the transmitter is changed to a new optimum value. Interruption occurs, of course, if ion concentration falls so low that the beam penetrates the ionosphere. Communication also fails if ion concentration increases markedly in regions of relatively high air density, for radio energy is absorbed by the ions if they are so close to neutral air molecules that collisions occur during their forced oscillation. In this case a shift to a high frequency which normally would penetrate the ionosphere may restore communication.

7.16   Acoustic Refraction

Upon combining Eq. (7.4) with Snell's law it may be recognized that refraction of sound waves must occur where temperature gradients exist. For example, the path of a sound "ray" may be calculated as follows. Imagine the atmosphere to be composed of a large number of horizontal layers each differing in temperature from the adjacent layers. Acoustic waves experience refraction at the boundaries between layers as shown in Fig. 7.21. From Snell's law for any two interfaces the following relation must hold

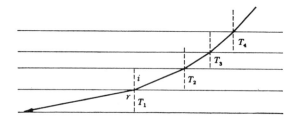

Fig. 7.21. Path of an acoustic ray in passing obliquely through a series of isothermal air layers for the case in which temperature (T) decreases with height.

$$\frac{\sin i_1}{c_1} = \frac{\sin r_1}{c_2} = \frac{\sin i_2}{c_2}, \qquad \text{etc.} \tag{7.37}$$

Thus, the ratio of the sine of the angle of incidence to speed is constant all along the ray, and this holds for continuous as well as for discontinuous variation of temperature. Substitution of Eq. (7.37) into Eq. (7.4) gives

$$\frac{\sin i}{\sqrt{T}} = \text{constant}$$

so long as $(c_p/c_v)R_m$ is constant along the ray. To evaluate the constant, T and i must be known at some point along the ray. For a ray which becomes horizontal at some point, this equation yields

$$\sin i = \sqrt{\frac{T}{T'}}$$

where T' represents the temperature at the point of horizontal incidence. It follows from the Pythagorean theorem that

$$\tan i = \sqrt{\frac{T}{T' - T}}$$

If it is assumed that the temperature is a linear function of height expressed by $T = T' - \gamma z$ where T represents the temperature at any reference height (usually, the ground) and if $\tan i$ is replaced by dx/dz

$$dx = \sqrt{\frac{T}{\gamma z}}\, dz \tag{7.38}$$

Equation (7.38) may be integrated (usually by numerical means) to give the path of the ray. An application is given in problem 6.

Equation (7.38) has been used in the original detection of the stratospheric temperature inversion. In the months following the First World War large quantities of explosives were detonated at various places in Germany. It was observed that several approximately concentric circles could be drawn separating alternate regions in which explosions were audible at the ground from those in which the explosions were inaudible. As shown in Fig. 7.22, within the central circle the "ground" wave was audible, but at greater radii attenuation and upward refraction made the ground wave inaudible. At still greater radii waves were apparently re-

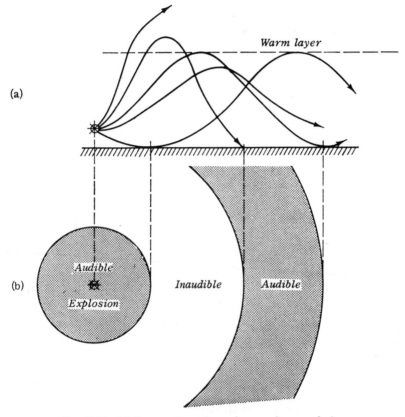

FIG. 7.22. (a) Propagation of sound waves from explosion.
(b) Distribution of audibility zones.

fracted toward the earth by a layer of warm air at considerable height above the ground. Measurement of the elapsed time between explosion and arrival of the sound waves permits an estimate of the height and temperature of the refracting layer. These calculations supplemented by the results of later experimental detonations show that the temperature at 50 kilometers above the earth in middle latitudes may be as high as 325°K during daytime. Usually the effect of refraction by windshear has also to be taken into account, which complicates the interpretation of the observations.

Further examples of acoustic refraction are discussed in Part IV of this chapter.

7.17　The Sonic Anemometer-Thermometer

Sound waves travel relative to the air in which they move. If the speed of sound waves is measured in opposite directions with respect to a fixed system, then the difference between the two speeds is twice the wind speed in the direction of measurement. The sum of the two speeds is equal to twice the speed of sound and therefore provides a measure of the temperature. Suomi* has developed on this basis an instrument which measures the rapidly fluctuating temperature and wind velocity components and which is illustrated in Fig. 7.23. An array of two sound-pulse emitters, E_1

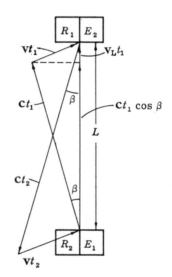

Fig. 7.23. The essential geometry of the sound rays emitted by a sonic anemometer-thermometer. The emitters are indicated by E and the receivers by R.

and E_2, and two receivers, R_1 and R_2, are placed so that E_1 opposes R_1 and E_2 opposes R_2. The distance between E_1 and R_1 and between E_2 and R_2 is L, whereas the distance between E_1 and R_2 and between E_2 and R_1 is very small. In order for a sound pulse emitted by E_1 to arrive at R_1 it has to be emitted in the direction ct_1 because the wind carries the pulse in the same time t_1 to R_1 along vt_1. Solving for t_1 yields

$$t_1 \doteq \frac{L}{c \cos \beta - v_L}$$

* V. E. Suomi and J. A. Businger, *Geophys. Research Papers* No. 59, Vol. III, 1 (1959).

Similarly, the transit time of a pulse from E_2 to R_2 may be expressed by

$$t_2 = \frac{L}{c \cos \beta + v_L}$$

Therefore, after realizing that $c^2 \cos^2 \beta = c^2 - v_n^2$ and $v_n^2 + v_L^2 = v^2$

$$t_1 - t_2 = \frac{2Lv_L}{c^2 - v^2}$$

For $v \ll c$, the component of wind speed in the L direction is approximately

$$v_L = \frac{(t_1 - t_2)c^2}{2L} \tag{7.39}$$

Similarly

$$t_1 + t_2 = \frac{2L(c^2 - v_n^2)^{1/2}}{c^2 - v^2} \approx \frac{2L}{c}$$

Upon combining this equation with (7.4)

$$T = \frac{4c_v L^2}{c_p R_m (t_1 + t_2)^2} \tag{7.40}$$

Equations (7.39) and (7.40) are accurate for dry air to within 1% for wind speeds up to 30 m sec^{-1}. The temperature determination using Eq. (7.40) is strictly valid only for dry air, and the extension necessary for humid air is considered in problem 7.

Simultaneous observation of the instantaneous temperature and the vertical wind component over the same path makes the sonic anemometer-thermometer suitable for the determination of the convective heat flux directly by use of Eq. (5.9).

PART IV: NATURAL SIGNAL PHENOMENA

Parts I, II, and III of this chapter provide the background from which many natural phenomena may be understood. In the following sections only a few of the naturally occurring signal phenomena are discussed. Many questions are left to the reader to answer or to wonder about. For example, why is the blue of the sky deepest in directions at right angles to the sun? Why are distant snow-covered mountains somewhat orange, whereas forest-covered mountains appear blue?

7.18 Refraction of Light by Air

Mirages

A very common form of mirage is observed when the earth's surface is much warmer than the air. The lapse rate then is very steep close to the surface and decreases rapidly with height; see, e.g., the temperature profile in Fig. 7.24. Equation (7.31) relates the vertical gradient of the speed of light to the lapse rate. From this equation it may be deduced that the curvature of the light ray is largest where the lapse rate is steepest. In

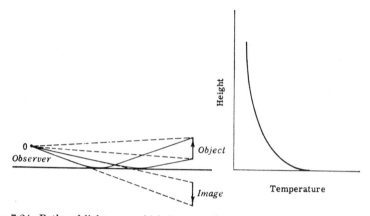

FIG. 7.24. Paths of light rays which form an inverted inferior mirage over a warm surface and typical temperature profile accompanying it.

Fig. 7.24 a number of nearly horizontal rays are drawn from a distant object to the observer to illustrate the development of an *inferior mirage* over a warm surface. This phenomenon may be observed almost any sunny day when driving on the highway and accounts for the sad experience of a pelican found on an asphalt highway near Wichita, Kansas.* "The miserable bird had obviously been flying, maybe for hours, across dry wheat stubble and had suddenly spotted what he thought was a long black river, thin but wet, right in the midst of the prairie. He had put down for a cooling swim and knocked himself unconscious." He was discovered by a local farmer and now enjoys the duckpond in the Wichita zoo.

The *superior mirage* may be observed when there is a strong inversion layer at a height of ten meters or so lying above a layer which is roughly isothermal. The light from distant objects is virtually reflected by the inversion layer. This phenomenon is rather rare because the required atmospheric conditions do not develop easily. There are many conditions under which complex mirages may occur, and it often is a challenge to explain the phenomenon one sees.

Shimmer and Twinkle of Stars

When the air is turbulent and eddy density gradients are large the light path from an object to the observer fluctuates rapidly, giving the object the

* C. A. Goodrum, *The New Yorker* **38** (8), 115 (1962).

appearance of continuous motion. When air flows over a very much warmer or very much colder surface this phenomenon is marked and is called *shimmer*. Less pronounced density fluctuations, occurring, however, through the depth of the atmosphere, are responsible for twinkling of stars.

Astronomical Refraction

The refraction of light from outside the atmosphere may be calculated with the help of Fig. 7.25. The angular deviation (δ) of the light ray in

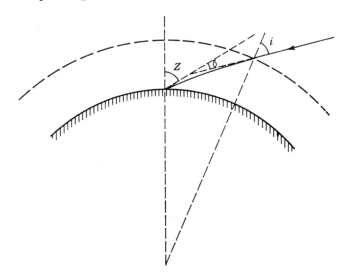

FIG. 7.25. Path through the atmosphere of a light ray from a star.

passing through the atmosphere is found by recalling from Snell's law that $(\sin i)/c$ is constant along the ray. Thus

$$\frac{\sin i}{c} = \frac{\sin Z}{c'}$$

The problem may be solved quite generally, but the trigonometry is somewhat tedious, so only the approximate solution which neglects the curvature of the atmospheric pressure surfaces will be discussed. Under these conditions the radii in Fig. 7.25 may be considered parallel and $i = Z + \delta$. Therefore

$$\frac{\sin Z \cos \delta + \cos Z \sin \delta}{\sin Z} = \frac{c}{c'}$$

Since δ is very small, $\cos \delta \approx 1$ and

$$\sin \delta = (n - 1) \tan Z \tag{7.41}$$

and because $n \approx 1.000293$, $\delta \approx 2.93 \times 10^{-4} \tan Z$. For zenith angle of 80° this gives a δ of $5' \, 48''$ which is within a half-minute of arc of the correct value. At larger zenith angles the curvature of the atmosphere must be considered. As a result of atmospheric refraction the disks of the sun and moon are visible when they are entirely below the horizon, and the disks may appear to be flattened.

Refraction depends on the wavelength, so that the atmosphere disperses light into its component colors in the same way as does a glass or quartz prism. For this reason, the stars may appear near the horizon as vertical lines containing the usual spectral distribution of color with blue and green at the top and red at the bottom. The *green flash* which sometimes is seen just as the limb of the sun vanishes below the horizon is due also to dispersion of light by the atmosphere.

7.19 Refraction by Ice Crystals

It has been shown in Chapter III that naturally occurring ice crystals are of several different forms with hexagonal basic structure. The great variety of shapes and orientations of the ice particles gives rise to a large number of refraction and reflection phenomena sometimes arresting and at other times hardly noticeable. Only a few of the most common phenomena will be discussed.

Haloes of 22° and 46°

Refraction occurring through hexagonal prisms, which are typical of cirrus clouds, is illustrated in Fig. 7.26a. If the crystal in Fig. 7.26a is ro-

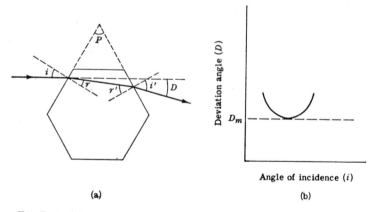

(a)   (b)

FIG. 7.26. (a) Path of a light ray refracted by a hexagonal crystal.
(b) Deviation angle as a function of the angle of incidence.

tated about its axis and the direction of the incident ray is kept constant, the angle of deviation (D) varies in a fashion as sketched in Fig. 7.26b. It is seen that as the angle of minimum deviation (D_m) is approached, an increment in angle of incidence brings about only a small change in angle of deviation. Therefore, a group of randomly oriented identical crystals transmit maximum intensity in the direction of minimum deviation. Now imagine an observer looking at the sun or moon through a cloud sheet of hexagonal needles with random orientation. Light reaches him by refraction at many different angles, but the intensity is greatest 'at the angle of minimum deviation. Thus he sees a ring of bright light (a *halo*) of angular radius equal to the angle of minimum deviation, surrounding a region which is darker than the region outside the halo.

From Fig. 7.26a it follows that the deviation angle may be expressed by

$$D = (i - r) + (i' - r') = i + i' - P$$

For minimum deviation

$$\frac{dD}{di} = 0 = 1 + \frac{di'}{di}$$

Because $r + r' = P$, also

$$\frac{dr'}{dr} + 1 = 0$$

The last two equations are satisfied when $i = i'$ and $r = r'$. Upon applying Snell's law and eliminating i and r

$$\sin \frac{D_m + P}{2} = n \sin \frac{P}{2} \qquad (7.42)$$

where n is the index of refraction for ice. An exercise is given in problem 8.

Since the index of refraction varies with wavelength, white light is dispersed into its component colors with red refracted least and blue refracted most. Notice that this order is the reverse of that observed in the corona, providing a means of distinction between ice crystals and water drops.

The most common halo has angular radius of 22° indicating refraction by hexagonal prisms. This halo can easily be identified because the angle of 22° is just the span of a hand at arm's length. The halo of 46°, which is produced by refraction by rectangular prisms, is less frequently observed.

Sundogs and Tangent Arcs

Horizontally oriented hexagonal plates are responsible for the appearance of *sundogs*. These are bright spots at the same elevation angle as the sun at an angular distance slightly greater than 22° left and right from the sun. The sundogs result from the fact that the light rays from sun to observer are not perpendicular to the vertical axes of the prisms. Therefore, the angle P appears to be somewhat larger than 60°, and consequently the angle of minimum deviation is somewhat larger than 22°. The vertical orientation of the prisms prevents the formation of a ring halo and causes bright spots to be visible on either side of the sun. Calculation of the locations of these spots is required in problem 9.

Tangent arcs on the 22° halo may appear at the highest and lowest points of the circle when sunlight is refracted by hexagonal needles with horizontal orientation. In this case the angle P is equal to 60° just above and below the sun, but is larger than 60° to the left and right of these points. Once the geometry of the problem is understood the computation of various points of the arcs is merely tedious.

There are many other phenomena caused by refraction and also by reflection by ice crystals which are discussed by Humphreys (see references).

7.20   Refraction by Water Drops

The angle of minimum deviation for light passage through a spherical water drop can be determined in a manner similar to that outlined in the previous section. However, the spherical shape of the droplets makes the geometry of the refraction phenomena somewhat different. In Fig. 7.27 the

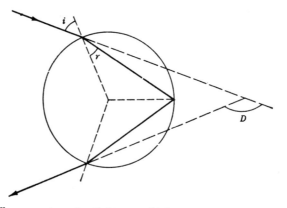

Fig. 7.27. The geometry of a light ray which contributes to the first order (42°) rainbow.

geometry of refraction and reflection of a light ray by a spherical drop is illustrated. From this figure it may be deduced that

$$D = 2(i - r) + \pi - 2r = \pi + 2i - 4r$$

This equation together with Snell's law and the condition for minimum deviation ($dD/di = 0$) provides three equations for the three unknowns, D_m, i, and r. Upon solving for the angles i and r

$$i = \text{arc cos} \; \frac{n^2 - 1}{3} \qquad r = \text{arc cos} \; \frac{2}{n} \frac{n^2 - 1}{3}$$

The corresponding solution for D_m yields for the primary rainbow an angular radius of about 42°. It is comparatively simple to extend this development to two or more internal reflections yielding second and higher order rainbows, as required in problem 10.

The appearance of the rainbow depends rather strongly on the drop size. Large drops (2 mm diameter) are responsible for very clear and colorful rainbows with distinct separation between the colors. Small drops (0.2–0.3 mm) produce less colorful rainbows, and very small drops (0.05 mm) produce a white primary bow (mist bow).

When the rainbow is very pronounced supernumerary bows often become visible. This is an interference phenomenon accompanying the refraction of the light in the drop which may be understood by following a wave which strikes the drop as shown in Fig. 7.28. The incident wave

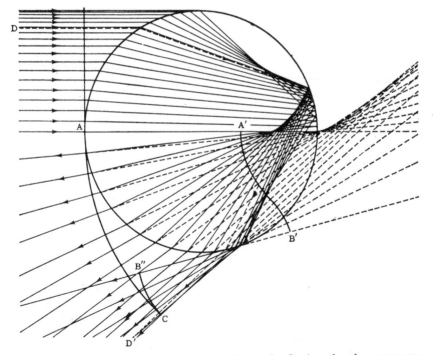

Fig. 7.28. Light rays representing the refraction and reflection of a plane wave surface AB by a spherical water drop (after W. J. Humphreys, "Physics of the Air," p. 481. McGraw-Hill, New York, 1940).

phase AB appears when emerging as ACB'', just as if it had come from the virtual phase $A'B'$. The section CB'' trails somewhat behind CA, and this gives rise to interference patterns. When the distance is a half-wavelength the two sections counteract each other, and when it is a full wavelength they reinforce each other, giving rise to maxima and minima inside the primary bow. These maxima correspond to the supernumerary bows.

7.21 Naturally Occurring Atmospheric Radiowaves

Lightning discharge generates a wide spectrum of electromagnetic waves extending from the ultraviolet to very low-frequency radiowaves. Near thunderstorms radio broadcasts may be seriously "jammed" by "sferics" produced in this way. At great distances from the thunderstorm, a remarkable phenomenon may be observed. Waves of radio audio frequency emitted by a lightning discharge may penetrate the ionosphere, travel approximately along the lines of force of the earth's magnetic field and return to the earth's surface at the opposite geomagnetic latitude. The higher frequencies travel faster than the lower frequencies because of dispersive propagation through the upper atmosphere. The result is that an observer with a detector for these low frequencies (e.g., a very long wire and an earphone) hears a whistling tone of steadily falling pitch, called a "whistler."

Whistlers were first observed in 1888 by Pernter and Trabert in Austria using a telephone wire 22 kilometers long as the antenna. The German scientist Barkhausen rediscovered the phenomenon while eavesdropping on Allied military telephone conversations during the First World War. Barkhausen also suggested that whistlers might arise from propagation of sferics over long dispersive paths, thus linking the phenomenon to lightning discharges; but he could not explain how radio waves of such low frequencies could possibly penetrate the ionosphere.

Understanding of whistlers depends upon the magneto-ionic theory, the theory of interaction between electromagnetic waves and charged particles in the earth's magnetic field. The basis for this theory has been mainly developed by Appleton and others (see, for instance, Goldstein*). As might be expected, the effect of an electron which spins about a magnetic line of force is similar in certain respects to that of a bound charge which oscillates in response to an incident polarized wave. Because the orientation of the electron orbit is determined by the geomagnetic field, the index of refraction has different values for radiation perpendicular and parallel to the magnetic lines of force. The plasma in the magnetic field is said to

---

* S. Goldstein, *Proc. Roy. Soc.* A121, 260 (1928).

be *doubly refracting*, and an incident ray is split into two parts, the "ordinary" and "extra-ordinary" rays. The gyro-frequency of the electron plays a role analogous to the resonance frequency of the oscillating dipole. Therefore, the index of refraction may be expected to be expressed by an equation similar to Eq. (7.16). The complete theory shows that the whistlers are caused by the extraordinary mode traveling approximately along the magnetic lines of force and that the index of refraction for this case is given by

$$n^2 = 1 + \frac{Ne^2}{\pi m \nu (\nu_G - \nu)} \tag{7.43}$$

where ν_G is the gyro-frequency, defined by Eq. (6.25) in the form

$$\nu_G = \frac{Be}{2\pi m}$$

The similarity of this result with Eq. (7.16) is striking.

As has been pointed out in Section 7.6, the index of refraction is the ratio of the speed of light in vacuum to the phase velocity. The signal, however, travels with the group velocity which is given by Eq. (7.7). By substituting Eq. (7.43) into (7.7) and assuming that $Ne^2/\pi m \gg \nu_G \nu$ and $\nu_G > \nu$, the group velocity may be expressed by

$$c_g = \frac{2c_0 \nu^{1/2} (\nu_G - \nu)^{3/2}}{\nu_p \nu_G} \tag{7.44}$$

where $\nu_p{}^2 \equiv Ne^2/\pi m$. This equation has a maximum for $\nu = \nu_G/4$. For smaller values of ν the group velocity decreases with decreasing frequency, giving rise to the dispersion observed in whistlers. An exercise is given in problem 11. Storey[*] has shown that the rays tend to follow the lines of magnetic force shown in Fig. 6.3, so that electromagnetic waves generated by a lightning flash at geomagnetic latitude $-\phi$ are observed as a whistler at geomagnetic latitude ϕ in the opposite hemisphere. Whistlers may be reflected from the earth's surface and travel back and forth several times between northern and sourthern hemispheres producing the typical form of dispersion illustrated in Fig. 7.29.

From the observed dispersion it is possible from Eq. (7.44) to make an estimate of the average electron density along the path of propagation. The electron density found in this manner is about 400 electrons per cubic centimeter, which is a much higher value than previously was believed to apply for heights of several earth radii.

Scarf[†] has been able to use observations of the cutoff frequency of

---

[*] L. R. O. Storey, *Phil. Trans. Roy. Soc. (London)* **A246**, 113 (1953).

[†] F. L. Scarf, *Phys. Fluids* **5**, 6 (1962).

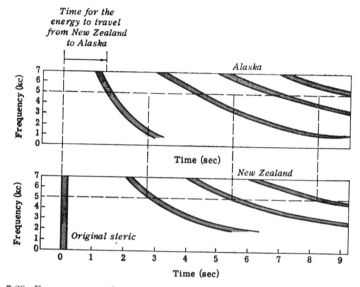

FIG. 7.29. Frequency as a function of time for whistlers traveling between Unalaska, Alaska, and New Zealand [after M. G. Morgan, *Ann. IGY* 3, 315 (1957)].

whistlers to compute the temperature at roughly four earth radii. His results indicate an electron temperature between 10^4 and 10^5 °K.

In addition to whistlers a wide variety of very low frequency (VLF) phenomena not associated with lightning flashes have been observed. Some of these phenomena have been given the suggestive name of "dawn chorus" describing a fuzzy sound of rapidly increasing pitch. Other fairly frequently observed phenomena are the "hooks" (falling pitch followed by a sharp rise in pitch) and "risers" (continuous rising pitch). A theory developed by Gallet† indicates that these noises are probably caused by clouds of charged solar particles entering the magnetosphere at speeds of the order of 10^4 km sec^{-1}. They also influence the earth's magnetic field and the aurora as described in Chapter VI.

7.22   Refraction of Sound Waves

The Audibility of Thunder

Sound waves produced by lightning above the earth's surface are refracted away from the earth under conditions of normal atmospheric lapse rate. For any lapse condition there is a "critical" ray which strikes the earth at grazing incidence as illustrated in Fig. 7.30; at greater distances

† R. M. Gallet, *Proc. I.R.E.* **47**, 211 (1959).

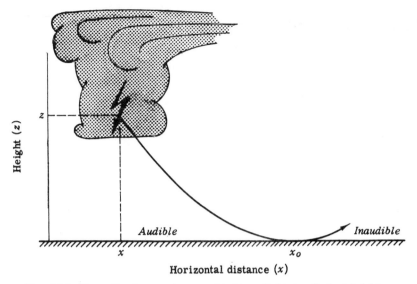

FIG 7.30. The path of a sound ray emitted by a lightning flash at height z.

the only sound reaching the ground results from scattering| by atmospheric inhomogeneities and is usually very weak. To determine the maximum range of the critical ray in the case of constant lapse rate, Eq. (7.38) may be integrated, treating T as a constant, with the result

$$x - x_0 \approx 2 \left(\frac{Tz}{\gamma} \right)^{1/2}$$

which is the equation of a parabola whose properties depend on the temperature (T) and the lapse rate (γ). For a lapse of 7.5°C km^{-1}, temperature of 300°K, and height of 4 kilometers, thunder should be inaudible beyond about 25 kilometers. Lightning which is visible at distances greater than the maximum range of audibility frequently is referred to as *heat* or *sheet* lightning.

Temperature Inversions near the Ground

Within the surface temperature inversion sound rays are refracted downward and may be partially focused with the result that sounds are heard with remarkable clarity. The phenomenon is particularly marked on mornings following clear calm nights and during widespread rain storms, for these conditions frequently are accompanied by strong temperature inversions.

The noise of jet aircraft or rockets taking off, of ground level explosions, and of other high-intensity sounds are heard at very great distances under inversion conditions. The sound of a plane rising through such an inversion may drop very suddenly as heard by an observer on the ground.

Effect of Wind Shear

Under normal conditions of increasing wind with height, waves moving in the direction of the wind are refracted downward, whereas waves moving against the wind are refracted upward as illustrated in Fig. 7.31. The

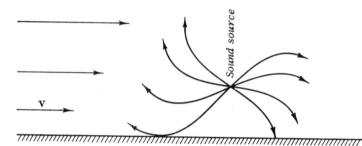

FIG. 7.31. Sound rays from an elevated source refracted by vertical wind shear.

usual effect of refraction by wind shear near the ground is that audibility in the direction of the wind is decreased. Vertical shear of 6 m sec⁻¹ km⁻¹ has an effect on refraction equal to vertical temperature gradient of 10°C km⁻¹. For this reason fog horns are audible at sea at great distances during offshore winds but may be inaudible at short distances during onshore winds.

Observations indicate that the normal wind shear near the base of mature thunderstorms is directed toward the storm. Consequently, wind shear is effective in refracting away from the earth thunder originating within the storm. In this case the rays are refracted symmetrically due both to temperature and to wind shear with the result that the audibility of thunder is markedly reduced. The countryman speaks of "the calm before the storm."

PART V: EFFECTS OF NUCLEAR EXPLOSIONS

The physical properties of the atmosphere determine how the enormous energies released by nuclear explosions are transmitted from the point of explosion. Energy is released as mechanical work done against atmospheric pressure, as electromagnetic radiation in the X-ray, ultraviolet, visible, and infrared regions, and as particulate radiation. The proportions of energy released in each of these forms varies with the design of the explosive and with height of the explosion. Total energies emitted by nuclear explosions vary over a range which includes "small" tactical weapons of 10^{20}

erg or less capacity to hydrogen bombs of 10^{25} erg but for which an upper limit does not exist. We shall discuss here the effects of the "nominal" bomb, the type tested at Alamagordo on July 16, 1945, and dropped on Hiroshima and Nagasaki that same summer. Detailed data have been published for the nominal bomb, and less complete data are available for the megaton bomb. The following discussion, except where noted, applies to the nominal bomb. Problem 7.12 provides an illustrative example.

The nominal bomb released 8×10^{20} erg in a time interval probably of the order of a microsecond. About half of this energy appeared as mechanical energy, about one-third as electromagnetic radiation, and the remaining energy as particulate radiation.

Rough scaling on the basis of total energy released is possible, but scaling differs for the various forms of energy. Particulate and very short wavelength electromagnetic radiation is strongly absorbed by air, so that the "radius of radiation danger" increases only fairly slowly with increase of total energy. Mechanical energy of the shock wave is absorbed much less strongly; the "radius of shock danger" has been observed to increase as the cube root of the total energy released. On the other hand, visible and infrared radiation is only weakly absorbed in the atmosphere, so that the "radius of fire danger" increases as the square root of the total energy. Thus, as the total energy of nuclear explosions increases, the danger from fire dominates increasingly the other dangers.*

For a bomb detonated outside the atmosphere mechanical energy is much less and radiation energy correspondingly greater. In this case absorption of high-energy radiation in the upper atmosphere may heat the ionosphere to incandescence over a large area; subsequent visible and infrared radiation may be more intense and extend over a larger area at the ground than would have been the case following an atmospheric burst.

7.23 Electromagnetic Radiation

Air in the immediate vicinity of an exploding nuclear bomb is heated to incandescence, forming what is known as the ball of fire. About 10^{-4} seconds after explosion the ball of fire is a sphere 15 meters in radius which has a surface temperature of 3×10^5 °K.

At this stage a wave of high pressure known as a shock wave forms the boundary of the ball of fire. Inside the ball of fire thermal radiation is

---

* In regions so remote from the explosion that the incident energy is negligible, danger remains in the damage to living cells produced by "fallout" of radioactive debris. Fallout danger depends strongly on the quantity of debris carried upward in the convection cell and on the distribution of wind velocity as well as on the quantity of fissionable material in the bomb.

transmitted in all directions from the incandescent air. It might be inferred that the ball of fire should expand with the speed of light until its temperature falls below the temperaure of incandescence; instead, after about 1/100 second the ball of fire expands more slowly than does the shock wave. This curious phenomenon may be understood if the processes which occur immediately after the explosion are visualized. The shock wave moves outward and heats the air adiabatically to incandescence. Air is opaque to much of the ultraviolet radiation emitted by this very hot air, with the result that the transfer of energy by radiation occurs in a series of steps separated by alternate absorption and emission in a random direction. Thus, destructive ultraviolet and X-ray radiation lags behind the less destructive visible and infrared. The process is analogous to the diffusion of a gas through another gas. Longer wave radiation (visible) is of course transmitted with the speed of light from the surface of the sphere.

The temperature of the shock wave falls due to expansion and loss of energy by radiation, and when the shock wave and the ball of fire have a radius of about 100 meters and a temperature of about 2000°K, the shock wave "breaks away" from the ball of fire—that is, no longer produces incandescence. The ball of fire continues to expand into the shock-heated (but not incandescent) air by the process of absorption and emission of radiation. A maximum radius of about 150 meters is reached after about one second when the temperature of the surface is about 6000°K.

The increase of temperature which follows the "break away" of the shock wave may be explained by the following considerations. Prior to break away the surface of the ball of fire has a temperature determined by the rapidly expanding shock wave. Following break away, expansion is much slower, giving the surface time to be heated more strongly than before by the very intense interior radiation.

The ball of fire radiates approximately as a black body; consequently, its surface temperature determines the proportion of the total radiation emitted at a particular wavelength. Also, it must be borne in mind that the ball of fire grows in surface area. Where F_0 represents the total radiation flux density emitted by the ball of fire at radius r_0 and temperature T_0, the total flux density F at radius r and temperature T may be expressed from Stefan-Boltzmann law in the form

$$\frac{F}{F_0} = \left(\frac{r}{r_0}\right)^2 \left(\frac{T}{T_0}\right)^4$$

It has been observed that at a particular radius r the flux density falls to a minimum about 10^{-2} seconds after the explosion (when the temperature is a minimum), then rises to a secondary maximum about 0.2 second after the explosion, and then gradually falls as expansion and radiation dissipate the thermal energy through a larger and larger volume.

Wien's law shows that as the surface temperature changes, the wavelength of maximum energy changes. Thus, the ball of fire emits mostly ultraviolet radiation in its first stage, and then emits predominantly infrared radiation which grows in intensity for several seconds after the explosion.

These variations in emitted radiation together with the complex absorption spectrum for air (Fig. 4.13) result in a complex variation in the radiant energy reaching a particular point. In the first stage of predominant ultraviolet radiation, the intense radiation of wavelengths less than about $0.2\,\mu$ is absorbed in the immediate vicinity of the ball of fire, and the wavelengths between 0.2 and 0.3 μ are absorbed within a few kilometers of the explosion. Visible and infrared radiation is not strongly absorbed, but its intensity is far less than that of the ultraviolet radiation, and therefore is not initially of great danger. Subsequently, however, when the ball of fire has grown in size and decreased in temperature, most intense radiation is emitted in the visible and, later, in the infrared region. At this time radiation passes through the atmosphere with little attenuation, and great damage to life and property by burning may extend over many miles.

The direct radiation incident on a unit area at a distance $r \gg r_0$ may be expressed, using Beer's law, in the form

$$F_d\, \tau = \frac{E}{4\pi r^2}\, e^{-\sigma r} \tag{7.45}$$

where τ represents the duration of the flash, E the total radiant energy, and σ the extinction coefficient of the atmosphere. The value of σ depends upon wavelength interval so that the energy delivered at a certain distance depends on wavelength. However, except for radiation in the ultraviolet and in the water vapor and carbon dioxide bands, σ reflects fairly closely the optical extinction coefficient used for computing the visual range [see Eq. (7.25)]. In addition to the direct radiation, scattered radiation from other directions may be large for distances less than about half the visual range.

The fact that the wavelength of maximum energy varies with time indicates that the intensity of radiation received at a particular place varies during the period of incandescence of the fire ball. The initial ultraviolet radiation is absorbed fairly close to the explosion, whereas a substantial amount of visible and infrared radiation, which reaches its maximum after several seconds, may penetrate over large distances. At distances greater than a few kilometers the maximum danger to life and the maximum possibility of setting fires occurs 0.3 to 3 seconds after the explosion. This time interval is sufficient for individuals to act to reduce the danger due to thermal radiation.

The effect on inanimate objects as well as on living tissue depends on the factors which have been discussed—intensity and distribution of radiation, distance, and absorption in air. It also depends on the properties of the receiving surface. Thus, surfaces which reflect are less affected than those which absorb, and surfaces of higher thermal diffusivity are less affected than those of low thermal diffusivity. The critical energy required for ignition (or for charring or evaporation) and the distance at which this may occur for a few common materials is illustrated in Table 7.1. The distances (particularly for large visual range) tend to be underestimates due

TABLE 7.1

Material	Critical energy (erg cm^{-2})	Distance from 20-kiloton bomb (visual range >50 km)	Distance from 20-kiloton bomb (visual range 2.5 km)
Skin	1.2 × 10^8 (moderate burn)	3.5 km	1.3 km
White paper	4.2 × 10^8 (ignites)	2.2	1.0
Wood (Douglas fir)	4.6 × 10^8 (ignites)	2.2	1.0
Gray cotton shirt	4.2 × 10^8 (ignites)	2.2	1.0
Bakelite	30 × 10^8 (chars)	1.0	0.5
Cloud 500 m in depth (0.02 g H$_2$O cm^{-2})	5 × 10^8 (evaporates)	2.2	1.0

to neglect of scattering. Shielding from thermal radiation is afforded by absorbing surfaces; thus, in contrast with the damage due to the shock wave, protection is provided by light screens, clothing, clouds, or other opaque objects.

Of the multiple dangers to life which nuclear explosions present, the absorption of electromagnetic radiation and the resulting setting of innumerable fires is perhaps the worst. A single one-megaton bomb may ignite clothing, paper, dry wood, and other similarly combustible materials at distances up to 15 kilometers, and present capabilities make it necessary to scale this range upward by an order of magnitude. The resulting fire storm would in many populated areas "escalate" until destruction of life and property would be virtually total. Rain or the presence of clouds might reduce the range of ignition greatly, but this amelioration is scarcely to be relied on.

7.24   The Shock Wave in Air

The shock wave created by the expanding gas is a compression wave and so has certain features in common with sound waves. The distinction lies in the fact that variation of pressure within the shock wave is comparable to atmospheric pressure or greater, whereas variation of pressure within the sound wave is several orders of magnitude smaller.

The development of the shock wave may be explained quite simply although thorough theoretical treatment requires advanced methods. The explosion releases energy which very rapidly heats the air surrounding the bomb. As has been described in Section 7.23 this results in expansion of the air and consequent increase of pressure along a spherical surface which moves radially outward. The pressure in the immediate neighborhood of the explosion may reach many hundred thousand atmospheres, and the corresponding adiabatic temperature change is sufficient to render the air incandescent. As the wave moves outward from the center of the explosion, the leading part of the wave heats the air through which it passes by compression. Because the speed of wave is proportional to the square root of the absolute temperature, the following portion of the wave gradually overtakes the leading portion as shown in Fig. 7.32. The pressure profile acquires a steeper and steeper "front" until a *shock* wave has been formed.

As the shock front moves outward from the point of explosion, the spherical wave surface becomes larger and larger; consequently, the temperature and pressure of the shock front must fall, and the speed of the front

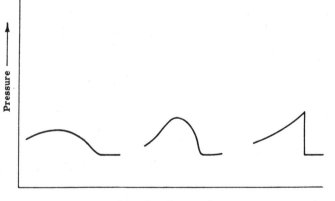

Direction of propagation   ⟶

Fig. 7.32. The distribution of pressure at three stages in the development of a shock wave.

drops toward the speed of sound in the original air. Behind the shock front a region of sub-atmospheric pressure develops as a result of the outward rush of air, and in this region the air which had rushed outward from the blast center rushes back (suction phase). The pressure at a point as a function of time or the pressure at a single time as a function of distance is shown in Fig. 7.33.

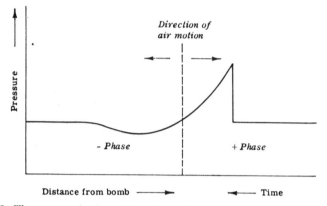

FIG. 7.33. The pressure in the neighborhood of a shock wave as a function either of distance or of time.

The temperature variation at a point as the shock wave passes is also of interest. As the shock front reaches a particular point, the temperature rises almost discontinuously to a maximum which depends upon the distance from the explosion and the quantity of energy released. Then, as the suction phase develops, the temperature falls adiabatically to a value below the original air temperature, and finally it rises again above the original temperature.

If the air through which the shock wave passes is nearly saturated, the fall in temperature which follows the shock front results in condensation of water vapor and the appearance of a ring-shaped or spherical cloud. At a particular point the following increase in temperature results in evaporation of the cloud. The sequence of temperature changes and the accompanying cloud formation and dissipation is called the "cloud-chamber effect" since it is identical in principle with the laboratory technique for detection of high-energy particles or condensation nuclei by cloud chamber. The duration of the cloud at a point is of the order of one second.

Reflection and diffraction of shock waves may change the *overpressure* (pressure in excess of atmospheric) of the waves several fold under certain conditions. In general, the shock wave is increased in amplitude by reflection and is decreased in amplitude in regions reached only by the diffracted wave. Diffracting obstacles of the size of houses have a negligible effect on the shock wave except in the region directly "behind" the house. Larger objects may have an important effect.

Reflection of sound waves doubles the overpressure at the reflecting surface, and the overpressure in the reflected shock wave may be as much as eight times the overpressure in the original shock wave. A familiar example is provided by the great amplification of water waves which may occur when reflection occurs at a concrete wall. The great increase in overpressure results from the sudden conversion of the kinetic energy of the particles in the shock wave to work done in compression at the rigid surface. Even for doubling of the pressure, which is appropriate to the acoustic case, it is easy to find arrangements of surfaces which result in very large overpressures. For example, a shock wave which approaches a 90° corner obliquely is reflected at the two surfaces as shown in Fig. 7.34. The

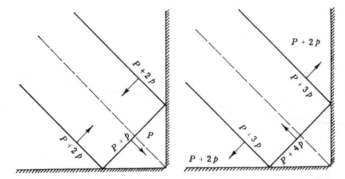

FIG. 7.34. Overpressures (p) developed behind a shock wave in the neighborhood of a 90° reflecting corner.

original shock wave has the pressure P (atmospheric) $+ p$ (overpressure) and is traveling as shown by the arrow. If, for simplicity, the gradual drop in pressure after the shock front has passed is not considered, then in the region crossed by the original wave and by the reflected wave, the total pressure is $P + 2p$. The two reflected waves meet along the dotted line and are reflected again as shown in Fig. 7.34. As a result, the pressure in the region crossed by three waves is $P + 3p$, and the pressure in the region crossed by four waves is $P + 4p$. Pressure at a point near the intersection of the two walls increases discontinuously with each passage of the shock wave. The distance of the point from each of the two surfaces determines the time interval between successive shock fronts. It follows that at a point on one of the surfaces the shock front is $2p$ in amplitude, and at the intersection of the two surfaces there is a sudden increase in pressure of $4p$.

Because the pressure begins to fall immediately after the passage of the shock front, the total pressure attained after reflection depends on distance from the reflecting surfaces. Under certain conditions the drop in pressure may exceed the pressure rise which occurs with the passage of the reflected shock wave.

In the case of an air burst above a flat plane it is convenient to think of the reflected wave as originating from a "virtual bomb" situated below the plane. Successive stages in the reflection of an acoustic wave then may be drawn very simply as shown in Fig. 7.35. It is evident from Fig. 7.35 that

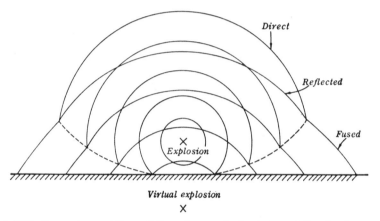

FIG. 7.35. Successive positions of the primary shock wave and the shock wave reflected from a horizontal boundary.

the reflected wave lags the original wave by a maximum distance of twice the height of the explosion above the ground. At the ground the two waves occur together resulting in an overpressure of $2p$ as stated earlier. At points above the ground two waves are experienced, the second slightly weaker than the first due to its greater distance from the explosion. Due to the fact that the reflected wave passes through air which has been heated by the original shock wave, it tends to catch up to the original wave. Since the reflected and incident waves are close to each other near the ground, the two quickly fuse in this region and then travel as a single shock. The shock wave formed in this way is called the "Mach front." As the spherical shock wave expands, fusion of the two waves occurs at greater and greater heights above the ground, and finally there is a single spherical wave whose overpressure is, in general, greater than the overpressure of the incident wave. Under proper conditions, the Mach front may be detected clearly in pictures of atomic bomb explosions. What makes the Mach front visible?

The preceding discussion has considered an atmosphere of uniform temperature and composition. Since temperature within the troposphere decreases upward, shock waves travel more slowly with increasing height; as a result the spherical wave is flattened as it moves upward. Vertical wind shear and temperature inversions throughout the atmosphere may be effective in focusing shock waves under the conditions discussed in Section 7.16. The propagation of sound or shock waves may be more thoroughly studied as a scattering problem, analogous to the Mie scattering of electromagnetic waves. In the case of shock waves appreciable energy may reach the observer by paths other than that of the simple ray analysis.

In concluding our discourse on so dismal a subject as the effects of nuclear explosions no subtle symbolism is intended. The all-encompassing atmosphere need not be destined to serve as the medium of transmission for man's final inhumanity to man.

List of Symbols

		First used in Section
a	Radius of scattering particle	7.7
A	Amplitude	7.1
B	Brightness	7.11
a, b	Constants	7.13
c	Phase speed	7.1
c_g	Group velocity	7.4
c_0	Speed of light in vacuum	7.6
c_p, c_v	Specific heat at constant pressure and constant volume, respectively	7.2
C	Contrast	7.11
D	Angle of deviation	7.19
e	Elementary electric charge	7.6
E	Radiant energy released by bomb	7.23
$\mathbf{E}$	Electric field strength	7.3
F	Force, energy flux density	7.2, 7.9
g	Force of gravity per unit mass	7.2
h	Depth of water column, difference between true and apparent height	7.2, 7.13
$\mathbf{H}$	Magnetic field strength	7.3
i	Angle of incidence	7.10
I	Intensity	7.11
k	Wave number, absorption coefficient	7.4, 7.6
L	Distance between emitter and receiver	7.17
m	Mass per unit length, mass of electrical charge	7.2, 7.6
n	Index of refraction, integer	7.4, 7.9
N	Number of scattering particles per unit volume	7.6
p	Pressure	7.2
P	Prism angle, atmospheric pressure	7.18, 7.24
$\mathbf{P}$	Polarization	7.6

List of Symbols (Continued)

		First used in Section
q	Specific humidity	7.12
r	Radius (polar coordinates), angle of refraction, radius of curvature	7.2, 7.10, 7.13
R_m	Specific gas constant for air	7.2
s	Path length	7.2
t	Time	7.1
T	Tension, temperature	7.2, 7.13
v	Wind speed	7.14
x_v	Visual range	7.11
x, y, z	Cartesian coordinates	7.1, 7.4
Z	Apparent zenith angle	7.18
α	Specific volume, polarizability, angular radius of diffraction band	7.2, 7.6, 7.9
β	Angle between sound ray and direct path	7.17
γ	Damping coefficient, temperature lapse rate	7.6, 7.13
δ	Angular deviation in astronomical refraction	7.18
ϵ	Permittivity	7.3
θ	Angle between dipole axis and direction of wave propagation, angle	7.7, 7.13
κ	Absorption index	7.6
λ	Wavelength	7.1
μ	Permeability	7.3
ν	Frequency	7.4
ν_G	Gyro-frequency	7.21
ρ	Density	7.2
ρ_w	Density of water vapor	7.13
σ	Surface area, extinction coefficient	7.7, 7.11
τ	Time interval of flash	7.23
ϕ	Longitude angle in spherical coordinates	7.7
ψ_λ	Sensitivity of the human eye	7.11
χ	Susceptibility	7.6
ω	Angular frequency	7.4

Problems

1. Show that the following differential equation which governs a wave on a stretched string may be written

$$\frac{\partial^2 z}{\partial t^2} = \frac{T}{m}\frac{\partial^2 z}{\partial y^2}$$

where T represents the tension in the string and m the mass per unit length.

2. Find from Eq. (7.22) the diameter of cloud droplets which produce a white corona of 4° angular radius. What may be concluded about the range of sizes of the cloud droplets?

3. For a particular object contrast at short range appears to be -0.5. What is the maximum distance that this object can be seen when the visual range is 10 km?

4. Find the temperature lapse rate required for light to travel around the earth at sea level. What value of g would result in curvature around a planet whose size is the same as the earth and whose atmosphere is isothermal, if the index of refraction is the same as that in our atmosphere?

5. Find the increase of specific humidity with height which would result in straight line propagation of radar waves if temperature is constant with height and equal to 290°K and the specific humidity is 10 g/kg. Find under the same conditions the change of specific humidity with height which results in curvature of a horizontal beam equal to the curvature of the earth.

6. Find the critical temperature lapse rate just sufficient to prevent the sound of a volcanic eruption at the top of a mountain from reaching an observer by a direct path for the following case: the mountain is 100 km from the observer, and the top is 4×10^3 meters higher than the observer. Neglect the curvature of the earth.

7. If the specific humidity is 10 g/kg, find the error in temperature made by using Eq. (7.40) for dry air.

8. Find the difference in angular radius of the red and violet rings of the haloes formed by prism angles of 60° and 90° if

$$\lambda(\mu) \qquad n$$
$$0.656 \text{ (red)} \qquad 1.307$$
$$0.405 \text{ (violet)} \qquad 1.317$$

9. Compute the angular distance between sun and sundog when the elevation angle of the sun is 30°.

10. Derive the angle of minimum deviation for the secondary bow. Describe the color sequence of the primary and secondary bow using the following indices of refraction and wavelengths.

$$\lambda(\mu) \qquad n$$
$$0.656 \qquad 1.332$$
$$0.405 \qquad 1.344$$

11. Compare Eq. (7.44) with the expression for the group velocity of radio waves to be derived from Eqs. (7.7) and (7.20).

12. If a nuclear explosion releases 4×10^{20} erg as thermal energy within a sphere of 1.0 kilometer radius, what is the average temperature within the sphere after the pressure has returned to its initial value? The initial temperature at the center of the sphere (1.0 kilometer above earth) is 7°C, and temperature decreases linearly with height. At a height of 11.0 kilometers the temperature is −55°C, and above this height temperature is uniform. Find the height at which the buoyant cloud would reach equilibrium with its environment under the above conditions assuming no mixing with the environment and no loss or gain of energy through radiation, turbulence, or condensation. How far above this height could the cloud penetrate under these idealized conditions?

General References

Sears, *Principles of Physics*, Vol. 3, gives an elementary but correct and clear account of optics including the ray treatment of diffraction and refraction.

Joos, *Theoretical Physics*, gives an excellent discussion of the physical basis of index of refraction, diffraction, and electromagnetic waves.

Unsöld, *Physik der Sternatmosphären*, gives a comprehensive account on an advanced level of the mechanism of absorption of radiation.

Panofsky and Phillips, *Classical Electricity and Magnetism*, gives an excellent treatment on an advanced level of dipole radiation and other fundamental topics concerned with electromagnetic waves.

Van de Hulst, *Light Scattering by Small Particles*, provides an advanced account of the problems associated with scattering.

Johnson, *Physical Meteorology*, rambles through a wide range of topics in atmospheric physics. His discussion of visibility and other topics related to scattering is particularly useful.

Battan, *Radar Meteorology*, gives a coherent, well-organized account of the application of radar to atmospheric observation and measurement. The theoretical background given is more brief than it needs to be.

Humphreys, *Physics of the Air*, presents the physical basis for a great range of atmospheric phenomena, especially those which are distinctly acoustic or optical.

Proceedings of the I.R.E., *Special Issue on the Nature of the Ionosphere*, provides data on whistlers from the IGY and a useful summary.

Glasstone, *Effects of Nuclear Weapons*, presents the most complete data available to the public on many aspects of nuclear explosions. It was first published in 1950 and in recent editions presents data on bombs in the megaton range.

Mathematical Topics

A. Partial Differentiation

The quantities which serve to describe the state of the atmosphere, for example, temperature, specific humidity, velocity, etc., are called dependent variables. They depend on four independent variables: time and the three space coordinates.

First consider a function represented by $z = f(x)$ which is continuous, has continuous derivatives and is single valued. The dependent variable is z, and the independent variable is x. The derivative is defined by

$$\frac{dz}{dx} \equiv \lim_{\Delta x \to 0} \frac{f(x + \Delta x) - f(x)}{\Delta x}$$

which describes the slope of the curve of z in the xz plane. Now, suppose that the function z represents the height of the surface of a mountain. The height must depend on both x and y, the two horizontal coordinates. It is clear that a derivative of z may be calculated with respect to x along any of an infinite number of directions depending on the direction chosen for x. If x is chosen along the line of steepest ascent, dz/dx is large; but if the direction chosen is along a contour, dz/dx vanishes. To avoid this ambiguity, derivatives of functions of two or more variables are defined holding all but one independent variable constant. Thus, for a function of two independent variables

$$\left(\frac{dz}{dx}\right)_y \equiv \frac{\partial z}{\partial x} \equiv \lim_{\Delta x \to 0} \frac{f(x + \Delta x, y) - f(x, y)}{\Delta x}$$

This is called the *partial derivative* of z with respect to x. The derivative with respect to y holding x constant is written

$$\left(\frac{dz}{dy}\right)_x \quad \text{or} \quad \frac{\partial z}{\partial y}$$

For a function of more than two variables, the partial derivative is formed by holding all but one independent variable constant. Higher partial derivatives may be formed in the same way; the order of differentiation may be interchanged without affecting the result. Thus

$$\frac{\partial^2 z}{\partial x\, \partial y} = \frac{\partial^2 z}{\partial y\, \partial x}$$

as may be demonstrated easily by choosing an elementary function of x and y and carrying out the differentiation in the indicated order.

In the examples z has been used as a dependent variable. In many problems in geophysics z also is used as an independent variable.

B. Elementary Vector Operations

A quantity which is described by a direction and a magnitude, force, for example, is called a *vector* quantity; whereas a quantity which is described by magnitude alone, temperature, or pressure, for example, is called a scalar quantity. Vectors may be represented by bold-faced letters, for example, **A**.

Vector Addition

The vectors **A** and **B** may be added by making their lengths proportional to their respective magnitudes and attaching the tail of **B** to the head of **A**. The vector connecting the tail of **A** and the head of **B** is defined as the vector sum; it is written **A** + **B**. Vectors are equivalent if they have the same direction and magnitude. It is clear from Fig. A-1 that the vector **A** may be expressed as the sum of three vectors taken along the three coordinate axes, x, y, and z. This sum may be expressed by the vector equation

$$\mathbf{A} = \mathbf{A}_x + \mathbf{A}_y + \mathbf{A}_z$$

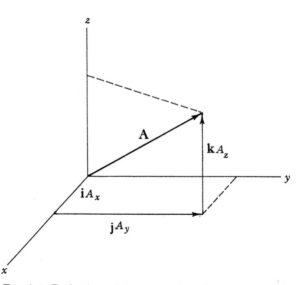

FIG. A-1. Projections of the vector **A** on the axes x, y, and z.

It is customary to define unit vectors in the x, y, and z directions by the equations

$$\frac{A_x}{A_x} \equiv i \qquad \frac{A_y}{A_y} \equiv j \qquad \frac{A_z}{A_z} \equiv k$$

It is important to realize that a unit vector describes a direction but has no dimensions and has unit magnitude in any system of units. The use of unit vectors permits expanding A in the form

$$A = iA_x + jA_y + kA_z$$

The Pythagorean theorem shows that

$$A^2 = A_x{}^2 + A_y{}^2 + A_z{}^2$$

Vector Multiplication

The product of a scalar and a vector is a vector with magnitude equal to the product of the magnitudes and the direction of the original vector. Multiplication by a negative number reverses the direction of the vector.

The product of two vectors is defined in two different ways, reflecting two physical relationships between vectors. First, the *scalar* or *dot* product of two vectors is defined by the equation

$$A \cdot B \equiv AB \cos (A, B)$$

and is read, A dot B. Because the cosine of an angle equals the cosine of the corresponding negative angle, $A \cdot B = B \cdot A$. An example of the use of the scalar, or dot, product is the definition of work

$$W \equiv \int F \cdot ds$$

where F represents force, s represents displacement and the integral is taken along a specified line or contour.

Second, the vector product of vectors A and B is defined as a vector normal to the plane of A and B in the direction which a right-hand screw would advance if rotated from A into B through the smaller angle and with a magnitude given by $AB \sin (A, B)$. The *vector* or *cross* product is written $A \times B$ and is read, A cross B. Because the order of the vectors determines the direction of the vector product, $A \times B = -B \times A$.

An example of the vector product is the definition of the angular momentum in the form, $mr \times v$. From Fig. A-2 it is seen that rotation of r into v generates a vector directed out of the paper as a right-hand screw would advance if rotated from r into v.

From Fig. A-2 it may also be recognized that the angular frequency vector ω may be defined by

$$\omega \equiv \frac{r \times v}{r^2}$$

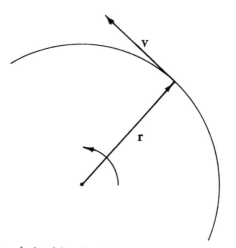

FIG. A-2. The velocity (**v**) and position vector (**r**) of a point in rotation.

It follows then that $\boldsymbol{\omega} \times \mathbf{r} = \mathbf{v}$ and that the angular momentum may be defined by $m\mathbf{r} \times (\boldsymbol{\omega} \times \mathbf{r})$. It may easily be shown that these relations hold whether or not the vector **r** lies in the plane normal to $\boldsymbol{\omega}$.

The Vector Operator ∇

It is convenient to introduce the *del* operator which is defined in Cartesian coordinates by

$$\nabla \equiv \mathbf{i} \frac{\partial}{\partial x} + \mathbf{j} \frac{\partial}{\partial y} + \mathbf{k} \frac{\partial}{\partial z}$$

It may operate on a scalar field $\phi(x, y, z)$ in which case the *gradient* of ϕ is expressed by

$$\nabla \phi = \mathbf{i} \frac{\partial \phi}{\partial x} + \mathbf{j} \frac{\partial \phi}{\partial y} + \mathbf{k} \frac{\partial \phi}{\partial z}$$

Or, ∇ may operate on a vector in either of two ways. The *dot* operation of ∇ into **A**, called the *divergence* of **A**, is defined as

$$\nabla \cdot \mathbf{A} \equiv \mathbf{i} \cdot \frac{\partial \mathbf{A}}{\partial x} + \mathbf{j} \cdot \frac{\partial \mathbf{A}}{\partial y} + \mathbf{k} \cdot \frac{\partial \mathbf{A}}{\partial z} = \frac{\partial A_x}{\partial x} + \frac{\partial A_y}{\partial y} + \frac{\partial A_z}{\partial z}$$

The *cross* operation of ∇ into **A**, called the *curl* of **A**, is defined as

$$\nabla \times \mathbf{A} \equiv \mathbf{i} \times \frac{\partial \mathbf{A}}{\partial x} + \mathbf{j} \times \frac{\partial \mathbf{A}}{\partial y} + \mathbf{k} \times \frac{\partial \mathbf{A}}{\partial z}$$

$$= \mathbf{i} \left(\frac{\partial A_z}{\partial y} - \frac{\partial A_y}{\partial z} \right) + \mathbf{j} \left(\frac{\partial A_x}{\partial z} - \frac{\partial A_z}{\partial x} \right) + \mathbf{k} \left(\frac{\partial A_y}{\partial x} - \frac{\partial A_x}{\partial y} \right)$$

C. Taylor Series

It is often important to expand a function in a series of terms about a particular point. Examples occur in the solution of differential equations (both by analytic and numerical methods), in the analysis of errors and in the evaluation of indeterminate forms of functions. The Taylor series provides a systematic method for such expansion. It may be illustrated by a geometrical example: suppose that it is desirable to evaluate a function $f(x)$ as shown in Fig. A-3 by measurements made in the neighborhood of

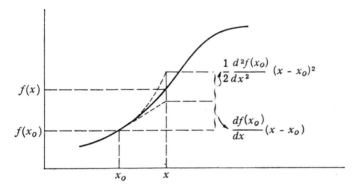

FIG. A-3. Geometrical representation of the expansion of $f(x)$ about the point x_0.

the point x_0. Taylor's theorem states that a function which has continuous derivatives in a certain interval may be expressed in that interval by an infinite series of the form

$$f(x) = f(x_0) + \frac{df(x_0)}{dx}(x - x_0) + \frac{d^2f}{dx^2}\frac{(x - x_0)^2}{2!} + \cdots$$

The first term is the value of the function at x_0, the second is the slope of the tangent at x_0 times the increment $x - x_0$, the third is proportional to the curvature at x_0 times $(x - x_0)^2$ for small increments. In the illustration the third term overshoots the point $f(x)$; but if the interval $x - x_0$ is not too large, the overshooting is counteracted by the next term which is proportional to the third derivative. It is important that the series converge and that enough terms of the infinite series be used to reduce the remainder of the series to negligible magnitude.

It is clear that because each succeeding term is multiplied by $x - x_0$ to an increasing power, the Taylor series converges more rapidly the smaller the interval, $x - x_0$. For differential quantities all differentials of order higher than the first vanish, so that the Taylor series becomes

$$f(x) = f(x_0 + dx) = f(x_0) + \frac{df}{dx} dx$$

which is a particularly useful expansion.

The Taylor series can easily be extended to a function of two or more independent variables. For a function of two variables for which all partial derivatives exist the expansion is

$$f(x + h, y + k) = f(x, y) + \left(h \frac{\partial}{\partial x} + k \frac{\partial}{\partial y} \right) f(x, y)$$

$$+ \frac{1}{2!} \left(h \frac{\partial}{\partial x} + k \frac{\partial}{\partial y} \right)^2 f(x, y) + \frac{1}{3!} \left(h \frac{\partial}{\partial x} + k \frac{\partial}{\partial y} \right)^3 f(x, y) + \cdots$$

D. The Total Differential

It is often important to be able to evaluate a differential increment taken in an arbitrary direction; that is, along neither the x nor y axis. It is easy to see, however, that because the variables x and y are independent, the increment can be represented by the sum of increments taken along each of the axes. Thus, for the function $z = f(x, y)$

$$dz = dz_x + dz_y = \frac{\partial z}{\partial x} dx + \frac{\partial z}{\partial y} dy$$

and the expansion of the differential of temperature as a function of space and time is

$$dT = \frac{\partial T}{\partial t} dt + \frac{\partial T}{\partial x} dx + \frac{\partial T}{\partial y} dy + \frac{\partial T}{\partial z} dz$$

This equation is quite general. It may be applied to many different problems by specifying the displacement in an appropriate manner. Where the increment is taken following the displacement of a particular element, dT/dt is called the *individual derivative* of temperature with respect to time.

When differentials having more than a single geometrical or physical meaning are employed, it is convenient to distinguish them by using different symbols. For example, dT usually represents the individual differential of temperature, whereas δT may be employed to represent a space differential of temperature.

E. The Exact Differential

Functions whose value is a unique function of position, like the height of a mountainside, have an important mathematical property. The line integral of the differential of such a function from point A to point B is independent of the path followed from A to B. Having reached B one might return to A by any other path, and he should have then found that the line integral around the closed path is zero. A differential which has this property is called an *exact differential*. Examples, in addition to height of a continuous surface, are the geopotential, or temperature distribution in a gas at a fixed time.

F. Gauss' and Stokes' Theorems

These two theorems are of great utility in transforming vector functions which are continuous and have continuous partial derivatives.

Gauss' Theorem (or the Divergence Theorem) expresses a general relation between surface and volume integrals which may be clearly visualized for the case of fluid flow. The mass of fluid passing a unit area per unit time, the fluid flux, is represented by $\mathbf{Q}$; and the component of flux normal to the surface is expressed by $\mathbf{Q \cdot n}$, where $\mathbf{n}$ represents the unit vector directed normally outward from the surface. The rate of outflow from an enclosed volume is then

$$\oiint \mathbf{Q \cdot n} \, d\sigma$$

Now divide the volume into differential rectangular elements as shown in Fig. A-4, and express the flux through each face by expansion in a Taylor

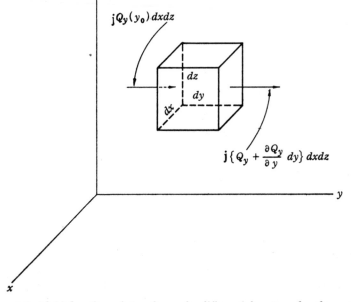

FIG. A-4. Fluid flux through two faces of a differential rectangular element.

series about the origin. Then, from Fig. A-4, the net outflow through the two xz faces is $(\partial Q_y/\partial y)\, dy\, dx\, dz$; and the net outflows through the xy and yz faces are, respectively, $(\partial Q_z/\partial z)\, dz\, dx\, dy$ and $(\partial Q_x/\partial x)\, dx\, dy\, dz$. If the contributions to net outflow from all the differential volume elements are added, the net outflow from the entire volume is expressed by $\iiint \nabla \cdot \mathbf{Q}\, d\tau$ where $d\tau$ represents the differential volume element. Upon equating the two expressions for net outflow, Gauss' theorem appears in the form

$$\oiint \mathbf{Q} \cdot \mathbf{n}\, d\sigma = \iiint \nabla \cdot \mathbf{Q}\, d\tau \qquad (F.1)$$

Stokes' theorem relates the area integral of the curl of a vector and the line integral of the tangential component of the vector taken around the area. To develop this theorem consider the component of $\nabla \times \mathbf{V}$ perpendicular to the xy plane as shown in Fig. A-5. Evaluation of the line integral

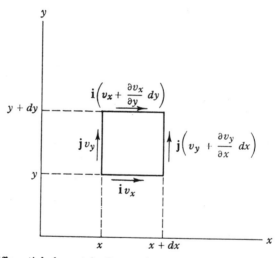

FIG. A-5. Differential element in the xy plane and the components of the vector $\mathbf{v}$ along the sides dx and dy.

around the figure in the positive or counterclockwise sense gives

$$\oint_{dx\, dy} \mathbf{V} \cdot d\mathbf{l} = \left(\frac{\partial V_y}{\partial x} - \frac{\partial V_x}{\partial y} \right) dx\, dy$$

Now consider a finite area enclosed by a curve as shown in Fig. A-6. Divide the area into differential elements and evaluate the line integral around each element. All contributions to the sum of the line integrals cancel except for the sides of the differential elements which lie along the bounding curve. Therefore, if equations for the enclosed area are summed

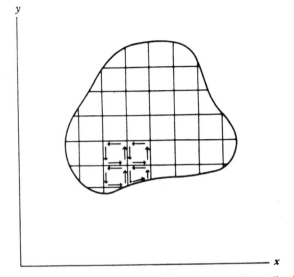

FIG. A-6. Differential area elements enclosed by a curve and contributions of several elements to the line integral taken around the boundary.

$$\oint \mathbf{V} \cdot d\mathbf{l} = \iint \left(\frac{\partial V_y}{\partial y} - \frac{\partial V_x}{\partial x} \right) dx\, dy$$

which is Stokes' theorem for the plane. Since orientation of the surface is arbitrary, a more general form is

$$\oint \mathbf{V} \cdot d\mathbf{l} = \iint \nabla \times \mathbf{V} \cdot \mathbf{n}\, d\sigma \tag{F.2}$$

where $\mathbf{n}$ represents the outwardly directed unit vector and $d\sigma$ represents the differential element of area.

G. The Potential Function

Irrotational fields, for which $\nabla \times \mathbf{V} = 0$, form an important class of vector fields. Such vectors can be represented by the gradient of a scalar "potential" function ϕ, as is evident if one substitutes the components of $\nabla\phi$ into the expansion of $\nabla \times \mathbf{V}$. The right-hand side of Stokes' theorem now vanishes and the left-hand side becomes

$$\oint \nabla\phi \cdot d\mathbf{l} = \oint d\phi = 0$$

Evidently, the existence of an exact differential implies that the gradient of the scalar function is irrotational. Also, it follows that a vector which is irrotational may be expressed by the gradient of a scalar function.

H. Solid Angle

In dealing with several physical problems an understanding of the concept of solid angle is required. Consider a cone with vertex at the origin of a concentric spherical surface as shown in Fig. A-7. The solid angle is defined as the ratio of the area of the sphere intercepted by the cone to the square of the radius. Thus

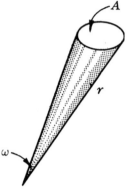

$$\omega \equiv \frac{A}{r^2}$$

The unit of solid angle is the steradian. The area cut out of a sphere by a steradian is equal to the square of the radius.

The solid angle encompassing all directions at a point is given by the total area of a circumscribed sphere divided by the square of the radius. Therefore, integration over all angles yields

Fig. A-7. A cone of solid angle ω which intercepts the surface A at distance r from the origin.

$$\omega = 4\pi \text{ steradians}$$

Special problems dealing with solid angle are the determination of the number of collisions at a surface area of a wall (Chapter II), and the determination of the radiation flux passing through a unit area (Chapter IV). In these problems the orientation of the area is important, and it is convenient to consider the solid angle with respect to the plane dA shown in Fig. A-8. An increment of solid angle can then be represented using spherical coordinates by

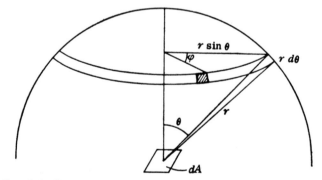

Fig. A-8. Geometry of solid angle with respect to a fixed plane A.

$$d\omega_{\theta\phi} = \sin\theta \, d\phi \, d\theta$$

In the special case that the vertical axis is an axis of circular symmetry, $d\omega$ may be integrated over ϕ, giving a solid angle increment

$$d\omega_\theta = 2\pi \sin\theta \, d\theta$$

Physical Topics

A. Units and Dimensions

Various internally consistent systems of units are used in scientific work. The system in widest use in physics is the centimeter-gram-second (CGS) system; it is based upon arbitrary units of length, mass, and time. The centimeter was originally defined as $1/10^9$ part of the arc length between the north pole and the equator. The *gram* was originally defined as the mass of a cubic centimeter of distilled water at the temperature of its greatest density, but since 1889 it has been defined as $1/1000$ of the standard kilogram, a platinum-iridium cylinder which is carefully preserved in its original state. The *centimeter* is now defined as 1,650,763.73 times the wavelength of the orange-red line of krypton-86, and the *second* is defined as $1/31,556,925.9747$ of the tropical year 1900.* Many other systems of units are used in technical fields, viz., the meter-kilogram-second (MKS) or the foot-slug-second. The CGS system is followed in this book except that in discussing electrical quantities the MKS system with the additional unit ampere (A) is used.

The quantities of length, mass, and time are referred to as *dimensions.* Other physical quantities, such as force, momentum, energy, etc., have dimensions which are made up of the three fundamental dimensions. The dimensions of all physical quantities may be determined from the mathematical expression which defines the quantity or from a physical relation between quantities. Thus, the dimensions of momentum and force are, respectively,

$$\text{mass} \times \text{velocity} = MLt^{-1}$$
$$\text{mass} \times \text{acceleration} = MLt^{-2}$$

where M represents mass, L length, and t time. An additional dimension, temperature K, is sometimes used.

B. Significant Figures†

It is important in calculation to distinguish between *mathematical* numbers which are known to any accuracy required and *physical* numbers whose accuracy is limited by errors of measurement. It follows that the

---

* Adopted by 11th General Conference on Weights and Measures in Paris, 14 October 1960. Definition of the second in terms of the period of atomic vibrations has been discussed.

† An excellent book on the subject is Yardley Beers, "Introduction to the Theory of Error," 2d ed., 66 pp. Addison-Wesley, Cambridge, Massachusetts, 1957.

last digit of a physical number is uncertain. The number of digits in such a number, after any zeros to the left of the first number different from zero, is called the number of significant figures. Thus, if we can read a distance as 98 km, we say that there are two significant figures. We are not permitted to say that this is 98,000 meters, for this erroneously gives the impression of a measurement accurate to the order of meters. Instead, we should say that the distance is 98×10^3 meters.

Mathematical numbers, which carry any number of significant figures, are treated as though they have at least as many significant figures as the largest number of significant figures in the problem.

Often one more figure may be used than is significant in order to prevent round-off error from influencing the result. The mean of more than 10 and less than 1000 numbers may contain one more significant figure than the individual observations. This may sometimes be used to advantage, as Chapman demonstrated in finding the lunar tide in high latitudes from barometric observations.*

In adding or subtracting all numbers should have the same number of significant *decimal places*. In order to achieve this, digits beyond the last significant decimal place should be dropped.

In multiplying and dividing the number of significant figures in the result is equal to the number of significant figures in that number which has the smallest number of significant figures.

Special rules may be developed for handling the number of significant figures of functions. The basic idea, however, is that the uncertainty of a function $y = f(x)$ is expressed by

$$dy = \frac{df}{dx} dx$$

where dx represents the uncertainty of the error of measurement of x.

---

* S. Chapman, *Quart. J. Roy. Meteorol. Soc.* **44**, 271 (1918).

C. ELECTROMAGNETIC CONVERSION TABLE[a]

Quantity	Symbol	Rationalized CGSe system	Rationalized Giorgi or MKSA system
Force	$\mathbf{F}$	10^5 dynes	$= 1$ newton (1 N)
Charge	Q	2.997930×10^9 Fr	$= 1$ coulomb (1 C)
Current	I	2.997930×10^9 Fr sec^{-1}	$= 1$ ampere (1 A)
Potential	V	3.33563×10^{-3} erg Fr^{-1}	$= 1$ volt (1 joule C^{-1})
Resistance	R	1.11265×10^{-12} erg sec Fr^{-2}	$= 1$ ohm (1 JA^{-2} sec^{-1})
Capacitance	C	8.98758×10^{11} Fr^2 erg^{-1}	$= 1$ farad (1 CV^{-1})
Electrical field strength	$\mathbf{E}$	3.33563×10^{-5} dyn Fr^{-1}	$= 1$ NC^{-1}
Magnetic induction[b]	$\mathbf{B}$	3.33563×10^{-7} dyn sec Fr^{-1} cm^{-1}	$= 1$ weber m^{-2} (1 NA^{-1} m^{-1})
Work	W	10^7 erg	$= 1$ joule (1 J)
Power	P	10^7 erg sec^{-1}	$= 1$ watt (1 J sec^{-1})

Permeability $\mu_0 = 4\pi \times 1.11265 \times 10^{-21}$ dyn sec^2 Fr$^{-2} = 4\pi \times 10^{-7}$ weber A^{-1} m^{-1} (NA^{-2})

Permittivity $\epsilon_0 = \dfrac{1}{4\pi}$ Fr^2 dyn^{-1} cm$^{-2} = \dfrac{1}{4\pi} \times 1.11265 \times 10^{-10}$ farad m^{-1} (A^2 sec^2 N^{-1} m^{-2})

[a] Calculated from Nederlandse Natuurkundige Vereniging ("Jaarboek"), p. 131 (1960).

[b] The magnetic induction is frequently expressed in the CGS electromagnetic unit, called the gauss, which is 10^{-4} weber m^{-2}.

D. TABLE OF PHYSICAL CONSTANTS

Velocity of light in vacuum[a]	(c_0)	$(2.997930 \pm 0.000003) \times 10^{10}$ cm sec^{-1}
Boltzmann constant[a]	(k)	$(1.38044 \pm 0.00007) \times 10^{-16}$ erg °K^{-1}
Mass of proton[a]	(m_+)	$(1.67239 \pm 0.00004) \times 10^{-24}$ g
Mass of electron[a]	(m_-)	$(9.1083 \pm 0.0003) \times 10^{-28}$ g
Charge of electron[a]	(e)	$(4.80286 \pm 0.00009) \times 10^{-10}$ Fr
Planck constant[a]	(h)	$(6.62517 \pm 0.00023) \times 10^{-27}$ erg sec
Wien's radiation constant[a]		(0.289782 ± 0.000013) cm °K
Stefan-Boltzmann constant[a]	(σ)	$(0.56687 \pm 0.00010) \times 10^{-4}$ erg cm^{-2} sec^{-1} °K^{-4}
Avogadro's number[a]	(N_0)	$(6.02295 \pm 0.00016) \times 10^{23}$ mol^{-1}
Standard molar volume of gas[a]	(V_m)	(22413.6 ± 0.6) cm^3 mol^{-1}
Molar gas constant[a]	(R)	$(8.31432 \pm 0.00034) \times 10^7$ erg mol^{-1} °K^{-1}
Atomic mass, carbon-12 nucleus[a]	(C^{12})	12 (exact)
Atomic mass, hydrogen[a]	(H^1)	1.007822 ± 0.000003.
Atomic mass, oxygen-16 nucleus[a]	(O^{16})	15.994915
Length of year[a]		365.24219878 days
Gravitation constant[b]	(G)	$(6.668 \pm 0.005) \times 10^{-8}$ dyn cm^2 g^{-2}
Standard sea level pressure[b]	(p_0)	$(1.013246 \pm 0.000004) \times 10^6$ dyn cm^{-2}
Ice point[b]	(T_0)	273.155 ± 0.015°K
Solar mass[b]		$(1.991 \pm 0.002) \times 10^{33}$ g
Solar radius[b]		$(6.960 \pm 0.001) \times 10^{10}$ cm
Earth mass[b]		$(5.977 \pm 0.004) \times 10^{27}$ g
Mean solar day[b]		1.00273791 sidereal days
Solar constant[c]	$(\overline{F}_s)$	1.98 cal cm^{-2} min^{-1} (138.2 mw cm^{-2})

[a] In Nederlandse Natuurkundige Vereniging ("Jaarboek"), p. 126, from E. R. Cohen, J. W. M. DuMond, J. W. Layton, and J. S. Rollett, *Revs. Modern Phys.* **27**, 363 (1955). The atomic mass scale based on the carbon-12 nucleus as 12 exactly has been used.
[b] C. W. Allen, "Astrophysical Quantities," p. 11. Univ. London, 1955.
[c] Adopted by the International Radiation Committee in 1957.

E. International Reference Atmosphere[a]

Height (km)	Density (g cm^{-3})	Pressure (dyn cm^{-2})	Temp. (°K)	Mol. mass[b]	Number density (cm^{-3})
0	1.23×10^{-3}	1.02×10^{6}	289.25	28.966	2.55×10^{19}
10	4.19×10^{-4}	2.68×10^{5}	222.36		8.72×10^{18}
12	3.16×10^{-4}	1.96×10^{5}	216.65		6.56×10^{18}
14	2.33×10^{-4}	1.43×10^{5}	214.21		4.84×10^{18}
16	1.70×10^{-4}	1.04×10^{5}	214.00		3.53×10^{18}
18	1.23×10^{-4}	7.59×10^{4}	215.18		2.55×10^{18}
20	8.89×10^{-5}	5.54×10^{4}	217.07		1.85×10^{18}
30	1.84×10^{-5}	1.21×10^{4}	228.28		3.83×10^{17}
40	4.07×10^{-6}	2.90×10^{3}	247.86		8.46×10^{16}
50	1.02×10^{-6}	7.91×10^{2}	269.56		2.13×10^{16}
60	3.04×10^{-7}	2.25×10^{2}	257.82		6.33×10^{15}
80	1.94×10^{-8}	1.03×10	184.60		4.03×10^{14}
90	3.12×10^{-9}	1.62	181.14	$\downarrow$	6.48×10^{13}
100	4.78×10^{-10}	2.92×10^{-1}	212.10	28.85	9.98×10^{12}
120	2.44×10^{-11}	2.44×10^{-2}	342.96	28.60	5.15×10^{11}
150	1.69×10^{-12}	5.08×10^{-3}	1014.53	28.09	3.63×10^{10}
200	3.61×10^{-13}	1.36×10^{-3}	1226.77	27.00	8.05×10^{9}
250	1.03×10^{-13}	4.34×10^{-4}	1301.45	25.55	2.42×10^{9}
300	3.34×10^{-14}	1.59×10^{-4}	1358.51	23.74	8.48×10^{8}
400	5.09×10^{-15}	3.04×10^{-5}	1436.16	20.00	1.53×10^{8}
500	1.17×10^{-15}	8.20×10^{-6}	1474.15	17.48	4.03×10^{7}
600	3.45×10^{-16}	2.70×10^{-6}	1474.15	15.70	1.32×10^{7}
700	1.19×10^{-16}	1.00×10^{-6}	1474.15	14.62	4.90×10^{6}
800	4.60×10^{-17}	4.05×10^{-7}	1474.15	13.88	1.99×10^{6}

[a] H. Kallmann-Bijl, R. L. F. Boyd, H. Lagow, S. M. Poloskov, and W. Priester, Committee on Space Research (COSPAR), Intern. Council Sci. Unions, 177 pp. North-Holland, Amsterdam, 1961.

[b] The molecular mass is based on the oxygen-16 nucleus as 16 exactly.

Bibliography

Advances in Geophys. **4**, 1 (1958): C. E. Junge, Atmospheric chemistry. (Chapter 3)

Advances in Geophys. **5**, 223 (1958): J. E. McDonald, The physics of cloud modification. (Chapter 3.)

Advances in Geophysics. **7**, 105–187 (1961): M. Siebert, Atmospheric Tides. (Chapter 1.)

Alfvén, H., "Cosmical Electrodynamics." Oxford Univ. Press, London and New York, 1950. (Chapter 6.)

Allis, W. P., and M. A. Herlin, "Thermodynamics and Statistical Mechanics." McGraw-Hill, New York, 1952. (Chapter 2.)

Battan, L. J., "Radar Meteorology." Univ. of Chicago Press, Chicago, Illinois, 1959. (Chapter 7.)

Berry, F. A., *et al.* (eds.), "Handbook of Meteorology," p. 283: J. Charney, Radiation. McGraw-Hill, New York, 1945. (Chapter 4.)

Brunt, D., "Physical and Dynamical Meteorology," 2nd ed. Cambridge Univ. Press, London and New York, 1941. (Chapter 5.)

Budyko, M. I., "The Heat Balance of the Earth's Surface." U. S. Weather Bureau, Washington, D.C., 1956 (translated by Nina A. Stepanova, 1958). (Chapter 5.)

Chalmers, J. A., "Atmospheric Electricity." Pergamon Press, New York, 1957. (Chapter 3.)

Chamberlain, J. W., "Physics of the Aurora and Airglow." Academic Press, New York, 1961. (Chapters 4 and 6.)

Chandrasekhar, S., "Radiative Transfer." Oxford Univ. Press, London and New York, 1950. (Chapter 4.)

Chapman, S., and J. Bartels, "Geomagnetism," Vols. 1 and 2. Oxford Univ. Press, London and New York, 1940. (Chapter 6.)

Compendium Meteorol., p. 101 (1951): O. H. Gish, Universal aspects of atmospheric electricity. (Chapter 3.)

Fletcher, N. H., "The Physics of Rainclouds." Cambridge Univ. Press, London and New York, 1962. (Chapter 3.)

Geiger, R., "The Climate Near the Ground," 2nd ed. (translated by M. N. Stewart and others). Harvard Univ. Press, Cambridge, Massachusetts, 1950. (Chapter 5.)

Glasstone, S., ed., "The Effects of Nuclear Weapons," U. S. Atomic Energy Commission. U. S. Govt. Printing Office, Washington, D.C., 1957. (Chapter 7.)

Goody, R. M., "The Physics of the Stratosphere." Cambridge Univ. Press, London and New York, 1954. (Chapter 4.)

Halliday, D., and R. Resnick, "Physics for Students of Science and Engineering." Wiley, New York, 1960. (Chapters 3, 4, and 6.)

Haltiner, G. J., and F. L. Martin, "Dynamical and Physical Meteorology." McGraw-Hill, New York, 1957. (Chapters 1 and 4.)

"Handbuch der Physik" (J. Bartels, ed.), Vol. 47, p. 498: S. K. Runcorn, The magnetism of the earth's body. Springer, Berlin, 1956. (Chapter 6.)

"Handbuch der Physik" (J. Bartels, ed.), Vol. 48, p. 155: F. Möller, Strahlung in der unteren atmosphäre. Springer, Berlin, 1957. (Chapter 4.)

Hirschfelder, J. O., C. F. Curtiss, and R. B. Bird, "Molecular Theory of Gases and Liquids." Wiley, New York, 1954. (Chapters 2 and 3.)

Holmboe, J., G. E. Forsythe, and W. Gustin, "Dynamic Meteorology." Wiley, New York, 1945. (Chapter 2.)

Humphreys, W. J., "Physics of the Air," 3rd ed. McGraw-Hill, New York, 1940. (Chapter 7.)

Johnson, J. C., "Physical Meteorology." M.I.T. Press and Wiley, New York, 1954. (Chapter 7.)

Joos, G., "Theoretical Physics," 2nd ed. (translated by Ira M. Freeman). Hafner, New York, 1950. (Chapters 1 and 7.)

Kallmann-Bijl, H. K., ed., "Space Research," Proc. 1st Intern. Space Sci. Symposium. North-Holland Publishing Co., Amsterdam, 1960. (Chapter 6.)

Kennard, E. H., "Kinetic Theory of Gases." McGraw-Hill, New York, 1938. (Chapter 2.)

Kourganoff, V., "Basic Methods in Transfer Problems." Oxford Univ. Press, London and New York, 1953. (Chapter 4.)

Lettau, H., and B. Davidson, eds., "Exploring the Atmosphere's First Mile," Vols. 1 and 2. Pergamon, New York, 1957. (Chapter 5.)

Massey, H. S. W., and R. L. F. Boyd, "The Upper Atmosphere." Philosophical Library, New York, 1959. (Chapter 6.)

Mason, B. J., "The Physics of Clouds." Oxford Univ. Press, London and New York, 1957. (Chapter 3.)

Meteorol. Monographs 4, No. 23 (1960): W. M. Elsasser with M. F. Culbertson, Atmospheric radiation tables. (Chapter 4.)

Mitra, S. K., "The Upper Atmosphere." Asiatic Society, Calcutta, 1952. (Chapter 4.)

Panofsky, W. K. H., and M. Phillips, "Classical Electricity and Magnetism." Addison-Wesley, Reading, Massachusetts, 1955. (Chapter 6.)

Pasquill, F., "Atmospheric Diffusion." D. Van Nostrand, London, New York, Princeton, Toronto, 1962. (Chapter 5.)

Priestley, C. H. B., "Turbulent Transfer in the Lower Atmosphere." Univ. of Chicago Press, Chicago, Illinois, 1959. (Chapter 5.)

Proc. I.R.E. 47, Chapters 4, 6, 7 (February, 1959).

Ratcliffe, J. A., ed., "Physics of the Upper Atmosphere," p. 1: S. Chapman; pp. 219, 269, 298: D. R. Bates; p. 378: J. A. Ratcliffe and K. Weekes; p. 471: E. H. Vestine. Academic Press, New York, 1960. (Chapters 2, 4, and 6.)

Ridenour, L. N., ed., "Modern Physics for the Engineer," p. 330: L. B. Loeb, Thunderstorms and lightning strokes. McGraw-Hill, New York, 1954. (Chapter 3.)

Sears, F. W., "An Introduction to Thermodynamics, The Kinetic Theory of Gases and Statistical Mechanics." Addison-Wesley, Reading, Massachusetts, 1953. (Chapter 2.)

Sears, F. W., "Principles of Physics," Vol. 1: Mechanics, Heat, and Sound. Addison-Wesley, Reading, Massachusetts, 1947. (Chapter 1.)

Sears, F. W., "Principles of Physics," Vol. 2: Electricity and Magnetism. Addison-Wesley, Reading, Massachusetts, 1947. (Chapters 3 and 6.)

Sears, F. W., "Principles of Physics," Vol. 3: Optics. Addison-Wesley, Reading, Massachusetts, 1947. (Chapter 7.)

Sommerfeld, A., "Thermodynamics and Statistical Mechanics." Academic Press, New York, 1956. (Chapter 2.)

Sutton, O. G., "Micrometeorology." McGraw-Hill, New York, 1953. (Chapter 5.)

Unsöld, A., "Physik der Sternatmosphären." Springer, Berlin, 1955. (Chapter 7.)

Van de Hulst, H. C., "Light Scattering by Small Particles." Wiley, New York, 1957. (Chapter 7.)

Wilkes, M. V., "Oscillations of the Earth's Atmosphere." Cambridge Univ. Press, London and New York, 1949. (Chapter 1.)

Index

A